AF849486

# Robotic Swarming Control via Voronoi Tessellations

With the help of a Voronoi diagram, this book systematically discusses how to solve the core cooperative control problem in a multi-robot system. From basic theory to cutting-edge applications, *Robotic Swarming Control via Voronoi Tessellations: Foundations and Applications* provides a comprehensive overview of the subject matter, offering readers a rigorous and practical knowledge system.

Central to this book's contribution is its transformation of centroidal Voronoi tessellations (CVT) theory into innovative multi-robot control algorithms. The effectiveness and superiority of these algorithms are verified through numerous simulations and real-environment experiments. The book demonstrates the deep integration of mathematical geometry and robotics, offering universal solutions for swarm intelligence in complex environments. It serves as an authoritative guide, bridging the gap between theoretical geometry and robotics engineering practice.

This text will be invaluable to scholars and students of robotics, automation, control engineering, and computer science.

**Kai Cao** is an associate professor at Xi'an Technological University and deputy director of the International Joint Research Center for Autonomous Systems and Intelligent Control in Shaanxi Province, China. His main research directions include multi-agent systems, multi-robot formation control and source localization, and collaborative multispectral remote sensing based on small drones.

**ChaoBo Chen** is a professor at Xi'an Technological University. He is currently the dean of the School of Electronic Information Engineering of Xi'an Technological University and the council member and deputy secretary-general of Shaanxi Provincial Automation Society. His current research focuses on intelligent control, fault diagnosis, and fault tolerance control.

**Song Gao** is a professor at Xi'an Technological University. Director of the International Joint Research Center for Autonomous Systems and Intelligent Control in Shaanxi Province, his research interests include autonomous intelligent unmanned systems, robot multi-mode intelligent inspection, automation control, and grid-connected inverter technology.

**YangQuan Chen**, a Clarivate Highly Cited Researcher, is a professor in the Department of Mechanical and Aerospace Engineering, School of Engineering, University of California, Merced. His research interests include mechatronics, smart control engineering via digital twins, and applied fractional calculus in STEM.

# Robotic Swarming Control via Voronoi Tessellations

## Foundations and Applications

Kai Cao, ChaoBo Chen, Song Gao,
and YangQuan Chen

CRC Press
Taylor & Francis Group
Boca Raton London New York

CRC Press is an imprint of the
Taylor & Francis Group, an **informa** business

MATLAB® and Simulink® are trademarks of The MathWorks, Inc. and are used with permission. The MathWorks does not warrant the accuracy of the text or exercises in this book. This book's use or discussion of MATLAB® or Simulink® software or related products does not constitute endorsement or sponsorship by The MathWorks of a particular pedagogical approach or particular use of the MATLAB® and Simulink® software.

First edition published 2027
by CRC Press
2385 NW Executive Center Drive, Suite 320, Boca Raton, FL 33431

and by CRC Press
4 Park Square, Milton Park, Abingdon, Oxon, OX14 4RN

*CRC Press is an imprint of Taylor & Francis Group, LLC*

© 2027 Kai Cao, ChaoBo Chen, Song Gao, and YangQuan Chen

Reasonable efforts have been made to publish reliable data and information, but the author and publisher cannot assume responsibility for the validity of all materials or the consequences of their use. The authors and publishers have attempted to trace the copyright holders of all material reproduced in this publication and apologize to copyright holders if permission to publish in this form has not been obtained. If any copyright material has not been acknowledged, please write and let us know so we may rectify in any future reprint.

Except as permitted under U.S. Copyright Law, no part of this book may be reprinted, reproduced, transmitted, or utilized in any form by any electronic, mechanical, or other means, now known or hereafter invented, including photocopying, microfilming, and recording, or in any information storage or retrieval system, without written permission from the publishers.

For permission to photocopy or use material electronically from this work, access www.copyright.com or contact the Copyright Clearance Center, Inc. (CCC), 222 Rosewood Drive, Danvers, MA 01923, 978-750-8400. For works that are not available on CCC, please contact mpkbookspermissions@tandf.co.uk

For Product Safety Concerns and Information, please contact our EU representative GPSR@taylorandfrancis.com. Taylor & Francis Verlag GmbH, Kaufingerstraße 24, 80331 München, Germany.

*Trademark notice:* Product or corporate names may be trademarks or registered trademarks and are used only for identification and explanation without intent to infringe.

ISBN: 978-1-041-19913-7 (hbk)
ISBN: 978-1-041-20003-1 (pbk)
ISBN: 978-1-003-71456-9 (ebk)

DOI: 10.1201/9781003714569

Typeset in LMRoman
by KnowledgeWorks Global Ltd.

*To Prof. YangQuan Chen, with special gratitude and enduring remembrance, your guidance and inspiration continue to light our path.*

*-K.C., C.C. and S.G.*

*To Rong, my family, & mentors,*
*without your support this accomplishment would not have been possible.*

*To Shiyu (Niu Zai),*
*through clam thinking, hardwork, and persistent effort, anything is possible.*

*To my readers,*
*I hope this book helps you implement robotic swarming control to make the world more efficient.*

*-K.C.*

*To our mentors and families.*

*-C.C., S.G., and Y.C.*

# Contents

# Preface

In the rapidly developing field of robotic technology, one of the most promising challenges is the coordination and collaboration of multiple robots to complete complex tasks. Robot swarm technology enables multiple robots (a multi-robot system or MRS) to work together to accomplish specific tasks. Compared to a single robot, a robot swarm can cover a broader task area and achieve faster task execution through parallel processing, especially in areas that require efficient spatial division, task allocation, and coverage control. Voronoi tessellations (VT), Voronoi diagrams (VD), or Dirichlet tessellations, powerful geometric tools, have emerged as a key method for solving these problems, enabling robots to partition space and collaborate effectively in dynamic and uncertain environments. A Voronoi diagram divides a space into regions based on proximity to a set of seed points; in multi-robot control, each robot's position acts as a seed, and its Voronoi cell consists of all points in the environment closer to that robot than to any other. In essence, each robot is in charge of its own Voronoi region.

This Voronoi partitioning possesses powerful theoretical properties that make it highly suitable for swarm robot control. It has distributed characteristics, where the partitioning of each robot's unit mainly depends on the position of its neighboring robots, and it can cover the entire target area without overlap or omission. According to classical coverage control theory, driving each robot toward the centroid of its Voronoi cell effectively optimizes overall coverage performance (similar to the Lloyd algorithm used for centroidal Voronoi tessellations). By assigning the corresponding Voronoi cells to each robot, spatial load balancing is achieved essentially creating roughly equitable task area, which guarantees comprehensive area coverage. Control laws based on Voronoi partitions typically require only local sensing or communication (each robot only needs position information from neighbors sharing a cell boundary), enabling dynamic updates to the partition. As a result, the approach offers strong scalability and robustness for large-scale swarm systems.

In Voronoi-based robotic swarm control, Voronoi neighbors (robots with adjacent units) typically define the communication or interaction topology. As robots move, the Voronoi boundaries dynamically change. This provides a formal framework for analyzing the global convergence of distributed control laws: since each robot optimizes only its own unit, under certain conditions, the entire swarm will converge to a balanced partition (usually the centroid Voronoi configuration used for coverage tasks). The current Voronoi methods have been extended beyond simple coverage tasks to include pursuit and evasion games, task allocation, and collision avoidance. For example, Voronoi diagrams have been used to define dominion areas in pursuit problems (where each pursuer restricts the evader within its own area) and to compute safe areas for each robot to achieve collision-free navigation.

In summary, Voronoi tessellation provides a framework for the coordinated control of robotic swarms that combines mathematical rigor with engineering practicality. By partitioning continuous space into well-defined regions of responsibility, it enables distributed, scalable, and efficient task allocation and area coverage for multi-robot systems. As robotic technology continues to advance, Voronoi diagrams and their derivative methods will remain core tools, driving progress in collaborative decision-making and autonomous operation capabilities of multi-robot systems in complex, dynamic environments.

This book provides a comprehensive exploration of the application of Voronoi tessellations in multi-robot systems, covering both foundational concepts and advanced techniques. The concept of Voronoi tessellations was first introduced by the German mathematician Peter Gustav Lejeune Dirichlet in the 19th century, who provided a simple proof of quadratic reducibility and proposed the method of partitioning spaces using Voronoi diagrams. Over the decades, Voronoi tessellations evolved from a theoretical concept in geometry to a powerful tool with a wide range of applications, particularly in robotics. The application of Voronoi diagrams in multi-robot systems has enabled the effective solution of complex problems such as task allocation, formation control, and collision avoidance. By leveraging the geometric properties of Voronoi tessellations, multi-robot teams can partition space, optimize their positions, and coordinate their actions in dynamic environments.

Over the past few decades, the integration of Voronoi-based methods with multi-robot systems has significantly advanced. The introduction of Centroidal Voronoi Tessellations (CVT), originally proposed by Professor Stuart P. Lloyd in 1982, marked a pivotal development, enabling robots to continuously adapt to changing environments and tasks. Professor Jorge Cortés and his team played a significant role in advancing the application of CVT in multi-robot systems, particularly in areas such as formation control and coverage algorithms in static spatial environments. Building on these foundations, Professor YangQuan Chen and his team contributed to the practical deployment of CVT by focusing on distributed control and sensor network applications, facilitating scalable and adaptive coverage strategies in dynamic evolving spatial environments. Today, Voronoi tessellations are widely used in multi-robot formation control, coverage algorithms, and safety strategies. These techniques have become integral in ensuring that robots can effectively collaborate, handle dynamic changes, and operate in unstructured environments. Moreover, the integration of Voronoi methods with modern robotic platforms, such as ROS, TurtleBot3, and Crazyflie, has further propelled their practical applications. In addition to these foundational applications, researchers have explored novel uses of Voronoi tessellations in areas such as source search, capture strategies, and autonomous systems.

Part of this text is extracted from the first author's PhD dissertation. Part of the book contents is also based on Professor Cao's one-year visiting stay with the Mechatronics Embedded Systems and Automation (MESA) Lab of the University of California Merced, CA, USA (2020-2021) focusing on multi-robot control, swarm robots, and source localization. The roadmap of this text is shown in Fig. 1.

The structure and chapter arrangement of this book are as follows:

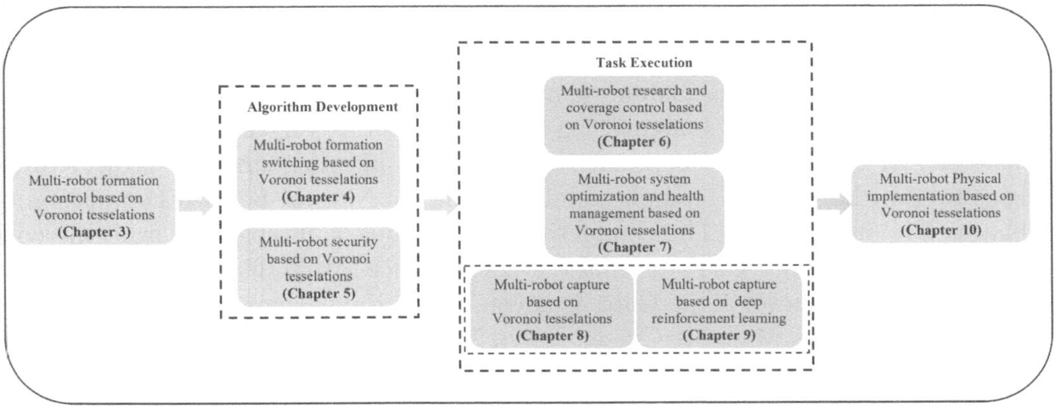


Figure 1: The monograph roadmap.

Chapter 1 provides an introduction to Voronoi tessellations, including their basic concepts, definitions, generation methods, and various application areas. Chapter 2 focuses on the geometric properties and computational methods of Voronoi tessellations, analyzing their fundamental features, properties, and their behaviors in different environments such as 2D, 3D, and non-convex spaces. Chapter 3 presents multi-robot cooperation and formation control strategies based on Voronoi tessellations, with emphasis on centroidal Voronoi tessellations and formation simulation in different dimensions. Chapter 4 introduces dynamic formation switching methods guided by density functions, along with related simulation analyses. Chapter 5 addresses multi-robot safety and collision avoidance strategies, exploring buffered uncertainty Voronoi cells and obstacle avoidance techniques. Chapter 6 investigates source localization and coverage control for multi-robot systems using Gaussian process regression and Voronoi-based exploration strategies. Chapter 7 explores optimization and health management of multi-robot systems through Voronoi-based methods, supported by simulation analyses. Chapter 8 discusses multi-robot capture strategies for targets with known or unknown locations, presenting algorithms and simulation results. Chapter 9 introduces multi-robot cooperative capture strategies based on deep reinforcement learning. By integrating the MADDPG algorithm with Voronoi tessellations, a multidimensional fusion framework is proposed to enhance coordination and control among agents. Simulation experiments in the PettingZoo environment verify the effectiveness of the method, demonstrating improvements in encirclement success rate and convergence stability. Chapter 10 covers practical implementation on ROS and multi-robot platforms with experiments on formation control, switching, source search, and capture. Finally, Chapter 11 concludes the book with a summary and outlook on future research directions and challenges.

Professor Cao would like to thank Professors Song Gao, ChaoBo Chen and his research team members from Xi'an Technological University for their support and help. Professor Cao also wishes to thank the funding from the International Joint Research Center for Autonomous Systems and Intelligent Control in Shaanxi Province. In addition, he wishes to thank Xi'an Technological University Study Abroad Fund for funding his one-year exchange PhD studentship at the MESA Lab of the University of California Merced, CA, USA. Professor Cao wishes to thank China Scholarship Council (CSC) for funding his one-year visiting scholar position at the Institute of Thin Films, Sensors and Imaging, Scottish Universities Physics Alliance (SUPA), University of the West of Scotland. He would also like to thank the project funding grant No. H20251091, and the funding from the Natural Science Foundation of Shaanxi Province (2026). The completion of this book is inseparable from the contributions of Professor Cao's graduate students since 2016. Here, he would like to express his gratitude to: Jiajia Gao, Xuemeng Yang, Hang Zhang, Muhan Cao, Zhao Wang, Haixin Dang, Manying Cheng, Haiyu Yang, Shuangchao Li, Qian Cheng, Jiahao Zhang, Shanshan Lang, Ge Zhang, Yunbo Wei, Kang Li, Fan Yang, Weichao Liu, Yufei Ding, Zhi Liu, Shibo Yang, Qilong Shi, Yige Liu, Lulu Zhao, Wenle Qi, Heyang Wang, Jiarui Zhang among many others. Together, we have refined the theory and applications of Voronoi Diagram-based methods in multi-robot control.

Professor YangQuan Chen wishes to thank the funding support by the Center for Methane Emission Research and Innovation (CMERI) through the Climate Action Seed Funds grant (2023–2026) at the University of California Merced, CA, USA.

Last but not least, the authors would like to thank Ms. Xiaoyin Feng, Books Editorial Assistant, Routledge and CRC Press, part of Taylor & Francis Group, for her professional coordination of the book project; Ms. Lian Sun, Publisher, Head of China Books Publishing and International Cooperation Taylor & Francis Advanced Learning for her effective peer review process of this book project proposal.

The simulation code and dataset supporting each chapter of this book have been uniformly hosted on the GitHub platform (Link: `https://github.com/storybox126/Robotic-Swarming-Control-via-Voronoi-Tessellations.git`) Readers can obtain complete resources through this repository.

The authors made judicious use of LLM for polishing of various versions of Chapter 11 of the book draft. The authors reviewed and revised all content and bear full responsibility for its accuracy and originality.

Xi'an, China, August 2025 — *Kai Cao, ChaoBo Chen, Song Gao*
Merced, CA, USA, August 2025 — *YangQuan Chen*

# Abbreviations

There are numerous abbreviations used throughout the book. Books from various fields may have abbreviations that use the same letters, but have different meanings. So to avoid confusion, we list in alphabetical order the abbreviations used.

| | |
|---|---|
| 2D | Two-Dimensional |
| 3D | Three-Dimensional |
| 4D | Four-Dimensional |
| APF | Artificial Potential Fields |
| BUVC | Buffered Uncertainty Voronoi Cell |
| BVC | Buffered Voronoi Cell |
| CMERI | Center for Methane Emission Research and Innovation |
| CRF | Centroid Repulsion Field |
| CSC | China Scholarship Council |
| CTDE | Centralized Training Decentralized Execution |
| CVT | Centroid Voronoi Tessellations |
| DDPG | Deep Deterministic Policy Gradient |
| DE | Distance Error |
| DGPR | Distributed Gaussian Process Regression |
| DQN | Deep Q-Networks |
| DRL | Deep Reinforcement Learning |
| DWA | Dynamic Window Approach |
| EE | Extremum Error |
| FISMO | Fuzzy Integral Sliding Mode Observe |
| FITC | Fully Independent Training Conditional |
| G-CVT | Greedy-Centroidal Voronoi Tessellation |
| GMM | Gaussian Mixture Model |
| GP | Gradient Projection |
| GPR | Gaussian Process Regression |
| GRF | Gibbs Random Field |
| HOM | Hybrid Optimization Method |
| HSNs | Heterogeneous Sensor Networks |
| HVCR | Hierarchical Virtual Communication Ring |
| IP | Internet Protocol |
| KL | Kullback-Leibler |
| MADDPG | Multi-Agent Deep Deterministic Policy Gradient |
| MADRL | Multi-Agent Deep Reinforcement Learning |

| | |
|---|---|
| MAPPO | Multi-Agent Proximal Policy Optimization |
| MASAC | Multi-Agent Soft Actor-Critic |
| MATD3 | Multi-Agent Twin Delayed Deep Deterministic Policy Gradient |
| MATRPO | Multi-Agent Trust Region Policy Optimization |
| MDP | Markov Decision Process |
| MESA | Mechatronics Embedded Systems and Automation |
| ML | Machine Learning |
| MLE | Maximum Likelihood Estimation |
| MRS | Multi-Robot System |
| MSNs | Mobile Sensor Networks |
| ODMA | Omni-Directional Minimum Area |
| ODMV | Omni-Directional Minimum Volume |
| PD | Proportional Derivative |
| PID | Proportional Integral Derivative |
| PITC | Partially Independent Training Conditional |
| PPO | Proximal Policy Optimization |
| PRM | Probabilistic Road Map |
| PSO | Particle Swarm Optimization |
| RL | Reinforcement Learning |
| RMSE | Root Mean Square Error |
| ROS | Robot Operating System |
| RRT | Rapidly-Exploring Random Trees |
| SGPR | Sparse Gaussian Process Regression |
| SLAM | Simultaneous Localization and Mapping |
| SUPA | Scottish Universities Physics Alliance |
| SVM | Support Vector Machine |
| TCP | Transmission Control Protocol |
| TD | Temporal Difference |
| UAV | Unmanned Aerial Vehicle |
| UCB-CVT | Upper Confidence Bound-Centroidal Voronoi Tessellation |
| VD | Voronoi Diagram |
| VFE | Variational Free Energy |
| VS-GPR | Variational Sparse Gaussian Process Regression |
| VT | Voronoi Tessellation |
| W-CVT | Weighted Centroidal Voronoi Tessellation |
| WPA | Wolf Pack Algorithm |
| WSNs | Wireless Sensor Networks |

CHAPTER 1

# Introduction

## 1.1 Motivation

In recent years, the rapid advancement of robotics has significantly expanded the application boundaries of robotic systems, extending from traditional industrial manufacturing to multiple domains such as services, healthcare, agriculture, and military operations. With their flexible mobility and autonomous decision-making capabilities, modern robots are increasingly able to perform tasks that would otherwise require human involvement in harsh environments, high-risk scenarios, or repetitive labor. Compared to fixed automation equipment, mobile robots equipped with onboard computing, sensing, and communication capabilities require no pre-installed infrastructure. They can dynamically perceive changes in their surroundings, adjust strategies in real time, optimize task execution and energy allocation, and continuously collect environmental data while responding accordingly. This exceptional adaptability enables robots to maintain operational effectiveness in unstructured scenarios, whereas traditional static automation systems often struggle in such environments.

Based on an analysis of real-world operational environments and task characteristics, robots are widely regarded as particularly well-suited for performing so-called "4D" tasks those that are "Dull, Dirty, Dangerous, and Deep." This classification not only highlights the advantages of robots in replacing humans but also underscores their unique value in complex, dynamic, and extreme environments. "Dull" primarily refer to highly repetitive, mechanical, and unchallenging work. Examples include daily sorting operations in warehouses and routine equipment inspections on production lines. Long-term performance of such tasks can lead to fatigue in human workers, resulting in loss of focus and increased error rates. In contrast, robots performing these operations can maintain consistent and stable efficiency without being affected by mental boredom or physical fatigue, thereby significantly enhancing overall operational stability and accuracy. "Dirty" involve environments with poor sanitary conditions or severe pollution and hazardous substances. Examples include inspecting high-radiation areas in nuclear power plants, handling accidents involving toxic chemical leaks, and maintaining sewer pipelines. Such environments often pose direct threats to human health, and prolonged exposure may even endanger lives. In these scenarios,

DOI: 10.1201/9781003714569-1

robots serve as effective substitutes that can perform high-risk operations, reduce human exposure, and fundamentally ensure personnel safety. "Dangerous" encompass various high-risk situations, such as fire rescue operations where secondary explosions may occur, mine operations prone to collapses or gas leaks, and the handling of live explosives by military and police units. These tasks present extreme threats to human life and safety. Robots, however, can be deployed into hazardous zones for real-time reconnaissance, monitoring, and intervention. On the one hand, this significantly reduces the likelihood of direct human exposure to danger; on the other hand, equipped with sensing and autonomous decision-making capabilities, robots can ensure both timeliness and reliability in performing high-risk operations. "Deep" refer to missions beyond the reach of humans or manned equipment, including deep-sea exploration, underground pipeline inspection, and outer space exploration. These areas are often inaccessible to humans due to extreme pressure, oxygen-deficient environments, or the physical constraints of remote space. When equipped with advanced sensors, computing, and communication modules, mobile robots can autonomously explore and collect data in such extreme or confined environments, performing high-precision monitoring and operational interventions. Thereby, they overcome human physiological limits and expand the boundaries of human activity and engineering. In summary, the application of robots in "4D" tasks not only effectively reduces the burden on humans in monotonous, hazardous, and unsuitable environments but also demonstrates unique advantages in accomplishing complex missions under extreme conditions, enabled by intelligent and autonomous technologies. This characteristic underscores the broad potential of robots across various fields, including modern industry, public safety, medical rescue, and deep-space exploration.

While a single robot is capable of performing many of these tasks, its limitations become apparent when confronted with complex scenarios involving large spatial scales, high variability, or significant uncertainty. Issues such as limited endurance, constrained field of view, and individual robot failures have driven a shift in robotic intelligence from monolithic intelligence to swarm intelligence in multi-robot teams. By coordinating multiple robots into an integrated system, it is possible not only to overcome the physical and operational constraints of a single robot but also to achieve more adaptive, robust, and scalable collective decision-making. Multi-robot teams can cover larger areas, handle multiple objectives simultaneously, and provide redundancy in case of individual failures. Essentially, collaboration within robot swarms enables higher resilience and intelligence.

In the research of multi-robot swarm, formation control plays a key role in enhancing the organization and collaborative efficiency of group behaviors. Formation control primarily addresses the problem of how robots arrange themselves in space so as to maintain a desired geometric structure during motion or cooperative tasks. According to the topological relationships among robots, existing studies can be broadly categorized into three main strategies: the leader–follower approach [1] [2] [3], the virtual structure approach [4] [5], and the behavior-based approach [6] [7]. In the leader–follower strategy, one or several robots act as leaders, while the remaining robots serve as followers that adjust their states based on relative positions with respect to the leaders, thereby maintaining the desired formation geometry. The virtual structure strategy treats the entire formation as if it were embedded within a rigid geometric framework. This framework translates and rotates as a whole, while each robot maintains a fixed relative position within the structure. By contrast, in the behavior-based strategy, each robot follows a set of local behavioral rules, and through interactions among individuals, the overall formation emerges spontaneously. In addition to classification based on topology, formation control can also be divided according to communication methods into distributed control [8] [9] and centralized control [10]. In distributed formations, each robot makes decisions autonomously based on local sensing or neighborhood communication, without any single node holding global authority. This yields strong robustness and scalability, and avoids the risk of single-point failure. In contrast, centralized formations rely on a central controller or a main node to collect global information and issue commands to the entire formation. While this can simplify coordination logic, the failure of the central node significantly increases the overall vulnerability of the system. Although these methods each have their own advantages in specific scenarios, they still face considerable challenges when dealing with collision avoidance and reorganization in dynamic environments.

Formation control methods based on Voronoi diagrams offer a novel approach to addressing the aforementioned challenges in multi-robot systems. Initially proposed by Cortes et al. [11] in 2004, this methodology represents a groundbreaking application of Voronoi diagram theory from computational geometry to multi-robot coordination. The core concept involves tesselatting the operational environment into multiple non-overlapping Voronoi cells according to robot positions, and utilizing the geometric properties of Voronoi partitions for formation control. This ensures that each robot operates exclusively within its designated Voronoi cell, thereby fundamentally preventing inter-robot collisions. Compared to various traditional methods discussed earlier, the Voronoi tessellations approach effectively avoids robot collisions and serves as a flexible, non-rigid formation solution particularly suitable for multi-robot swarm exploration. Consequently, this book will provide a detailed elaboration of both theoretical foundations and practical applications of Voronoi diagrams in multi-robot swarm control, systematically demonstrating how Voronoi tessellation strategies can achieve efficient, adaptive, and safe multi-robot coordination. It aims to serve as a valuable reference for both researchers and practitioners in the field.

## 1.2 Basic Concepts and Application Areas of Voronoi Tessellations

The origin of Voronoi diagrams can be traced back several centuries. As early as the 17th century, the French philosopher and mathematician René Descartes implicitly used the concept of Voronoi-like spatial partitioning in his work *Principia Philosophiae* describe the distribution of matter in the solar system. He envisioned cosmic space as filled with contiguous vortices, as shown in Fig. 1.1, essentially dividing space into convex polygonal regions centered around stars. Within each region, all matter rotates around the central star of that region. This concept of "each star

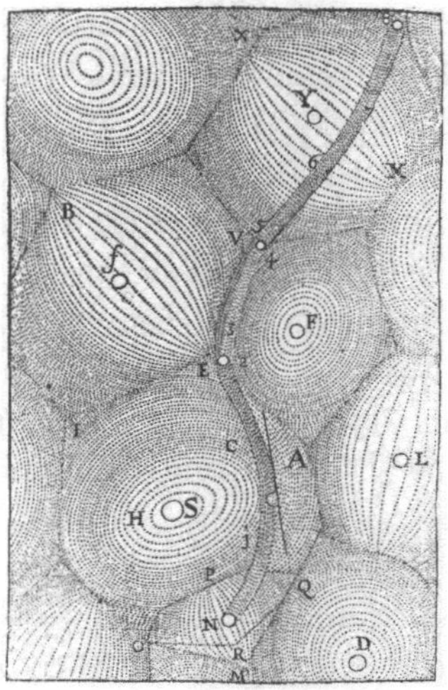

Figure 1.1: Descartes decomposition of space into vortices.

influencing its own spatial region" can be regarded as a precursor to the formal definition of Voronoi tessellation.

In the 19th century, the mathematical foundations of this theory were further developed. In 1840, Gauss discovered that positive quadratic forms could be associated with spatial lattice points, with their minima corresponding to spatial partitions resembling Voronoi diagrams of lattice points. However, he did not explicitly specify the formal definition of Voronoi regions. It was not until 1850 that the German mathematician Dirichlet [12], by providing a concise proof of quadratic form reduction, formally proposed the Voronoi tessellation method for spatial partitioning. After Voronoy extended this method to higher-dimensional spaces, Voronoi tessellations found widespread applications in fields such as geographic information mapping and biology. To honor these foundational contributions, the spatial partitioning method now known as Voronoi diagrams is also referred to in literature as "Dirichlet tessellations" or "Tyson polygons." Over time, Voronoi diagrams have been extensively applied in numerous fields including geography, meteorology, biology, and beyond. In 2004, Cortes et al. [11] pioneered the application of Voronoi partitioning to the field of multi-robot swarms. Building upon these historical developments, this section will begin with the basic concepts of Voronoi diagrams (VD), analyze their definitions and generation methods, and then introduce the application value of VD in various fields. The section will first review the fundamental definition of Voronoi tessellations and common construction methods, followed by an in-depth discussion of the significance of Voronoi tessellations by highlighting their value and typical applications across various engineering and scientific disciplines.

### 1.2.1 Definition and Generation Methods of Voronoi Tessellations

In simple terms, a Voronoi diagram is a method of dividing a space into regions where each region consists of all points that are closer to one specific designated point than to any other designated point. We will now formally define this concept. Setting a two-dimensional Euclidean plane as the task space $\mathbb{R}^2$, and a set of task points $\{p_i\}_{i=1}^n \in \mathbb{R}^2, \quad 3 \leq n < \infty$, within the task area, the Voronoi cell $v_i$ in two-dimensional space can be expressed as:

$$v_i = \left\{ \boldsymbol{q} \in \mathbb{R}^2 \mid \|\boldsymbol{q} - \boldsymbol{p}_i\| \leq \left\|\boldsymbol{q} - \boldsymbol{p}_j\right\|, i, j = 1, 2, \ldots, n, j \neq i \right\}, \tag{1.1}$$

where $\boldsymbol{q}$ is a point within the Voronoi cell, and the task area has certain limitations and must be strictly convex. The number of generating points in the environment determines the number of Voronoi cells. The Euclidean distance between the generating point and other points in the area is expressed as formula (1.2), with points with equal Euclidean distances as the boundaries of adjacent Voronoi cells. The distance from any point in the current Voronoi cell to the current generating point is the closest to other generating points. The complete set of Voronoi cells forms a partition of the task space. With $n$ generating points, there are exactly $n$ Voronoi cells, and these cells cover the entire task area without any overlap.

$$\|\boldsymbol{q} - \boldsymbol{p}_i\| = \sqrt{\left(\boldsymbol{q}_1 - \boldsymbol{p}_{i_1}\right)^2 + \left(\boldsymbol{q}_1 - \boldsymbol{p}_{i_2}\right)^2 + \cdots + \left(\boldsymbol{q}_n - \boldsymbol{p}_{i_n}\right)^2}. \tag{1.2}$$

The number of Voronoi cells is closely related to the number of generating points. When there are $n$ generating points, these points partition the space into $n$ Voronoi cells. The boundary of each Voronoi cell is formed by the perpendicular bisectors of the lines connecting adjacent generating points. Therefore, to determine the boundaries, it is necessary to first calculate the Euclidean distances between the generating points and all other points, and then delineate the cell boundaries based on geometric relationships of equidistance. According to the defining property of Voronoi diagrams, for any given Voronoi cell, the distance from the generating point within that cell to any point inside the cell is always less than the distance from any other generating point to that same point within the cell. The Voronoi tessellations in two-dimensional and three-dimensional environments are shown in Fig. 1.2, where circles represent the generating point within the Voronoi cell, while irregular polygons and polyhedra represent the Voronoi cell corresponding to different generating points. The Voronoi tessellations in two-dimensional environment only considers two directions, while in three-dimensional space, it is more complex, first manifested in the formula for Voronoi tessellations, setting a three-dimensional Euclidean plane as the task space $\mathbb{R}^3$, as shown below:

$$\begin{aligned} V_i^3 = \Bigg\{ q \in \mathbb{R}^3 \mid & \sqrt{\sum_{N=x,y,z} (q_n - p_{in})^2} \leq \sqrt{\sum_{m=x,y,z} (q_m - p_{jm})^2}, \\ & i, j = 1, 2, \ldots, N, \quad \forall j \neq i \Bigg\} \end{aligned} \tag{1.3}$$

From Fig. 1.2, it can be observed that in both two-dimensional and three-dimensional environments, each generating point exclusively occupies its corresponding Voronoi region, and these regions do not overlap with each other. The union of all

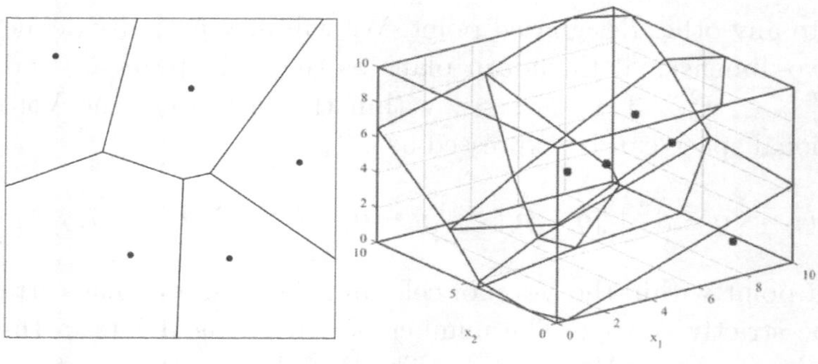

(a) (b)

Figure 1.2: Voronoi tessellations in two-dimensional and three-dimensional environment: (a) Voronoi tessellations in two-dimensional; (b) Voronoi tessellations in three-dimensional.

Voronoi cells precisely covers the entire task area $\mathbb{R}^3$. Further geometric properties of Voronoi tessellations, particularly their characteristics in planar environments, will be discussed in detail in Chapter 2.

There exist numerous methods for generating Voronoi diagrams, which can be broadly categorized into two classes: direct generation and indirect generation. The direct method refers to an algorithmic system that constructs Voronoi diagrams directly from spatial seed points through geometric computation, with its core principle being the direct calculation of Voronoi cell boundaries for each generating point based on mathematical definitions. The indirect method, on the other hand, first generates the dual structure of the Voronoi diagram—the Delaunay triangulation, and then converts it into the corresponding Voronoi diagram. Within the direct generation framework, commonly used algorithms primarily include the following three types: the incremental method, the divide-and-conquer method, and the parallel method, among others. Detailed descriptions are as follows:

1. Incremental method

Assuming there are $n$ growth points in two-dimensional space, and sorted in a certain way such as from $P_1$ to $P_n$ or from $P_n$ to $P_1$. Let $V_m$ represent the Voronoi cells of the first $m$ growth points $P_1, \ldots, P_m$. Starting from $V_3$, for each additional growth point $P_m \quad (m \leq n)$, $V_{m-1}$ is locally redivided to obtain $V_m$, as shown in Fig. 1.3.

In the process of expanding from $V_{m-1}$ to $V_m(n = 3, 4, 5, \ldots)$, there are two key steps: First, search for the nearest neighbor, from the growth points $P_1, \ldots, P_{m-1}$ corresponding $V_{m-1}$, find the growth point $P_{m(m)}$ that is closest to the newly added growth point $P_m$; Next, to perform local updates, starting from the perpendicular bisector of line segment $P_m P_{m(m)}$, find the intersection point between the perpendicular bisector and Voronoi cell $V_{m-1}\left(P_{m(m)}\right)$, and determine its adjacent Voronoi cell $V_{m-1}\left(P_{m1(m)}\right)$. Then, draw the perpendicular bisector of edge $P_m P_{m1(m)}$ and

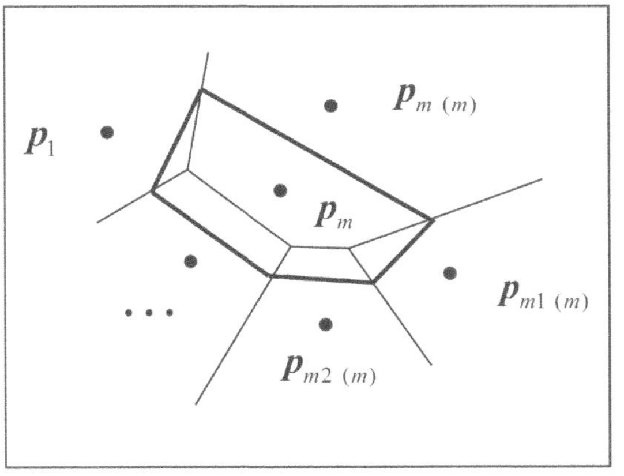


(a)

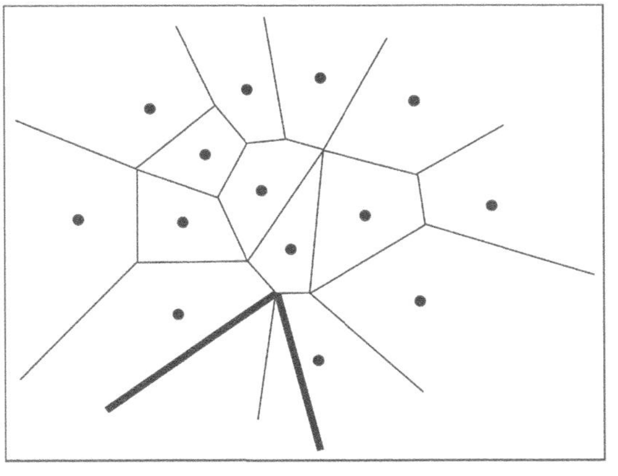

(b)

Figure 1.3: Incremental method: (a) The process of generating Voronoi cells using incremental method; (b) Voronoi diagram generated by incremental method.

find the other intersection point with Voronoi cell $V_{m-1}\left(P_{m1(m)}\right)$ and the adjacent polygon $V_{m-1}\left(P_{m2(m)}\right)$; Repeat this process for a circle to obtain the Voronoi cell $V_m$ of $P_m$.

2. Indirect method

This method first generates a Delaunay triangulation of a discrete growth point set, and then based on the duality property between the Delaunay triangulation and the Voronoi diagram, draws the perpendicular bisectors of each triangle edge. The intersection of all perpendicular bisectors forms the Voronoi diagram of the point set as shown in Fig. 1.4, indirectly generating the Voronoi diagram of the discrete point set.

Fig. 1.4a shows a randomly generated discrete growth point set, and Fig. 1.4b shows the Delaunay triangle network obtained by connecting the point sets. By drawing the perpendicular bisectors of each line segment, the Voronoi diagram shown in

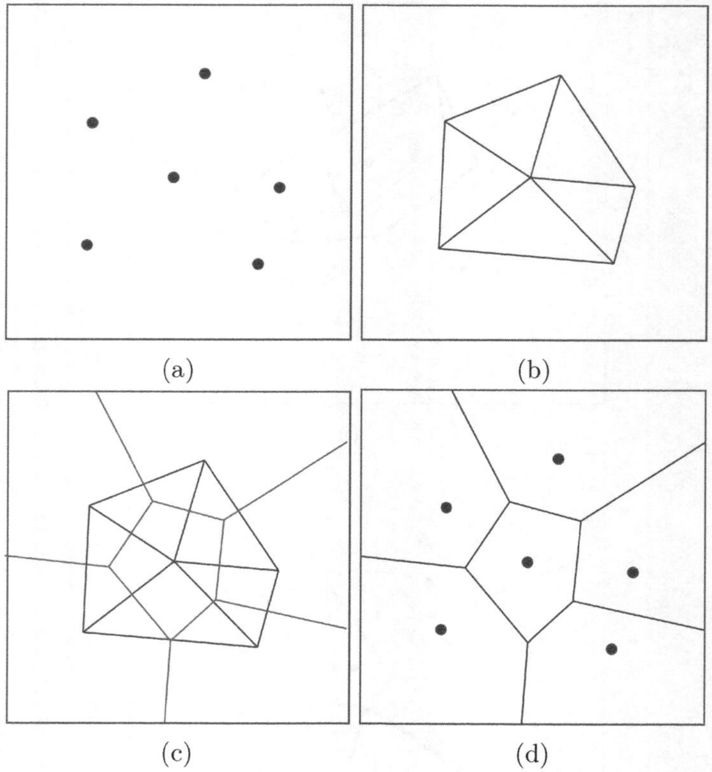

Figure 1.4: Indirect method for generating Voronoi diagram: (a) Discrete growth point set; (b) generating Delaunay triangulation; (c) generating Voronoi diagram from Delaunay triangulation; (d) finally generating Voronoi diagram.

Fig. 1.4c can be obtained. Removing the triangle network yields the Voronoi diagram where the growth point set is located, as shown in Fig. 1.4d.

The construction of Delaunay triangulation is the core of indirect method for generating Voronoi diagram. In order to construct a Delaunay triangulation, it is often required to follow criteria such as maximizing the maximum empty circle, minimizing the minimum angle of the triangle, and minimizing the total side length. The maximum empty circle requires that the circumcircle of each triangle in the Delaunay triangulation does not contain any vertices of other triangles.

3. Divide-and-conquer method

The divide-and-conquer method is another efficient approach for constructing Voronoi diagrams, conceptually similar to divide-and-conquer algorithms used in sorting or Delaunay triangulation. The core idea of this algorithm is to recursively divide a set of $n$ generator points into smaller subsets, compute the Voronoi diagram for each subset, and then merge these sub-diagrams to form the Voronoi diagram for the entire point set, as illustrated in Fig. 1.5. However, the implementation of the divide-and-conquer method is more complex than incremental or indirect approaches, particularly in handling the merging of sub-diagrams and maintaining data structures

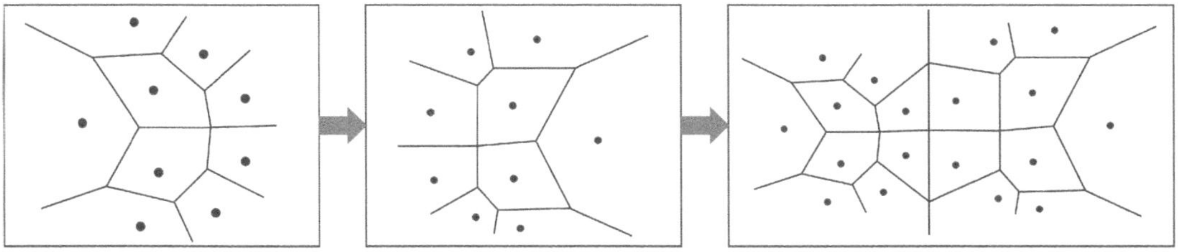

Figure 1.5: The convexity of the Voronoi diagram.

for the boundaries. Furthermore, if points are moved, added, or deleted, updating a Voronoi diagram constructed using the divide-and-conquer method is challenging, as the algorithm inherently requires processing the entire dataset at once. Therefore, this method is valued for its theoretical optimality and is typically applied in static scenarios where all data is known in advance.

4. Parallel method

After the 1980s, with the advancement of parallel processing technology, the Voronoi diagram parallel algorithm also emerged. Its basic idea is to simultaneously calculate all points and finally complete the construction of the entire Voronoi diagram. As for the evaluation of Voronoi diagram generation algorithms, the main considerations are their correctness and efficiency. In the worst-case scenario, the algorithm complexity of the incremental method is $O\left(n^2\right)$, and in general, its running time is $O\left(n^{3/2}\right)$. If further optimized in algorithm, the average time can reach $O(n)$. But its structure is simple and easy to implement dynamically, although its time efficiency is poor, it is still widely used. The time complexity of the divide and conquer method in the worst-case scenario is $O(n \log n)$, but its algorithm is complex and difficult to dynamically update during application due to the implicit need to construct the entire dataset at once.

### 1.2.2 Application Areas of Voronoi Tessellations

The Voronoi tessellation is a classical spatial partitioning technique that uses distance as the optimal criterion to address the question of "which generator governs which area." This method divides space into multiple regions based on a set of generator points, with each region comprising all locations closest to the corresponding generator. It thereby addresses problems related to spatial allocation and nearest-neighbor influence across various scenarios. Characterized by a clear geometric structure, it offers advantages such as high computational efficiency [13], strong scalability in the number of generators [14], and applicability to high-dimensional spaces [15]. With the rapid advancement of technologies in intelligent systems [16], spatial data analysis [17], and computational geometry [18], Voronoi tessellations have demonstrated growing practical value in solving spatial organization and resource allocation problems, establishing themselves as a fundamental tool across multiple research and engineering disciplines. To better appreciate the wide-ranging applicability of

Voronoi tessellations, the following provides a brief overview of their main application scenarios:

(1) Multi-robot Collaborative Control and Task Allocation

Efficient task allocation and coordinated operation are crucial for enhancing the overall performance and intelligence level of multi-robot systems. Voronoi tessellations provide a natural and efficient decentralized spatial task allocation mechanism for such systems. By using the current position of each robot as a generator point for spatial partitioning, the environment can be rationally divided into multiple subregions, with each subregion managed by the nearest robot. This approach explicitly addresses the question of "who is responsible for where" through the Voronoi diagram: each robot corresponds to its assigned Voronoi cell. The method ensures complete, non-overlapping coverage of the task area while preventing redundant operations by multiple robots in the same region. Moreover, the Voronoi boundaries dynamically adjust in real-time as robots move, making this approach particularly suitable for distributed robotic applications that operate without a central controller. Typical applications include environmental monitoring, dynamic patrol, material distribution, agricultural plant protection, and similar tasks [19] [20] [21].

In more complex scenarios, Voronoi diagrams can be extended into weighted or dynamically updated forms to accommodate heterogeneous robotic capabilities or changing environmental requirements. For instance, when employing a weighted Voronoi diagram for, the partition boundaries can be dynamically adjusted not only based on distance but also by incorporating specific parameters such as remaining battery power, movement speed, and load capacity. When a robot operates with low battery or moves at a slower pace, its "influence weight" decreases accordingly, leading it to cede portions of its jurisdictional area to neighboring robots. This enables fair and efficient task allocation based on actual operational capabilities [22]. In dynamic environments, as robots move, new members join, or existing robots depart, the system can update the Voronoi tessellation in real time, enabling continuous adaptive spatial partitioning [23]. This allows the system to automatically respond to changes such as unexpected obstacles, emerging task hotspots, or robot failures by dynamically reassigning coverage areas. The method demonstrates high scalability and significant application potential in scenarios involving frequent disturbances, complex terrain, or real-time task variations. Furthermore, by integrating machine learning (ML) techniques, an adaptive collaborative framework can be established, allowing robots to adjust partition weights or motion strategies based on empirical data, thereby further enhancing the intelligence level and operational efficiency of multi-robot systems [24].

(2) Deployment and Optimization of Wireless Sensor Networks

In wireless sensor networks (WSNs), the placement of sensor nodes directly impacts network coverage quality, energy efficiency, and overall lifecycle. Voronoi diagrams provide an intuitive mathematical tool for analyzing and optimizing such spatial deployment issues: by using the physical location of each sensor as a generator point, the monitoring area can be partitioned into non-overlapping Voronoi cells, with each cell representing the area of responsibility of its nearest sensor. This approach carries clear geometric significance—any point within a given cell is closest to the sensor corresponding to that cell. Designing the network layout based on Voronoi diagrams helps minimize coverage gaps and reduce monitoring redundancy. If sensors are overly densely distributed, their Voronoi cells will shrink and create overlapping coverage, leading to resource wastage; if too sparse, some cells will become excessively large, potentially resulting in coverage voids. Therefore, Voronoi diagrams can assist in initial deployment planning by identifying coverage blind spots and overlapping areas, guiding the adjustment of sensor positions or the deployment of additional nodes, thereby enhancing the overall balance and robustness of the deployment.

Voronoi tessellations also hold significant value in Mobile Sensor Networks (MSNs). Sensor nodes in such networks can dynamically adjust their positions to optimize coverage effectiveness. The adaptive strategy based on dynamic Voronoi tessellations guides sensor nodes to optimize their distribution according to environmental changes, target movements, or node failures. Each node periodically calculates its Voronoi cell relative to neighboring nodes—if a cell is found to be excessively large or if monitoring blind spots are detected within the cell, the node can autonomously move toward the uncovered area. Conversely, if multiple sensors cover overlapping areas, some nodes may reposition themselves or reduce their operational frequency to conserve energy. This approach ensures that sensors adaptively fill coverage gaps while avoiding resource clustering, achieving near-optimal area coverage and resource utilization in real time. This mechanism is particularly suitable for application scenarios requiring high spatial coverage and frequent environmental changes, such as environmental monitoring, disaster response, and smart cities [25] [26]. Meanwhile, Voronoi tessellations is often used in conjunction with power control, routing selection, and data fusion algorithms, forming a coupled mechanism that integrates spatial perception with communication collaboration. In recent years, researchers have also introduced weighted Voronoi models in Heterogeneous Sensor Networks (HSNs), enabling nodes to manage regions based on sensor type, remaining energy, or task priority while considering distance factors, thereby further enhancing system performance and intelligence levels.

(3) Robot Path Planning and Navigation

Voronoi tessellations are applicable not only to static spatial allocation but also to path planning and roadmap construction in robot navigation. A classic application in robotics involves constructing the corresponding Voronoi regions in environments containing obstacles. By using obstacle boundaries as generating elements and computing the Voronoi diagram of the obstacle boundary set, a path skeleton that traverses free space while maximizing obstacle avoidance distance can be obtained. Robots can follow this skeletal path, thereby maximizing their clearance from all obstacles. Such paths are referred to as "maximum clear path" or "safe path," as every point along the path is farthest from the nearest obstacle. This is highly valuable for applications with stringent safety requirements, such as autonomous driving [27], hazardous area inspection [28], and underground space navigation [29]. The Voronoi path inherently possesses obstacle avoidance capabilities, as its generation mechanism ensures that the path maintains the maximum possible distance from all obstacle boundaries, thereby enhancing the reliability and stability of robot navigation in complex environments.

Furthermore, the application of Voronoi tessellations in path planning demonstrates excellent scalability and environmental adaptability. In dynamic environments, when obstacles or target positions change, the Voronoi region can be updated in real time through local reconstruction or global updates, thereby providing mobile robots with timely and reliable motion path information. Compared to traditional grid-based algorithms such as A* or Dijkstra, the sparse topological graph structure constructed by Voronoi tessellations significantly reduces the complexity of the search space, particularly exhibiting higher computational efficiency and path planning performance in large-scale environments. This structure not only effectively captures the topological features of the environment but also avoids redundant computations associated with dense grids, thereby improving planning speed while ensuring path rationality. Additionally, Voronoi tessellations are often embedded as auxiliary structures in other path planning algorithms to optimize the quality and efficiency of generated paths. such as introducing it to guide the path generation direction in sampling algorithms such as RRT [30] and PRM [31], improving convergence efficiency and path quality. In multi-robot path coordination, combining Voronoi tessellation with path skeleton can achieve collision avoidance and local collaboration between robots.

(4) Computational Geometry and Image Segmentation

In the field of computational geometry, Voronoi tessellations serve as a fundamental and powerful spatial partitioning tool, widely employed in critical tasks such as spatial modeling, nearest-neighbor queries, point set reconstruction, spatial interpolation, and shape analysis. For instance, within Delaunay triangulation, the Voronoi area generated by Voronoi tessellations are often used in conjunction with their dual structures to construct efficient spatial indexes and topological maps [32]. Moreover, Voronoi tessellations demonstrate broad applicability in practical domains such as geographic information systems (GIS), urban planning, and meteorological simulation, where they can be utilized to model influence areas, resource distributions, or service regions, effectively addressing geometric optimization problems like "region ownership" and "nearest-neighbor distribution" [33]. Furthermore, Voronoi tessellations possess strong visualization capabilities. Their clear structure and well-defined boundaries allow for an intuitive representation of spatial data distribution characteristics and regional partitioning effects, making them highly suitable for various scenarios including spatial analysis, interactive design, and scientific visualization. By translating abstract spatial relationships into concrete geometric units, Voronoi tessellations make complex data structures more intuitively understandable and computationally tractable.

In the field of image processing and computer vision, Voronoi tessellations provides an image partition method based on spatial distance, widely applied to tasks such as image segmentation, feature extraction, texture synthesis, and color quantization. Overall, as a versatile tool in computational geometry and computer vision, Voronoi tessellations offer clear solutions to proximity assessment and spatial partitioning problems, while significantly aiding in the analysis and visualization of complex spatial data.

In addition to the application scenarios discussed in detail above, Voronoi diagram, as a basic and powerful spatial segmentation tool, its value and potential are also widely reflected in many other frontier fields and interdisciplinary subjects. For example in biology, Voronoi diagrams (VD) have significant applications in understanding the spatial organization of natural systems, especially in studying canopy segregation in forests. Canopy segregation refers to the spatial separation of tree crowns, which influences competition for resources like sunlight and water. By modeling each tree as a generating point, Voronoi diagrams partition the surrounding space into regions that represent the tree's influence. This spatial partitioning helps reveal how different tree species are distributed and segregated by their crown boundaries, providing insights into interspecies competition and how trees optimize resource use by maintaining physical distance [34].

In the case of multi-species forests, Voronoi diagrams can help reveal how different species of trees are spatially distributed and segregated by the physical boundaries of their crowns. This spatial partitioning aids in understanding the mechanisms of interspecies competition, as well as how trees reduce competition by maintaining physical distance through their crown spread. Researchers use this method to study the dynamics of tree growth and resource optimization at the community level, illustrating how trees organize themselves within the forest to optimize access to light and reduce shading on neighboring trees. Through these insights, Voronoi diagrams provide valuable tools for forest management and ecological studies, helping to optimize forest composition and biodiversity.

## 1.3 Contributions and Outline of This Book

### 1.3.1 The Contributions of This Book

This book systematically investigates the application of Voronoi tessellations in intelligent control of multi-robot systems, aiming to provide researchers and engineering practitioners with a systematic framework spanning from theory to practice. Compared to existing studies that focus solely on individual application scenarios or algorithmic analysis, this book achieves a comprehensive integration of theoretical foundations, algorithmic design, and experimental validation, with its main contributions summarized as follows:

(1) **Theoretical and system construction:** Starting from geometric foundations, this book systematically analyzes the existence, uniqueness, and construction methods of Voronoi tessellations in two-dimensional, three-dimensional, and non-convex environments, providing theoretical support and practical reference for the design and application of Voronoi tessellation strategies in multi-robot tasks.

(2) **Multi-robot formation and collaborative algorithm design:** By introducing a collaborative control framework based on centroidal Voronoi tessellations, a density function guided method for multi-robot formation and formation switching is proposed. During task execution, the system can recalculate the Voronoi tessellations and their centroids in real-time based on updated density functions, enabling the robot formation to achieve smooth and coordinated formation transitions. This

approach significantly enhances the overall flexibility and stability of the formation during operation.

(3) **Security modeling and obstacle avoidance strategy design:** A multi-robot formation safety strategy has been developed based on buffer zones, Buffered Uncertainty Voronoi Cells (BUVC) and a separation hyperplane mechanism. By integrating Voronoi tessellations, this approach fundamentally optimizes the obstacle avoidance performance of multi-robot systems in non-convex environments. Even when confronted with complex configuration spaces formed by irregular obstacles or walls, the system maintains superior obstacle avoidance capabilities.

(4) **Application extension of task oriented Voronoi tessellations:** Based on the traditional multi-robot cooperative control, this book further extends the Voronoi tessellations technology to multi-robot task fields such as source search, coverage control, target capture, and system health optimization. The construction of adaptive algorithm frameworks for different tasks and the systematic processing of these tasks have fully demonstrated the universality and scalability of the Voronoi-based control method in heterogeneous task scenarios.

(5) **Integration verification of physical robotic platform and ROS:** This book conducts a comprehensive physical platform integration and verification of the proposed algorithm through the Robot Operating System (ROS). it details the development of a multi-robot testbed built upon platforms such as TurtleBot3 and Crazyflie, along with the establishment of a complete simulation and experimental system within the ROS environment. The system accomplished tasks including collaborative formation control, formation switching, obstacle avoidance, source seeking, and capture tasks, thereby providing engineering level reproducibility and practical demonstration for the research results.

### 1.3.2 The Outline of This Book

This book consists of 11 chapters, each chapter progressively exploring the Voronoi tessellations theory and its applications in multi-robot systems. The main contents of each chapter are as follows:

Chapter 1 outlines the research background and motivations of this work, highlighting the importance of multi-robot swarm control. It provides a systematic description of the fundamental definitions and generation methods of Voronoi tessellations, reviews their broad applications across various engineering and scientific fields, and establishes a theoretical foundation for subsequent chapters.

Chapter 2 focuses on the geometric properties and computational methods of Voronoi tessellations. It provides an in-depth mathematical analysis of the essential characteristics of Voronoi tessellations, including proofs and discussions on the existence and uniqueness conditions given a set of points. The chapter also examines the construction algorithms and applicability of Voronoi cells in two-dimensional, three-dimensional, and non-convex environments, thereby establishing a geometric foundation for algorithm design and theoretical derivations in subsequent chapters.

Chapter 3 focuses on how to effectively integrate Voronoi tessellations with multi-robot systems to achieve collaborative control and formation generation. By constructing an algorithm model based on centroid Voronoi tessellations, each robot can self organize into formations according to spatial distribution, and the effectiveness of the algorithm in two-dimensional and three-dimensional environments can be verified through simulation, providing support for subsequent dynamic formations and obstacle avoidance mechanisms.

Chapter 4 is about multi-robot that formation needs to be flexibly adjusted according to task changes. This chapter proposes a formation switching strategy based on dynamic density function guidance, which combines the density interpolation and centroid guidance mechanism of the Voronoi cell to achieve smooth and efficient formation transition. Simultaneously verify the stability and adaptability of this method in complex environments through various simulation scenarios.

Chapter 5 proposes a series of safety strategies, including weighted buffer zone Voronoi tessellation, BUVC model, and buffer interval segment, to prevent collisions between robots and between robots and obstacles. Furthermore, it analyzes obstacle avoidance strategies in non-convex environments by integrating methods such as the dynamic window approach, separation hyperplanes, and artificial potential fields, thereby comprehensively enhancing the overall security of the system.

Chapter 6 focuses on the application of multi-robot systems in pollution source search and area coverage tasks. It proposes a pollution source search algorithm based on sparse Gaussian process regression, which is integrated with confidence upper bound guided centroidal Voronoi tessellation to achieve efficient source localization and estimation. For coverage control, a distributed exploration strategy is designed to enable the robot team to balance uniform coverage and adaptive focus in unknown environments. Simulation results validate the effectiveness and robustness of the proposed methods, demonstrating that the multi-robot system can successfully accomplish both pollution source localization and area coverage even under uncertainties and dynamic changes.

Chapter 7 proposes an optimization and health management framework for multi-robot systems, which integrates traditional control methods with system-level considerations such as resource allocation and fault tolerance. This framework combines a health management mechanism with weighted centroidal Voronoi tessellation, enabling robots to autonomously adjust their task roles or coverage areas based on real-time health status, thereby achieving load balancing and enhanced fault tolerance. The chapter also investigates coverage control strategies under different density functions. By adjusting the density distribution, the framework optimizes long-term mission performance and maintains coverage stability in the event of robot failures or the addition of new robots, ensuring sustained performance and robustness of the multi-robot system in complex tasks.

Chapter 8 focuses on the problem of target capture in multi robot systems, exploring capture strategies for both known and unknown target positions. In the case where the target position is known, the Omni-directional minimum volume (ODMV) is proposed, combined with Voronoi tessellations to achieve efficient capture and convergence guarantee based on multi-robot systems. In a scenario where the target location is unknown, a perception driven capture mechanism was designed to achieve dynamic tracking and uniform encirclement of unknown capture targets through target perception and collaborative decision-making. The effectiveness of the method was verified through simulation analysis.

Chapter 9 combines deep reinforcement learning with Voronoi diagrams and proposes a multi-robot cooperative capture strategy based on the MADDPG algorithm. This chapter constructs a multi-dimensional fusion framework, organically integrating the spatial partitioning characteristics of Voronoi diagrams with the decision-making optimization capabilities of deep reinforcement learning. Through a centralized training and decentralized execution architecture, agents can learn complex cooperative behavior strategies while maintaining the advantages of distributed control during execution. Simulation experiments in the PettingZoo environment demonstrate that this method significantly improves the success rate of encirclement capture and the stability of system convergence, providing a new solution for multi-robot cooperative control in dynamic and uncertain environments.

Chapter 10 is based on the Robot Operating System (ROS) and real-world robotic platforms. Building upon the simulation and theoretical research of previous chapters, it details the deployment process of multi-robot collaborative control algorithms in physical environments. Physical experiments for tasks such as formation switching, dynamic obstacle avoidance, source search, and target capture were conducted, validating the engineering usability and system stability of the proposed algorithms in real-world scenarios.

Chapter 11 provides a comprehensive summary of the book's content, reviewing the complete process of Voronoi tessellation theory from geometric modeling and algorithm design to engineering implementation in multi-robot systems. It also delves into the limitations and challenges of current research, discussing issues such as dynamic environment modeling, intelligent gaming, multi-task collaboration, and human-machine integration. By outlining these development trends, this final chapter aims to inspire further innovation based on the foundation laid in this book and point the way forward for advancing research in multi-robot swarm intelligence.

# Bibliography

[1] J Cruz. Leader-follower strategies for multilevel systems. *IEEE Transactions on Automatic Control*, 23(2):244–255, 2003.

[2] Jernej Čamernik, Rebeka Kropivšek Leskovar, and Tadej Petrič. Leader–follower dynamics in complex obstacle avoidance task. *International Journal of Social Robotics*, 15(1):59–70, 2023.

[3] Ziwen Yang, Shanying Zhu, Cailian Chen, Gang Feng, and Xinping Guan. Leader-follower formation control of nonholonomic mobile robots with bearing-only measurements. *Journal of the Franklin Institute*, 357(3):1628–1643, 2020.

[4] Kar-Han Tan and M Anthony Lewis. Virtual structures for high-precision cooperative mobile robotic control. In *Proceedings of IEEE/RSJ International Conference on Intelligent Robots and Systems. IROS'96*, volume 1, pages 132–139. IEEE, 1996.

[5] Guanqun Liu, Naifeng Wen, Feifei Long, and Rubo Zhang. A formation control and obstacle avoidance method for multiple unmanned surface vehicles. *Journal of Marine Science and Engineering*, 11(12):2346, 2023.

[6] Yinshuang Sun, Zhijian Ji, Yang Shi, and Yungang Liu. Event-based finite time stabilizability and formation control of multi-agent systems. *IEEE Transactions on Automation Science and Engineering*, 2024.

[7] Liang Cao, Zhijian Cheng, Yang Liu, and Hongyi Li. Event-based adaptive NN fixed-time cooperative formation for multiagent systems. *IEEE Transactions on Neural Networks and Learning Systems*, 2022.

[8] Shijie Dai, Zhengxing Wu, Pengfei Zhang, Min Tan, and Junzhi Yu. Distributed formation control for a multi-robotic fish system with model-based event-triggered communication mechanism. *IEEE Transactions on Industrial Electronics*, 70(11):11433–11442, 2023.

[9] Jianye Gong, Bin Jiang, Yajie Ma, and Zehui Mao. Distributed adaptive fault-tolerant formation control for heterogeneous multiagent systems with communication link faults. *IEEE Transactions on Aerospace and Electronic Systems*, 59(2):784–795, 2022.

[10] Kun Wang, Chih-Min Yu, Meng-Lin Ku, Li-Chun Wang, and Wen-Kang Jia. Joint shortest chain and fair transmission design for energy-balanced PEGASIS in WSNs. *IEEE Internet of Things Journal*, 10(8):6803–6817, 2022.

[11] Jorge Cortes, Sonia Martinez, Timur Karatas, and Francesco Bullo. Coverage control for mobile sensing networks. *IEEE Transactions on Robotics and Automation*, 20(2):243–255, 2004.

[12] René Descartes. *Principia philosophiae*, volume 1. ex typographia Blaviana... Sumptibus Societatis, 1905.

[13] Prathu Bajpai and Jagdish Chand Bansal. Voronoi Tessellations based simple optimizer. *Information Sciences*, 676:120795, 2024.

[14] Andreas Andreou, Constandinos X Mavromoustakis, Jordi Mongay Batalla, Evangelos K Markakis, and George Mastorakis. UAV-assisted RSUs for V2X connectivity using Voronoi diagrams in 6G+ infrastructures. *IEEE Transactions on Intelligent Transportation Systems*, 24(12):15855–15865, 2023.

[15] Bhagyashri Telsang and Seedik M Djouadi. Computation of centroidal Voronoi tessellations in high dimensional spaces. *IEEE Control Systems Letters*, 6:3313–3318, 2022.

[16] He-Lin Gong, Han Li, Dunhui Xiao, and Sibo Cheng. Reactor field reconstruction from sparse and movable sensors using Voronoi tessellation-assisted convolutional neural networks. *Nuclear Science and Techniques*, 35(5):43, 2024.

[17] Amy Richter, Kelvin Tsun Wai Ng, and Nima Karimi. Meshing centroidal Voronoi tessellation with spatial statistics to optimize waste management regions. *Journal of Cleaner Production*, 295:126465, 2021.

[18] Xinqiao Duan, Lin Li, Yong Ge, and Bo Liu. Exact Voronoi diagram for topographic spatial analysis. *GIScience & Remote Sensing*, 60(1):2171703, 2023.

[19] Alaa Z Abdulghafoor and Efstathios Bakolas. Motion coordination of multi-agent networks for multiple target tracking with guaranteed collision avoidance. *Journal of Intelligent & Robotic Systems*, 107(1):5, 2023.

[20] Po-wei Harn, Ji Zhang, Ting Shen, Wenlu Wang, Xunfei Jiang, Wei-Shinn Ku, Min-Te Sun, and Yao-Yi Chiang. Multiple ground/aerial parcel delivery problem: a weighted road network Voronoi Diagram based approach. *Distributed and Parallel Databases*, pages 1–21, 2023.

[21] Shailendra Tahilyani, Surabhi Saxena, Dimitrios A Karras, Shashi Kant Gupta, Chandra Kumar Dixit, and Bhadrappa Haralayya. Deployment of autonomous vehicles in agricultural and using Voronoi partitioning. In *2022 International Conference on Knowledge Engineering and Communication Systems (ICKES)*, pages 1–5. IEEE, 2022.

[22] Hao Cheng, Bailing Tian, Xuewei Zhang, and Hongming Shen. A distributed algorithm for multi-robot task allocation via weighted buffered Voronoi partition. *IEEE Transactions on Industrial Electronics*, 2024.

[23] Lei Zuo, Yang Shi, and Weisheng Yan. Dynamic coverage control in a time-varying environment using Bayesian prediction. *IEEE Transactions on Cybernetics*, 49(1):354–362, 2017.

[24] Hongyang Zhao, Yanan Guo, Yi Liu, and Jing Jin. Multi-robot unknown environment exploration and obstacle avoidance based on a Voronoi diagram and reinforcement learning. *Expert Systems with Applications*, 264:125900, 2025.

[25] Dharmendra Singh, Meenakshi Dahiya, Rahul Kumar, and Chintan Nanda. Sensors and systems for air quality assessment monitoring and management: A review. *Journal of Environmental Management*, 289:112510, 2021.

[26] Omer Melih Gul, Aydan Muserref Erkmen, and Burak Kantarci. UAV-driven sustainable and quality-aware data collection in robotic wireless sensor networks. *IEEE Internet of Things Journal*, 9(24):25150–25164, 2022.

[27] Chengke Xiong, Danfeng Chen, Di Lu, Zheng Zeng, and Lian Lian. Path planning of multiple autonomous marine vehicles for adaptive sampling using Voronoi-based ant colony optimization. *Robotics and Autonomous Systems*, 115:90–103, 2019.

[28] Bo Li, Pingfa Feng, Long Zeng, Chao Xu, and Jianfu Zhang. Path planning method for on-machine inspection of aerospace structures based on adjacent feature graph. *Robotics and Computer-Integrated Manufacturing*, 54:17–34, 2018.

[29] Ran Zhang, Guangbo Hao, Kong Zhang, and Zili Li. Unmanned aerial vehicle navigation in underground structure inspection: A review. *Geological Journal*, 58(6):2454–2472, 2023.

[30] Jiankun Wang and Max Q-H Meng. Optimal path planning using generalized Voronoi graph and multiple potential functions. *IEEE Transactions on Industrial Electronics*, 67(12):10621–10630, 2020.

[31] Ivan Palacios Serrano, Christyan Cruz Ulloa, and Manuel Barraza Rodríguez. Análisis de los algoritmos de planificación de trayectorias RRT, PRM y Voronoi en la solución de un laberinto modular controlado por una plataforma de dos gdl. *Ingeniare. Revista chilena de ingeniería*, 30(1):157–170, 2022.

[32] Rivka Gitik and Leo Joskowicz. Voronoi diagram and Delaunay triangulation with independent and dependent geometric uncertainties. *International Journal of Computational Geometry & Applications*, 31(02n03):75–121, 2021.

[33] Alessandro Bosisio, Alberto Berizzi, Marco Merlo, Andrea Morotti, and Gaetano Iannarelli. A GIS-based approach for primary substations siting and timing based on Voronoi diagram and particle swarm optimization method. *Applied Sciences*, 12(12):6008, 2022.

[34] Begoña Abellanas Oar, Miren del Río, and Fernando Montes. A forest simulation approach using weighted Voronoi diagrams: An application to mediterranean fir abies pinsapo boiss stands. *Forest Systems*, 25(2):6, 2016.

CHAPTER 2

# Geometric Properties and Computational Methods of Voronoi Tessellations

## 2.1 Introduction

THIS chapter explores the geometric properties and computational methods of Voronoi tessellations, with a focus on their application in multi-robot systems. Building upon the fundamental concepts of Voronoi diagrams introduced in Chapter 1, this chapter focuses on the geometric properties and computational methods of Voronoi tessellations, which are pivotal in dividing space into distinct regions of responsibility for multi-robot systems. Voronoi tessellations offer a natural framework for organizing robots in space, ensuring non-overlapping task areas, and enabling effective collaboration. The chapter begins by exploring essential geometric properties such as convexity, boundary consistency, and the empty circle property [1], while analyzing how these properties influence the performance of multi-robot systems. Furthermore, the dual relationship between Voronoi diagrams and Delaunay triangulations is examined, which provides a deeper understanding of how these structures complement each other in spatial analysis, robot coordination, and network connectivity. Additionally, various methods for generating Voronoi diagrams—both direct and indirect—are discussed, highlighting the computational efficiency and scalability of these methods, especially in large-scale, dynamic environments. The chapter provides a comprehensive foundation for understanding Voronoi tessellations and sets the stage for their practical application in multi-robot control.

## 2.2 Geometric Features and Shape Analysis of Voronoi Tessellations

As a region at tessellations method, the geometric characteristics of Voronoi tessellations determine its practical application effect in multi robot systems [2]. From the mathematical essence of spatial partitioning, the Voronoi diagram not

DOI: 10.1201/9781003714569-2

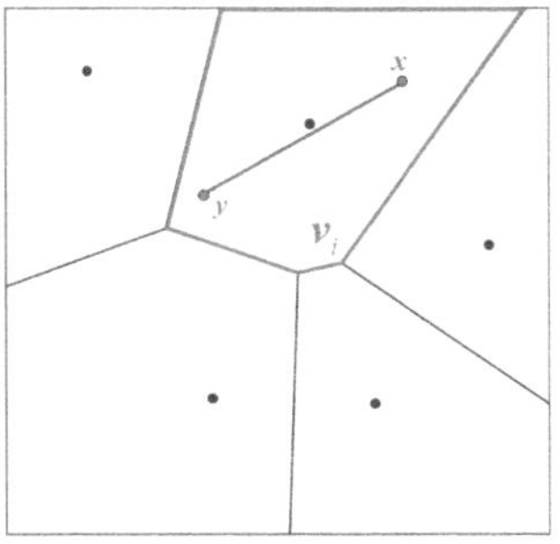


Figure 2.1: The convexity of the Voronoi diagram.

only provides an efficient method for environmental partitioning but also its unique geometric structure, such as convexity, circularity, adjacency topology, etc. [3] [4]. Throughout this book, we will show that VD directly determines the performance of robots in tasks such as coverage control, path planning, and collaborative obstacle avoidance. This section will discuss in detail the geometric properties of the Voronoi tessellations:

1. Convexity of cells

Convexity refers to the fact that in a Voronoi tessellations, each Voronoi cell is convex. The mathematical explanation is: for Voronoi cells $v_i$, any two points $x, y \in v_i$, their connection $\lambda x + (1 - \lambda)y$, that ($\lambda \in [0, 1]$) still belongs to $v_i$, as shown in Fig 2.1, where a schematic diagram of the convexity of the Voronoi diagram is illustrated. Among them, the blue line is the line connecting any two points $x$ and $y$ within the Voronoi cell.

Let $\{X_i \mid i \in I\}$ be an arbitrary collection of convex sets, and let $X$ denote their intersection. Consider arbitrary two points $p$ and $q$ in $X$. By the definition of intersection, these points lie in every set $X_i$ because every $X_i$ is convex, the entire segment $pq$ lies in every set $X_i$ and therefore in their intersection $X$ too. Thus, $X$ is convex.

In Voronoi tessellations, each cell is convex, meaning the straight line between any two points within it lies entirely inside the cell. This geometric property is crucial for collision avoidance in multi-robot systems. By ensuring each robot operates within a convex task area, the system guarantees that regardless of how robots move inside their designated regions, their paths will not cross the cell boundaries, thereby preventing intersections with adjacent cells and avoiding potential collisions. This significantly reduces coordination complexity, as robots only need to avoid the boundaries within their own convex region.

2. Empty circle property

For each Voronoi cell's vertex $\boldsymbol{w}_i$ of a Voronoi diagram, there exists a unique circle $C_i$ centered on $\boldsymbol{w}_i$, which satisfies the following conditions: passing through the three generated points $\boldsymbol{p}_{i1}$, $\boldsymbol{p}_{i2}$, and $\boldsymbol{p}_{i3}$ corresponding to that edge; $C_i$ does not contain any other generated points. In the two-dimensional case, this circle is called an "Empty Circle" or "Maximal Empty Circle." The schematic diagram of Voronoi's empty circle

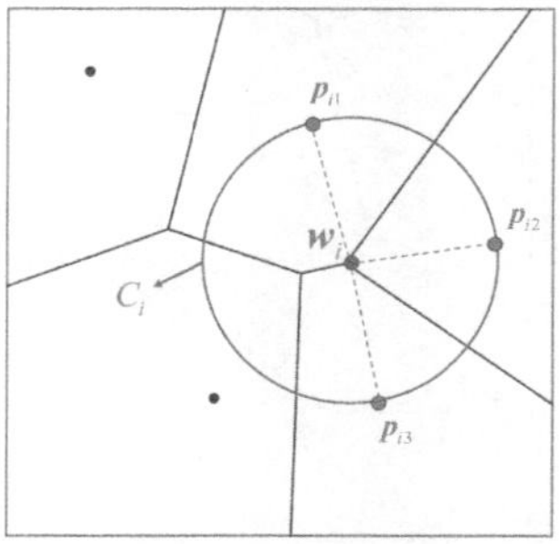


Figure 2.2: The empty circle property of the Voronoi diagram.

property is shown in Fig 2.2. The blue dots in the figure represent the vertices of the Voronoi cell, which is the center of circle $C_i$, while the red dots represent the three Voronoi cell generation points that circle $C_i$ passes through.

Due to a Voronoi vertex $\boldsymbol{w}_i$ consisting of three or more Voronoi cells $V_{i1}, \ldots, V_{ik}, k \geq 3$, together they form three or more Voronoi edges $\boldsymbol{e}_{(1,2)}, \ldots, \boldsymbol{e}_{(k,1)}$ that intersect at point $\boldsymbol{w}_i$. According to the definition of the Voronoi edge, each edge $e\left(p_{ij}, p_{ij+1}\right)$ is located on the perpendicular bisector $b\left(p_{ij}, p_{ij+1}\right)$ between the generation points $p_{ij}$ and $p_{ij+1}$. Therefore, the distance from vertex $\boldsymbol{w}_i$ to these generated points is the same, i.e. $\overline{q_i p_{ij}} = \overline{q_i p_{ij+1}}$. This means generating points $\boldsymbol{p}_{i1}, \ldots, \boldsymbol{p}_{ik}$ form a circle, denoted as $C_i$.

Assuming there is another generation point $p_i \in \mathbf{P} \backslash \{p_{il}, \ldots, p_{ik}\}$ located inside the circle $C_i$, then: $\overline{q_i p_i} < \overline{q_i p_{ij}}$ for all $j = 1, \ldots, k$. According to the definition of the Voronoi diagram, $\boldsymbol{q}_i$ should belong to the Voronoi cell of $\boldsymbol{v}_i$, but this contradicts the premise that $q_i$ is only shared by $V_{i1}, \ldots, V_{ik}, k \geq 3$. Therefore, the circle $C_i$ cannot contain any other generated points, meaning $C_i$ is an empty circle.

The empty circle property states that for any Voronoi cell's vertex, there is a unique circle centered at the vertex that passes through the three nearest neighboring robots, with no other robots inside this circle. This property optimizes path planning by maximizing the distance between robots. By guiding each robot toward the centroid of its Voronoi cell, robots maintain optimal spacing, avoid clustering, and maximize coverage, ensuring the entire area is covered without overlap.

Based on the analysis of the geometric properties of the Voronoi diagram above, it can be concluded that convex hull ensures that each Voronoi cell is a convex region. This means that in multi-robot formation tasks, each robot only needs to focus on its adjacent region, simplifying the design of path planning and collision avoidance strategies. Moreover the circularity ensures the geometric symmetry and accuracy at the Voronoi boundary, allowing the robot to maintain good balance and communication coverage in the boundary distribution. These two properties provide a strong geometric foundation for multi-robot systems, helping to achieve efficient area allocation, autonomous coverage, and collaborative control. In addition to these basic geometric characteristics at the Voronoi tessellations, there are many rich features that will be described in the subsequent sections.

## 2.3 Basic Properties of Voronoi Tessellations

In multi-robot systems, a reasonable division of task space has a critical impact on system performance [4], and Voronoi tessellations is a spatial segmentation method based on geometric distance, with a clear mathematical structure, naturally suitable for local perception and collaborative control in multi-robot scenes [5]. This section will delve into the five fundamental properties of Voronoi tessellations in multi-robot systems: locality, uniqueness and existence, boundary consistency, robustness and dynamic adaptability, and scalability, providing theoretical support for subsequent multi-robot systems to perform tasks in different scenarios.

### 2.3.1 Locality of Voronoi Tessellations

The most prominent feature of the Voronoi tessellations is its locality. The construction of each Voronoi cell relies solely on the geometric relationship between its corresponding generation point and adjacent generation points. Given task environment $Q \in \mathbb{R}^d$, let us use $\boldsymbol{P} = \{\boldsymbol{p}_1, \boldsymbol{p}_2, \cdots \boldsymbol{p}_n\}$ to represent the generation point of the Voronoi cells, and $V = \{v_1, v_2, \cdots v_n\}$ to represent the set of Voronoi cells. As shown in Fig. 2.3, the spatial tessellations area $\boldsymbol{v_i}$ that any robot $i$ in the system is responsible for is determined only by its generation point $\boldsymbol{p_i}$ and nearby generation points $\boldsymbol{p_j}$, without the global coordinates or information from other robots. This locality is particularly important in distributed multi-robot control. This approach ensures that robots only need to rely on local information, such as the positions of neighboring robots, to update their task areas. In a multi-robot system, this property allows robots to adjust to changes in their environment without needing global information, making it suitable for decentralized control. For example, in distributed coverage control tasks, robots can dynamically update their Voronoi cells by communicating with adjacent robots, in response to changes in the target area or movements of other robots. In addition, the locality of Voronoi tessellations also means that multi-robot systems have a high degree of adaptability when encountering node failures or topological changes. When a robot fails or goes offline, other robots can automatically expand their task area and fill the gap, without need to recalculate. Therefore, locality is not only an abstract attribute of Voronoi geometric tessellations but also a

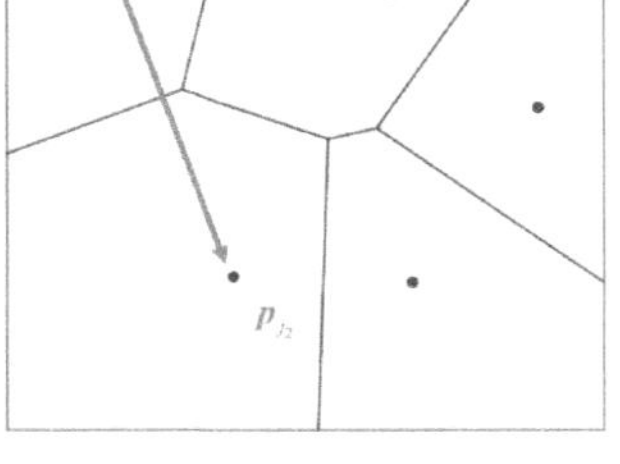

Figure 2.3: Locality of Voronoi tessellations.

key support for achieving task decoupling, computation distribution, communication minimization, and system resilience improvement in the actual deployment process of multi-robot. The multi-robot formation control, pollution source search, target encirclement, and other task drivers mentioned in subsequent chapters are all based on this property and serve as the foundation for the structural design of multi-robot systems.

### 2.3.2 Existence and Uniqueness of Voronoi Tessellations

As a distance based spatial tessellation method, Voronoi tessellations have a good mathematical structure. Its existence and uniqueness not only lay the foundation for geometric theory but also provide stable theoretical guarantees for spatial task tessellations in multi-robot systems. First, for any set of finite and non-overlapping Voronoi cells generating points $\boldsymbol{P} = \{\boldsymbol{p}_1, \boldsymbol{p}_2, \cdots \boldsymbol{p}_n\}$, a set of non-empty Voronoi cells $V = \{v_1, v_2, \cdots v_n\}$ can always be constructed in the Euclidean space, where each Voronoi cell $v_i$ is a closed convex body composed of the intersection of a series of hyperplane half spaces, as shown in Fig. 2.4. Due to the continuous and strictly monotonic nature of the distance function in the Euclidean space, for non-overlapping point sets, each corresponding $v_i$ is non empty, and the partition of the entire space is complete, satisfying:

$$\bigcup_{i=1}^{n} v_i = \boldsymbol{R}^d, \quad v_i \cap v_j = \varnothing \text{ if } i \neq j. \tag{2.1}$$

Therefore, the Voronoi tessellations always exist in theory, even in high-dimensional spaces. This property ensures that multi-robot systems can construct stable tessellation structures at any initial position, providing strong mathematical support for system initialization, self-organizing deployment, and task balancing.

The uniqueness of Voronoi is based on the premise that the generating points do not coincide. As long as $\left\| \boldsymbol{p}_i - \boldsymbol{p}_j \right\| > 0, \forall i \neq j$ is satisfied, the generated Voronoi tessellations structure is uniquely determined in terms of geometric structure. In the sense of distance, each point $\boldsymbol{q}$ belongs to the nearest generating point corresponding to the Voronoi cell, and therefore the boundary of each unit is uniquely determined

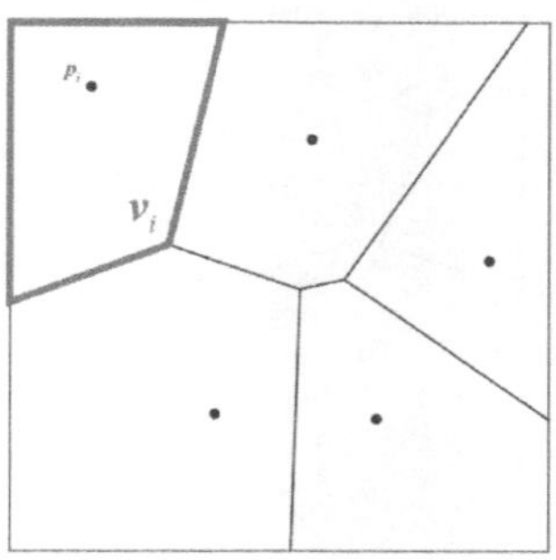


Figure 2.4: Existence and uniqueness of Voronoi tessellations.

by the perpendicular bisector hyperplane between the generating points. Under ideal conditions without floating-point errors or metric sharpening, the boundary of each Voronoi cell is stable and irreplaceable. For multi-robot systems, the existence and uniqueness of Voronoi tessellations mean that the task partitioning has clear and stable boundaries. Regardless of the initial state of the system, partitioning always exists and does not cause "area drift" or "task overlap" issues due to differences in the number or initial distribution of robots. Each robot can independently calculate its corresponding area based on its current position, ensuring the repeatability and invariance of system operations.

Voronoi tessellations are guaranteed to exist and be unique as long as the generator points are distinct. This property ensures that each robot has a clear, non-overlapping area of responsibility. In complex environments, such as those with irregular obstacles or varying terrain, Voronoi diagrams provide a stable partitioning of the space that adapts to robot movements and environmental changes without ambiguity. This guarantees the system's overall stability, even as the environment or robot configuration evolves. The uniqueness and existence of Voronoi tessellations ensure that task areas are well-defined, eliminating conflicts in multi-robot systems. Robots can confidently operate within their designated areas, knowing the partition will remain consistent and adaptable to environmental changes.

### 2.3.3 Boundary Consistency of Voronoi Tessellations

In multi-robot systems, the accuracy of boundary tessellations directly determines the effectiveness and safety of task collaboration. Boundary consistency is an important property inherent in the geometric construction of Voronoi tessellations, which refers to the shared boundaries between adjacent Voronoi cells being completely consistent in geometric definition, unambiguous, redundant, and seamlessly connected. In the Voronoi tessellations, if there is a shared boundary between two generating points $\boldsymbol{p}_i$ and $\boldsymbol{p}_j$, the boundary is defined by their mid vertical plane $H_{ij}$ instead:

$$H_{ij} = \left\{ \boldsymbol{q} \in Q \mid \|\boldsymbol{q} - \boldsymbol{p}_i\| = \left\|\boldsymbol{q} - \boldsymbol{p}_j\right\|, \quad i, j = 1, 2, \cdots n, \quad i \neq j \right\}. \tag{2.2}$$

Among them, the vertical plane is defined by the line connecting $\boldsymbol{p}_i$ and $\boldsymbol{p}_j$ as the normal vector. Through the hyperplane of its midpoint, all points in space are strictly divided into two regions near $i$ and near $j$. The specific effect is shown in Fig. 2.5. Due to the symmetry and continuity of Euclidean distance, this boundary has the same definition for both $v_i$ and $v_j$. Therefore, the adjacent boundaries between the VD naturally coincide in geometric structure, and there is no problem of "boundary misalignment" or "multiple definitions" between the two units.

In multi-robot systems, boundary consistency has multiple engineering implications. First, it ensures that the shared boundaries between adjacent Voronoi cells are well-defined and unambiguous. This is vital for multi-robot systems to avoid task allocation conflicts. Each robot is clearly aware of its designated area, preventing overlapping tasks. When robots dynamically adjust their positions or when conditions

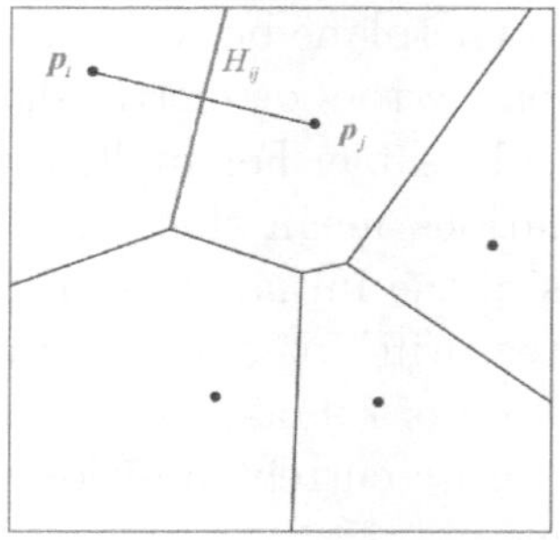


Figure 2.5: Boundary consistency of Voronoi tessellations.

change, their Voronoi cells are updated while maintaining consistent boundaries, ensuring smooth coordination. In addition, this consistency also makes collaborative tasks on the boundary possible. For example, in the construction of communication links, boundary consistency can be used to generate physical communication topology diagrams. At the same time, in our engineering implementation, boundary consistency also brings good visualization characteristics of Voronoi tessellations. The boundary of each cell is a clearly defined straight line segment or plane, so the division results can be displayed clearly in the graphical interface, which is convenient for debugging and verifying the system status. In practical deployment, robots can also send their Voronoi boundaries to central monitoring entity or other robots for consistency verification, ensuring the topological stability of the system during dynamic operation.

Boundary consistency allows robots to coordinate seamlessly without interference, improving task efficiency. For instance, in coverage or monitoring tasks, robots can operate without redundancy, ensuring that each area is covered without overlap. Furthermore, in real-time dynamic environments, boundary consistency facilitates quick updates to task areas as robots move or when the environment changes, ensuring smooth re-tasking and continuous cooperation.

### 2.3.4 Robustness and Dynamic Adaptability of Voronoi Tessellations

Multi-robot systems typically operate in dynamic, unstructured, or partially known environments, and must cope with a series of unpredictable factors such as generation point movement, communication interruptions, and changes in obstacles. In this context, the robustness and dynamic adaptability of Voronoi tessellations are particularly crucial, becoming an important geometric foundation for ensuring system stability and sustained operation capability. The response of Voronoi tessellations to small-scale disturbances is continuous and stable. Assuming a small displacement occurs at the generating point $\boldsymbol{p}_i$:

$$\boldsymbol{p}_i(t + \Delta t) = \boldsymbol{p}_i(t) + \delta \boldsymbol{p}_i. \tag{2.3}$$

Then, its corresponding Voronoi cell $v_i$ will undergo continuous deformation, and the new Voronoi cell can be represented as:

$$v_i(t+\Delta t) = \Big\{ \boldsymbol{q} \in Q \mid \|\boldsymbol{q} - \boldsymbol{p}_i(t+\Delta t)\| \leq \Big\|\boldsymbol{q} - \boldsymbol{p}_j(t+\Delta t)\Big\|, \quad i,j = 1,2,\ldots,n, \quad i \neq j \Big\}. \tag{2.4}$$

Due to the continuous partial derivatives of Euclidean distance with respect to the generating point, the deformation of the Voronoi cell in space is differentiable and will not cause structural mutations due to small perturbations of the generating point. This robustness has positive implications for path updates, obstacle avoidance adjustments, and adaptive deployment in multi-robot systems, ensuring that the system maintains consistency and effectiveness in dividing boundaries when facing uncertainty. This means that even in the presence of robot failures or environmental changes, the tessellation can adapt without causing significant disruptions. For example, if a robot fails or is repositioned, the remaining robots can expand or shift their task areas without needing a complete reorganization of the system. This adaptability is crucial for maintaining system performance in dynamic environments.

Multi-robot systems often face dynamic changes in task objectives, generating point density, or obstacle states. Voronoi tessellations have a good dynamic adaptation mechanism that can self-adjust boundaries with the movement of generating points, achieving dynamic evolution of spatial structure. Considering the robot system's execution of target point set tracking tasks, when the target point or robot itself moves, each Voronoi cell will evolve over time into:

$$v_i(t) = \Big\{ \boldsymbol{q} \in Q \mid \|\boldsymbol{q} - \boldsymbol{p}_i(t)\| \leq \Big\|\boldsymbol{q} - \boldsymbol{p}_j(t)\Big\|, \quad i,j = 1,2,\ldots n, \quad i \neq j \Big\}. \tag{2.5}$$

The geometric boundary evolves into a function $\partial v_i(t)$ over time, and the boundary is manifested graphically as edge reconstruction, vertex displacement, and even element merging or splitting, making the Voronoi diagram a natural structure for handling dynamic spatial tasks. Meanwhile, dynamic adaptability makes it easy to integrate Voronoi tessellations with sensor feedback mechanisms to construct closed-loop control systems. Robots can periodically obtain environmental changes and update their own generating point position information. The prediction mechanism is used to predict the future boundary change trend, and the Voronoi tessellations can be adjusted in advance to pre-adapt to the upcoming changes and improve the system response efficiency.

### 2.3.5 Scalability of Voronoi Tessellations

In the context of the increasing development of multi-robot systems toward large-scale, high-density, and cross dimensional, the scalability of the system has become one of the key factors in measuring its engineering feasibility and long-term operational potential. The scalability of Voronoi tessellations in multi-robot systems is mainly reflected in its local computing features and distributed implementation mechanism. Constructing a complete Voronoi diagram in two-dimensional environment has a complexity of $O(n \log n)$ [6], where $n$ is the number of robots. However, in actual distributed systems, each robot only needs to communicate with its neighboring robots and construct its own Voronoi cell based on local neighborhoods, whose computational complexity depends on the neighborhood size $k \ll n$. Therefore, even in large-scale systems, the computational and communication burden of each node remains reasonable. Scalability ensures that as a multi-robot system grows, it can maintain efficiency without excessive computational overhead. Whether it's adding new robots or adapting to larger task areas, Voronoi tessellations provide a flexible framework that can scale effectively, allowing large groups of robots to cooperate and adapt in real-time. This excellent scalability makes Voronoi tessellations particularly suitable for large-scale cluster systems composed of dozens to thousands of robots, such as warehousing and logistics, agricultural inspection, drones swarm, etc.

In addition, the scalability of Voronoi tessellations is also reflected in task load allocation and structural adaptation. The system can estimate the local workload through the area or density function of each Voronoi cell, which can be expressed as:

$$W_i = \int_{\mathcal{V}_i} \phi(\boldsymbol{q}) d\boldsymbol{q}. \tag{2.6}$$

Based on this, we can dynamically adjust the position of the generating points to achieve real-time balance in the task area. When a new robot is added to the system, it only needs to be added to the set of generating points, and the original tessellation structure can be automatically adjusted without the need to reconstruct the global path or task map. The specific effect is shown in Fig. 2.6. This feature endows the system with good fault tolerance and resilience, especially suitable for use in environments with dynamic topology or frequent node changes.

In summary, the Voronoi tessellations provide a solid theoretical foundation and engineering feasibility support for spatial partitioning and collaborative control of multi-robot systems, thanks to its excellent geometric and system level characteristics such as locality, uniqueness, boundary consistency, robustness, and scalability. These properties not only ensure the clarity and stability of the system partition

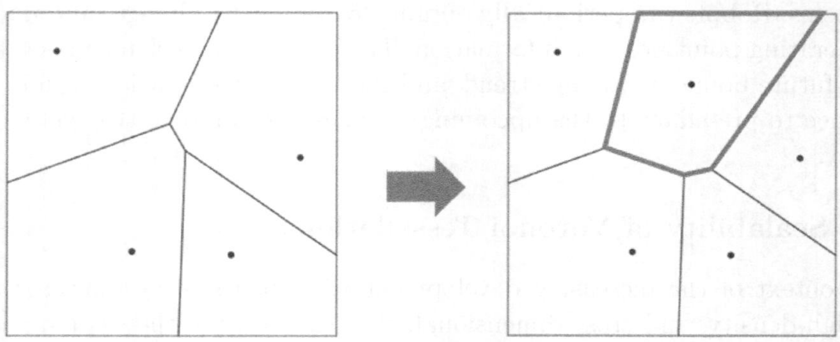

Figure 2.6: Scalability of Voronoi tessellations.

structure but also enable robot clusters to achieve efficient collaboration in dynamic, distributed, and even partially unknown environments. Especially in the face of large-scale deployment, frequent task changes, or resource constraints, Voronoi tessellations exhibits strong adaptability and structural flexibility. In the subsequent chapters, we will further explore its behavioral evolution in different geometric environments, as well as how to embed these basic properties into core algorithm designs such as coverage, division of work, and path planning for multi-robot systems.

## 2.4 Duality Between Voronoi Diagrams and Delaunay Triangulation

Voronoi diagram and Delaunay triangulation are a pair of inseparable basic concepts in computational geometry. They describe the spatial structure information contained in the same set of points from different perspectives. The Voronoi diagram divides space and generates an area for each point set, while the Delaunay triangulation defines spatial relationships by connecting the generated points into triangles. Understanding their duality relationship is the key to mastering spatial division, grid generation, and many related algorithm core ideas.

### 2.4.1 Review of Basic Concepts

We have introduced the basic concepts of Voronoi diagrams earlier, and now we will focus on Delaunay triangulation and their dual relationships. Delaunay Triangulation (DT) is one of the most fundamental and important concepts in computational geometry, which describes an optimal triangulation method for a set of planar discrete points. It is known for its Empty Circle Property and Max Min Angle optimization criteria. This structure is not only the core of many spatial analysis algorithms such as interpolation, mesh generation, and surface reconstruction but also has a beautiful dual relationship with Voronoi diagrams, which together form the cornerstone of spatial partitioning theory. Given a discrete point set $P = \{p_1, p_2, \ldots, p_n\}$ in the two-dimensional Euclidean space $\mathbb{R}^2$, its Delaunay triangulation $\mathcal{D}(P)$ satisfies the following core conditions:

The Delaunay triangulation $\mathcal{D}(P)$ of a discrete point set $P = \{p_1, p_2, \ldots, p_n\}$ in the two-dimensional Euclidean space $\mathbb{R}^2$ is defined by the following fundamental properties:

1. Empty Circle Property
   For any triangle $T$ in $\mathcal{D}(P)$ (whose vertices are from $P$), the interior of its circumcircle contains no other points from the set $P$. Formally:

   $$\forall T \in \mathcal{D}(P),\ \forall p \in P \setminus \text{vertices}(T),\ p \notin \text{interior of Circumcircle}(T).$$

   This property is both a necessary and sufficient condition for a triangulation to be a Delaunay triangulation, constituting its most fundamental definition.

2. Max-Min Angle Criterion

Among all possible triangulations $\mathcal{T}$ of the point set $P$, the Delaunay triangulation $\mathcal{D}(P)$ maximizes the minimum interior angle across all triangles. This criterion ensures that the resulting triangles are as "well-shaped" as possible (avoiding skinny triangles), thereby providing superior **stability** and **convergence** in numerical computations.

3. Uniqueness
   The Delaunay triangulation $\mathcal{D}(P)$ is unique when the point set $P$ is in general position, meaning no four points are cocircular. If four or more points are cocircular, a degeneracy occurs, leading to multiple valid triangulations that satisfy the empty circle property. In such cases, uniqueness can be enforced by applying additional rules (e.g., selecting the triangulation based on a lexicographical ordering of point indices).

### 2.4.2 The Mathematical Connotation of Dual Relationship

The concept of "duality" here means that one structure can be constructed from the other through specific geometric rules. Specifically:

#### 1. From Delaunay Triangulation to Voronoi Diagram

The Voronoi diagram is the dual graph of the Delaunay triangulation. The construction rules are as follows:

- Voronoi Vertex: The circumcenter of each triangle in the Delaunay triangulation is a vertex of the Voronoi diagram.
- Voronoi Edge: Connecting the circumcenters of two adjacent triangles that share a common Delaunay edge forms a Voronoi edge. This Voronoi edge is exactly the perpendicular bisector of the common Delaunay edge of these two triangles.

#### 2. From Voronoi Diagram to Delaunay Triangulation

Conversely, if the Voronoi diagram is given first, the Delaunay triangulation can be recovered as follows:

- For each Voronoi edge, which is constituted by the perpendicular bisector of two generator points $p_i$ and $p_j$, connect all such pairs of generator points $(p_i, p_j)$ to obtain the Delaunay triangulation.
- In the Voronoi diagram, each Voronoi vertex is the intersection point of three Voronoi cells, meaning that the vertex is equidistant to the three generator points. The triangle formed by connecting these three generator points is a Delaunay triangle, and this Voronoi vertex is precisely the circumcenter of that triangle.

Specifically, the Voronoi diagram is constructed by extracting the circumcircle centers of each triangle from the Delaunay triangulation, which is constructed by connecting the perpendicular bisectors of each edge in the Voronoi diagram. The two are interdependent and closely connected to each other through geometric properties and topological structures.

### 2.4.3 Equivalent Properties and Common Features

Due to the duality between the Voronoi diagram and the Delaunay triangulation, they share several equivalent properties:

- Uniqueness: Under general position (i.e., no four points are cocircular), the Delaunay triangulation of a point set is unique. Correspondingly, its dual Voronoi diagram is also unique.
- Convex Hull Relationship: The outermost boundary of the Delaunay triangulation constitutes the convex hull of the point set $P$. Similarly, in the Voronoi diagram, the Voronoi cells corresponding to the generating points located on the convex hull boundary are infinitely open.
- Local Optimization Property: The Delaunay triangulation possesses the characteristic of "maximizing the minimum angle," meaning that among all possible triangulations, it best avoids the appearance of sharp, skinny triangles, thereby optimizing mesh quality. This property is intrinsically unified with the spatial equilibrium of the structure of its dual, the Voronoi diagram.

### 2.4.4 Applications and Algorithmic Significance in Computational Geometry

This duality relationship is not only an elegant mathematical theory but also possesses strong practical value, providing significant convenience for algorithm design:

- Computational Mutual Transformation: Due to duality, one structure can be obtained by computing the other. In practice, first computing the Delaunay triangulation and then constructing its dual Voronoi diagram is usually a more efficient algorithmic path.
- Navigation and Proximity Queries: The Voronoi diagram naturally expresses the "nearest-neighbor" relationship, which can be used for robot path planning (e.g., navigating through the center of channels) or nearest-facility queries. Its corresponding Delaunay triangulation provides the topological connection relationship between points, making it suitable for surface reconstruction and finite element analysis, among other applications.
- Application in This Section: In the multi-robot cooperative pursuit strategy discussed later in this book, we will see how to utilize the Buffered Voronoi Cell (BVC) to assign independent collision-avoidance responsibility areas for

each agent. Understanding the Voronoi diagram and its relationship with the Delaunay triangulation is the geometric foundation for understanding how these areas are dynamically generated and evolved, and how agents make coordinated decisions based on them.

The dual relationship between Voronoi diagrams and Delaunay triangulation has wide applications in multiple fields such as multi robot collaboration, mesh generation, and path planning. For example, in multi robot systems, Voronoi diagrams are often used to allocate task areas, while Delaunay triangulation helps define the connection relationships between robots. The combination of the two can effectively avoid collisions between robots and improve the collaborative efficiency of the system. In short, mastering the dual relationship between Voronoi diagrams and Delaunay triangulation is the core of understanding algorithms in fields such as spatial segmentation, mesh generation, and multi robot collaboration.

## 2.5 Voronoi Tessellations in Different Environments

In the previous section, this book systematically introduced the geometric characteristics, existence, and uniqueness of Voronoi tessellations. However, in reality, the task space of multi-robot systems often has diverse geometric characteristics, such as dimensional differences, boundary constraints, obstacle distribution, etc. These factors can have a significant impact on the structure and behavior of Voronoi tessellations.

### 2.5.1 Voronoi Tessellations in 2D Environments

In two-dimensional Euclidean plane, the Voronoi tessellations exhibit a clear geometric structure, with its basic cell being a convex polygon surrounded by a set of straight boundaries. According to Voronoi's definition in Chapter 1, given a two-dimensional environment $Q \subseteq \mathbb{R}^N$, there are $n$ generating points for Voronoi cells in the environment. Using $V^2 = \{v_1^2, v_2^2, \cdots v_n^2\}$ to represent the set of Voronoi cells in the two-dimensional environment, the definition of the two-dimensional Voronoi tessellations can be expressed as:

$$v_i^2 = \left\{ \boldsymbol{q} \in Q \mid \|\boldsymbol{q} - \boldsymbol{p}_i\| \leq \left\|\boldsymbol{q} - \boldsymbol{p}_j\right\|, \quad i, j = 1, 2, \ldots n, \quad i \neq j \right\}, \tag{2.7}$$

where $\boldsymbol{q}$ represents the position of the point in the environment, $\boldsymbol{p}_i$ and $\boldsymbol{p}_j$ are two adjacent Voronoi cell generation points, $\|\boldsymbol{q} - \boldsymbol{p}_i\|$ can be expressed as:

$$\|\boldsymbol{q} - \boldsymbol{p}_i\| = \sqrt{(\boldsymbol{q}_1 - \boldsymbol{p}_{i1})^2 + (\boldsymbol{q}_2 - \boldsymbol{p}_{i2})^2}, \tag{2.8}$$

where $\boldsymbol{q} = (\boldsymbol{q}_1, \boldsymbol{q}_2)$ represents the coordinates of point $q$ in three-dimensional environment, $\boldsymbol{p}_i = (\boldsymbol{p}_{i1}, \boldsymbol{p}_{i2})$ represents the coordinates of point $\boldsymbol{p}_i$ in two-dimensional environment. When the distribution of generating points is uniform, the resulting Voronoi cells tend to be arranged uniformly, with spatial compactness and symmetry, which is conducive to balanced distribution of task load. The specific effect is

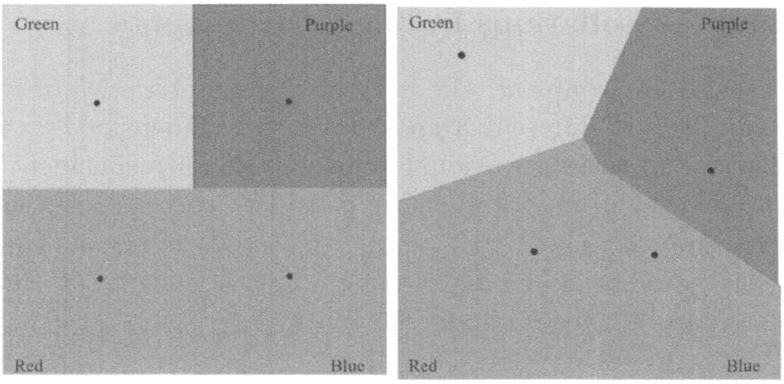


(a) (b)

Figure 2.7: Voronoi tessellations in two-dimensional environment: (a) Two-dimensional regular Voronoi cells; (b) 2D irregular Voronoi cells.

shown in Fig. 2.7a, the black dots in the figure represent the positions of the robots, and the geometric regions of different colors represent the Voronoi cells where each robot is located, such as the green, purple, and other areas indicated in the figure. Each boundary of the Voronoi cell is a segment of the perpendicular bisector between two points, and the intersection point is the geometric center of gravity of three points that are collinear. This tessellation method naturally has locality and boundary consistency, providing clear geometric basis for robots to perform tasks in a two-dimensional environment [7].

The Voronoi tessellations in two-dimensional environment rely on the Euclidean distance metric and can adapt to various forms of point set distributions in most planar task scenarios. It has a solid theoretical foundation, guarantees uniqueness and existence, and can adapt to changes in static and dynamic point sets. In addition, the Voronoi cells generated by two-dimensional Voronoi tessellations have good visualization characteristics, making them easy to analyze and debug. They are particularly suitable for scenarios such as multi-robot deployment, task tessellations, and communication range modeling in the primary stage. Its local update characteristics also make it easy to embed into distributed control and self-organizing algorithms.

As the non-uniformity of the generating point distribution increases or the environmental boundaries become more limited, the Voronoi tessellations in two-dimensional environment will gradually exhibit complex geometric evolution characteristics, and the edge Voronoi cells may exhibit irregular shapes. The specific effect is shown in Fig. 2.7b. Meanwhile, in dynamic task scenarios, the movement of generated points will cause the continuous evolution of the Voronoi cell, manifested as the reconstruction changes of the Voronoi cell vertices. The evolutionary ability of this structure enables it to have strong environmental adaptability and can cope with the challenges brought by the temporal changes in task objectives, obstacles, and member positions in multi-robot systems.

### 2.5.2 Voronoi Tessellations in 3D Environments

In three-dimensional environment, the geometric characteristics of the Voronoi tessellations are significantly different from those in two-dimensional environment. The Voronoi tessellations in two-dimensional environment only considers two directions, while in three-dimensional environment it considers three directions. Given three-dimensional environment $Q \subseteq \mathbb{R}^3$, $V^3 = \{v_1^3, v_2^3, \dots v_n^3\}$ represents the set of Voronoi units in the three-dimensional environment. The definition of three-dimensional Voronoi partition can be expressed as:

$$v_i^3 = \left\{ \boldsymbol{q} \in Q \mid \|\boldsymbol{q} - \boldsymbol{p}_i\|_3 \leq \left\|\boldsymbol{q} - \boldsymbol{p}_j\right\|_3, \quad i, j = 1, 2, \dots n, \quad i \neq j \right\}, \tag{2.9}$$

where $\|\boldsymbol{q} - \boldsymbol{p}_i\|_3$ can be represented as:

$$\|\boldsymbol{q} - \boldsymbol{p}_i\|_3 = \sqrt{(\boldsymbol{q}_1 - \boldsymbol{p}_{i1})^2 + (\boldsymbol{q}_2 - \boldsymbol{p}_{i2})^2 (\boldsymbol{q}_3 - \boldsymbol{p}_{i3})^2}, \tag{2.10}$$

where $\boldsymbol{q} = (\boldsymbol{q}_1, \boldsymbol{q}_2, \boldsymbol{q}_3)$ represents the coordinates of the point $\boldsymbol{q}$ in three-dimensional environment, and $\boldsymbol{p}_i = (\boldsymbol{p}_{i1}, \boldsymbol{p}_{i2}, \boldsymbol{p}_{i3})$ represents the coordinates of generating point $\boldsymbol{p}_i$ in three-dimensional environment. Different with the convex polygons of two-dimensional Voronoi, three-dimensional Voronoi cells are convex polyhedra, where each Voronoi cell is formed by the intersection of several planes. The distribution of generated points and the topological structure of the area will directly affect the shape and size of these polyhedra. For uniformly distributed generation points, the resulting Voronoi cells are usually close to a cubic structure, as shown in Fig. 2.8a. In three-dimensional environment, the boundaries delineated by Voronoi exhibit more diverse geometric shapes, and the surface of each cell may be composed of multiple planes interwoven [8].

In three-dimensional environment, the Voronoi tessellations also rely on the Euclidean distance metric and have properties of uniqueness and existence. However, compared to two-dimensional environments, the Voronoi tessellations in three-dimensional environment is more complex in terms of theoretical adaptability, and

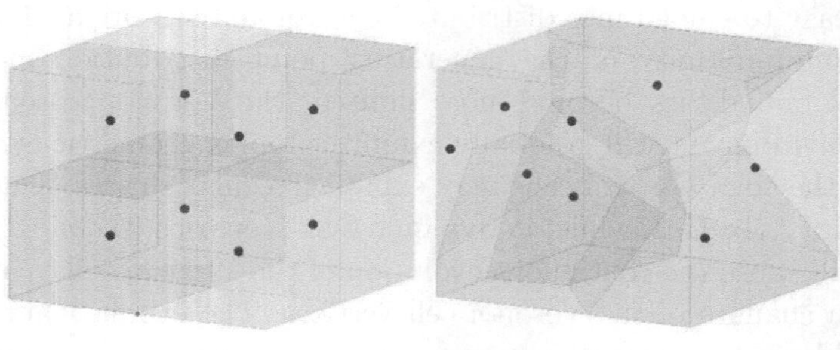

(a) (b)

Figure 2.8: Voronoi tessellations in three-dimensional environment: (a) Three-dimensional regular Voronoi cells; (b) 3D irregular Voronoi cells.

three-dimensional Voronoi tessellations also require handling more geometric dimensions and computational challenges. For example, the arrangement of generated points and the influence of environmental boundaries may lead to overlap or significant geometric deformation between Voronoi cells. In addition, obstacles and dynamic changes in three-dimensional space bring higher computational complexity, so generating and updating Voronoi tessellations require more efficient algorithms and data structure support.

As the distribution of generating points and the complexity of the environment increase, the structural evolution of three-dimensional Voronoi tessellations will exhibit stronger complexity. When adding obstacles or changing the position of generating points in the environment, the shape of the VD will undergo significant changes, especially in non-uniform generated point distribution or highly complex three-dimensional environments. The shape of the generated Voronoi cell will become more irregular, which may cause some units to shrink or expand excessively. The specific effect is shown in Fig. 2.8b. With the continuous changes in task requirements, it is necessary for three-dimensional Voronoi tessellations to flexibly adapt to the collaborative needs of robots in different task dimensions, so as to be widely applicable to tasks in various three-dimensional environments such as sea, land, or air. The various applications of Voronoi tessellations in two-dimensional and three-dimensional environments will be detailed in the subsequent chapters.

### 2.5.3 Voronoi Tessellations in Non-Convex Environments

According to regional characteristics, regions can usually be divided into convex regions and non-convex regions. If any two points can be taken within a region and each point on their connecting line is also within that region, then this region is called convex. Its mathematical definition is as follows. If for any two points $x, y \in D$, the region $D \subseteq \mathbb{R}^n$, and the real number $a (0 \leq a \leq 1)$ they all satisfy:

$$ax + (1 - a)y \in D. \tag{2.11}$$

However, in many cases, there are often obstacles within the region or concavity on the external boundaries. In contrast to convex regions, the line connecting any two points within such regions may not have all the points on the line segment still within that region. Therefore, this region is called a non-convex region. The schematic diagram shows the convex and non-convex regions as in Fig. 2.9.

In the non-convex environments, the geometric properties of Voronoi tessellations become more complex, especially in the presence of obstacles or irregular boundaries. Among them, the relevant formulas for Voronoi tessellations in non-convex environments will be introduced in detail in Chapter 5. Voronoi cells in non-convex environments often exhibit more irregular and tortuous shapes, as the Voronoi cells generating points may need to bypass obstacles or be constrained by irregular boundary conditions. These environments often result in boundary bending or interleaving between the VD, which is vastly different from Voronoi tessellations in convex environments. Due to the possible irregularity of the boundaries in non convex environments,

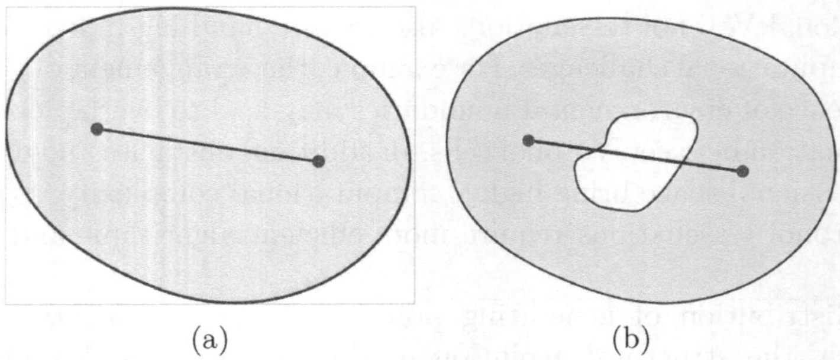


Figure 2.9: Schematic diagram of convex and non-convex regions: (a) Convex region; (b) non-convex region.

the shapes of Voronoi cells exhibit more diverse geometric forms, especially in narrow spaces or areas with obstacles [9].

In the non-convex environments, the theoretical adaptability of the Voronoi tessellations requires additional consideration. First, the boundary conditions of non-convex environments have a significant impact on the shape of the Voronoi cell, so it is necessary to extend the algorithm to ensure that the Voronoi tessellations can correctly generate these boundaries. Second, the presence of obstacles may prevent some VD from being calculated using simple distance metrics, forcing algorithms to adopt more complex collision detection and constraint handling methods. In addition, in dynamic environments, the Voronoi tessellations in non-convex environments needs to consider the relative position changes between robots, as well as the appearance and disappearance of obstacles, which increases the complexity of theoretical models.

With the changes in obstacle distribution and task requirements in non-convex environments, the structural evolution of Voronoi tessellations tends to be more flexible and dynamic. During the execution of tasks in a multi-robot system, the shape and size of the Voronoi cell will constantly change to adapt to new environmental challenges. For example, in certain areas, obstacles may cause VD to be compressed or stretched, while in other areas, Voronoi cells may expand to ensure sufficient operating space. The specific effects of Voronoi tessellations in two-dimensional and three-dimensional non-convex environments are shown in Figs. 2.10a and 2.10b, respectively. As the complexity of the environment increases, Voronoi tessellations may exhibit higher computational requirements, but its flexibility and adaptability also enable it to better meet the needs of multi-robot systems in non-convex environments.

This section analyzes the structural evolution of Voronoi tessellations in two-dimensional, three-dimensional, and non-convex environment from the perspectives of spatial dimensions and environmental geometry. With the increasing complexity of the environment, Voronoi tessellations gradually present richer geometric forms and higher computational challenges, providing complex but flexible mathematical support for subsequent multi-robot task execution.

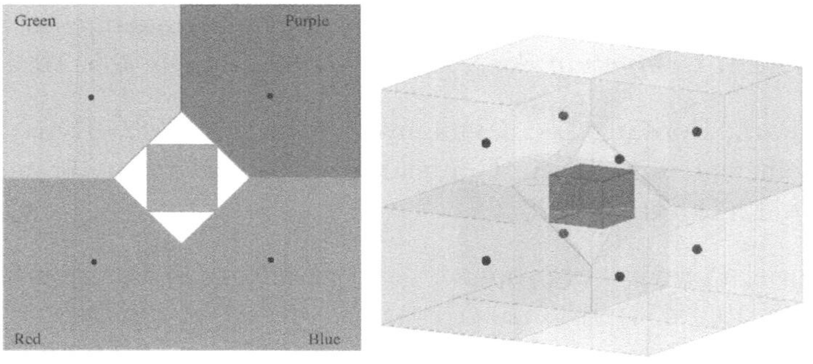


(a) (b)

Figure 2.10: Voronoi tessellations in non-convex environment: (a) Voronoi tessellations in a two-dimensional non-convex environment; (b) Voronoi tessellations in a three-dimensional non-convex environment.

## 2.6 Chapter Summary

This chapter provides a comprehensive examination of the geometric properties and computational methods of Voronoi tessellations, elucidating their core advantages in multi-robot systems. First, through analysis of the convex cell properties, the empty circle characteristics, and the adjacency topology, it clarifies how Voronoi tessellations offer robots safe motion spaces, optimal coverage control, and efficient communication topologies. Second, an in-depth investigation of fundamental properties—such as locality, uniqueness and existence, and boundary consistency—demonstrates how these features ensure the robustness and dynamic adaptability of Voronoi tessellations in multi-robot systems. Another key aspect of the chapter was the exploration of the duality between Voronoi diagrams and Delaunay triangulations, demonstrating how these two concepts are mathematically intertwined and how their duality is applied in tasks like path planning, network connectivity, and spatial optimization. Finally, for different environmental conditions (e.g., 2D/3D and non-convex environments), this chapter systematically analyzes the evolutionary patterns of Voronoi tessellations' geometric properties, providing theoretical support for algorithm design in complex scenarios. This Chapter aims to help better as well as to understanding of the mathematical essence of Voronoi tessellations establish a solid theoretical foundation for multi-robot cooperative control algorithms in the subsequent chapters.

## Bibliography

[1] Marjorie Senechal. Mathematical structures: Spatial tessellations. concepts and applications of Voronoi diagrams. *Science*, 260(5111):1170–1173, 1993.

[2] Atsuyuki Okabe, Toshiaki Satoh, Takehiro Furuta, Atsuo Suzuki, and Kyoko Okano. Generalized network Voronoi diagrams: Concepts, computational methods, and applications. *International Journal of Geographical Information Science*, 22(9):965–994, 2008.

[3] Franz Aurenhammer. Voronoi diagrams a survey of a fundamental geometric data structure. *ACM Computing Surveys (CSUR)*, 23(3):345–405, 1991.

[4] Jee Yong Park, Hoosang Lee, Changhyeon Kim, and Jeha Ryu. Voronoi tessellation for efficient sampling in Gaussian process based robotic motion planning. *Electronics*, 12(19):4122, 2023.

[5] Adam Dobrin. A review of properties and variations of Voronoi diagrams. *Whitman College*, 10(1.453):9156, 2005.

[6] Franz Aurenhammer. Voronoi diagrams—a survey of a fundamental geometric data structure. *ACM Computing Surveys (CSUR)*, 23(3):345–405, 1991.

[7] Joshua S Adams, Wei Sun, and YangQuan Chen. Formations with decentralized centroidal Voronoi tessellation algorithm. *IFAC Proceedings Volumes*, 42(22):13–18, 2009.

[8] Kai Cao, Yang-Quan Chen, Song Gao, Kun Yan, Jiahao Zhang, and Di An. Omnidirectional capture for multi-drone based on 3D-Voronoi tessellation. *Drones*, 7(7):458, 2023.

[9] Kai Cao, Yangquan Chen, Song Gao, and Manying Cheng. CVT-based swarming techniques for convex and non-convex cases. *International Journal of Advanced Robotic Systems*, 21(6), 2024.

CHAPTER 3

# Multi-Robot Cooperation and Formation Based on the Voronoi Tessellations

## 3.1 Introduction

MULTI robot systems (MRS) represent a critical aspect of modern autonomous robotics, enabling task parallelization, robustness to individual failure, and distributed intelligence. Compared to a single robot, MRS can cover a wider operational area, adapt to more complex task environments, and demonstrate immense potential in applications such as search and rescue, environmental monitoring, and warehouse logistics. Among the many challenges in coordinating such systems, achieving and maintaining a desired formation is of central importance. Traditional formation control techniques—such as leader-follower approach [1], virtual structure based methods [2], and behavior-based method [3]. While effective in structured environments, these methods face limitations in dynamic, uncertain, or decentralized contexts.

Leader-follower approach method is a method of achieving collaborative action among multi-robot formations through the leader follower relationship. In this method, one or more robots are designated as leaders, while other robots act as followers to follow the leader's movements. The goal of this control method is to enable followers to closely follow the leader, form and maintain a predetermined formation, and thus collaborate to complete the formation task. The navigation following method is mainly implemented by $l - \varphi$ control and $l - l$ control [4]. As shown in Fig. 3.1, formation control is achieved between robots through a master-slave relationship.

Under this control structure, numerous scholars have proposed various methods for formation control. Boo et al. [5] proposed a navigation following formation control method based on Fuzzy Integral Sliding Mode Observer (FISMO), which utilizes a ceiling mounted camera to perform in the absence of motion disturbances and velocity information of the leading and following robots. Only use the posture information of the guiding robot and the following robot obtained from the camera sensor to make the following robot follow the trajectory of the introduced target robot. Zhou et al. [6]

DOI: 10.1201/9781003714569-3

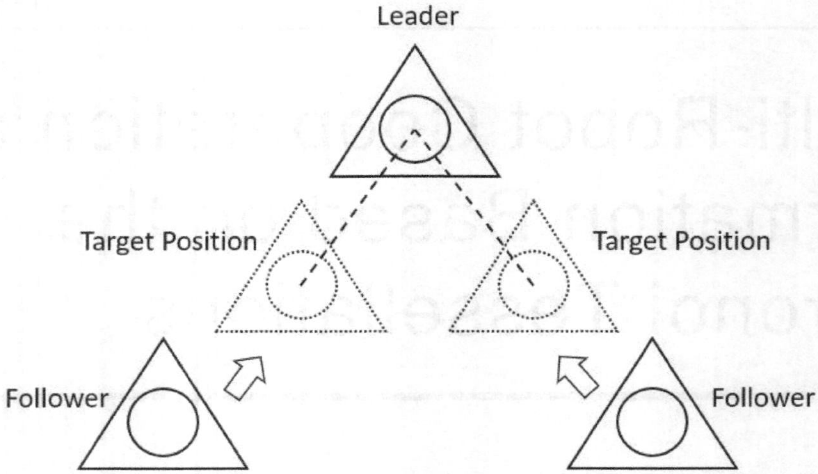


Figure 3.1: Leader-follower approach method.

considered the unknown control input of the navigator in the navigation following structure, relaxed the requirement of constant velocity for the navigator, introduced an estimator, designed a relative velocity based action function, and completed a distributed time-varying formation tracking control protocol. Tan Yao et al. [7] achieved formation control of robot formation based on navigation and following method using only azimuth information. Shen Guofeng et al. [8] proposed a fog computing enabled robot following system for navigation following formations. Using the leader's speed information to calculate the expected follow-up trajectory, real-time tracking error is obtained through visual measurement methods, and trajectory tracking control is achieved by combining speed replay and PD type iterative learning control. Liu et al. [9] transformed the second-order navigation following formation control problem into a stability problem of system error to achieve the desired formation tracking. A distributed adaptive control protocol based on local output information from adjacent agents as feedback is proposed, and an adaptive control strategy based on output feedback is proposed.

Behavior based approach is a multi-robot control method derived from the formation behavior of group organisms in nature, which achieves multi-robot formation control by defining and combining robot behaviors [10]. This method typically treats each robot as an individual with a series of basic behaviors, combining these behaviors to achieve complex tasks and collaborations. Due to its distributed nature, this method has strong flexibility and scalability in multi-robot control. At present, existing works mainly includes methods such as finite state machines [11], virtual potential field principles [12], consistency principles [13] for implementation.

Among them, Zhou et al. [14] studied the cluster event triggered tracking cooperation and formation control of a multi robot system with nonlinear vehicle dynamics under switch oriented communication topology and safety tasks. A cluster event triggered tracking cooperative control law with extended amplification region conditions is proposed based on limited information exchange between neighbors under the extended feasible domain of control parameters. Santilli et al. [15] introduced a robot formation control framework based on limited field of view sensors, which implements an adaptive decentralized switching control mechanism. Based on the switching mechanism of finite state machines, the formation control method adaptively switches between fully decentralized and fully centralized, and the formation encounters the use of a centralized planner in unstable situations. Wu et al. [16] proposed the concept of virtual kernel based on local potential fields between unmanned aerial vehicles to optimize the calculation of attraction fields between them; A method of artificial potential field backtracking filling has been proposed to solve the problem of local minima caused by compact concave obstacles in artificial potential fields. A virtual kernel based formation control algorithm for drone swarms has been proposed, which improves the effectiveness of swarm control. Feola et al. [17] focused on the distributed formation of minimalist robots, changing the duration of random walks based on the number of detected neighbors, simulating other collective behaviors of micro and nano molecules in active colloids, and using adaptive heuristic mechanisms to promote formation self-organization aggregation and task execution.

The virtual structure method defines the formation as a virtual structure, usually a geometric or topological structure, in which each robot is assigned to a node or a region in the virtual structure, and the robots perform corresponding tasks based on the task allocation and coordination information in the virtual structure. At the same time, the virtual structure of the formation can be dynamically updated, reassigned, or the robots' tasks or positions can be adjusted based on changes in task execution or task environment. Fig. 3.2 shows a simple virtual structure formation, where three robots are placed at the three vertices of a triangular virtual structure. When the virtual structure formation moves along a specified trajectory, the robots follow the formation to ensure that their relative positions remain unchanged in the formation

The virtual structure method was initially proposed by Anthony Lewis [18] and validated in a multi robot system consisting of three robots. This method can accurately control the member robots to move to the target position. Wu et al. [19] proposed a virtual structure of Hierarchical Virtual Communication Ring (HVCR), which reduces the boundaries of group communication and minimizes the unsafe range. The combination of collaborative control and secure communication ensures secure communication in multi-drone collaborative control flight scenarios. Introduced flight control factors to accelerate the convergence speed of multi-drone formations. Toumieh et al. [20] proposed a formation virtual structure for time aware

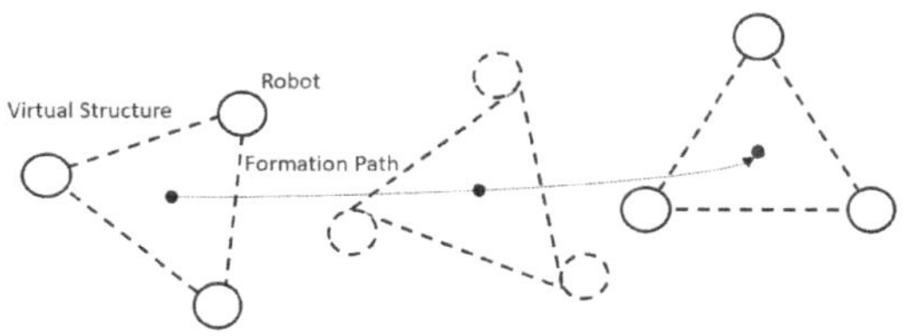


Figure 3.2: The virtual structure method.

safe corridors, based on which centralized formation control is implemented using model predictive control and mixed integer quadratic programming formulas. Zhang et al. [21] proposed a density based virtual structure, in which robots move toward the edges of a formation through edge following behavior and surround it into a group of robots. After the encirclement is completed, the group is compressed into a tight formation using a contraction mechanism, achieving large-scale robot formation generation. Yin et al. [22] proposed a dual event triggered formation control strategy based on virtual linkages to address the formation maintenance problem of a group of robots in confined spaces. Based on the idea of virtual linkages and event triggers, a dual event trigger is proposed to ensure formation maintenance and collision avoidance in multi-robot systems.

Based on the cobweb model proposed by in [93], a radar chart for the evaluation of robot formation control level is adopted. Starting from the five dimensions shown in Table 3.1, we can analyze the level of the three formation control methods mentioned above, as shown in Fig. 3.3.

Table 3.1: Evaluation criteria for multi-robot formation control methods

| **Evaluation dimension** | **0 Points** | **1 Point** | **2 Points** | **3 Points** |
|---|---|---|---|---|
| Autonomy (Control Method) | Robot fully relies on the main core for control | Main core participates in primary control, robot assists | Robot participates in primary control, main core assists | Robot is fully autonomous |
| Scalability (Number of Controlled Robots) | Controls a single robot | Controls 2–5 robots | Controls 5–20 robots | Controls more than 20 robots |
| Coordination (Communication with Other Robots) | No communication established with other robots | Can communicate with some robots | Can communicate with nearby robots | Can communicate with all robots in the formation |
| Robustness (Types of Faults or Disturbances Handled) | No robustness design | Handles one type of fault or disturbance | Handles two types of faults or disturbances | Handles more than three types of faults or disturbances |
| Obstacle Avoidance (Obstacle Avoidance Capability) | No obstacle avoidance design | Only considers collisions between robots | Formation can handle environmental obstacles and inter-robot collisions | Can predict obstacles while avoiding them |

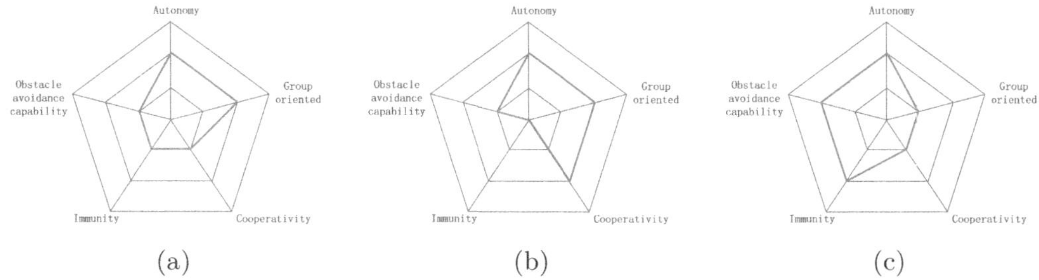


Figure 3.3: Multi-robot formation method evaluation radar chart: (a) Leader-follower approach; (b) behavior-based approach; (c) virtual structure approach.

## 3.2 Combination of Voronoi Tessellations and Multi-robot Systems

Through the introduction of the previous two chapters on Voronoi tessellations, it can be concluded that the method based on Voronoi partitioning can not only divide the task space into several regions but also has good robustness and scalability. It can effectively reduce redundancy and conflicts in the task allocation process, improve task execution efficiency, with the advantage of natural application in multi robot systems [23]. First, the definition of Voronoi tessellations in a multi-robot system is introduced. Given the task space $Q \in \mathbb{R}^d$ of the multi-robot system, the points in the environment are represented as $\boldsymbol{q}$, and the swarm system contains $n$ robots. The coordinate set of the robots is represented as $\boldsymbol{P} = \{\boldsymbol{p}_1, \boldsymbol{p}_2, \ldots \boldsymbol{p}_n\}$. The Voronoi cell $v_i$ generated by robot $i$ can be represented as:

$$v_i = \left\{ \boldsymbol{q} \in Q \mid \|\boldsymbol{q} - \boldsymbol{p}_i\| \leq \left\|\boldsymbol{q} - \boldsymbol{p}_j\right\|, \quad i, j = 1, 2, \ldots n, \quad j \neq i \right\}. \tag{3.1}$$

Here, $V = \{v_1, v_2, \ldots v_n\}$ represents the set of Voronoi cells generated by the multi-robot system. Each Voronoi cell $v_i$ contains all the points in the task space that are closest to the robot $\boldsymbol{p}_i$ [24]. In this way, robots can autonomously determine the areas they should be responsible for without the need for centralized coordination, achieving automation and localization of task allocation.

Based on the theory of Voronoi tessellations mentioned above, it can be concluded that Voronoi tessellations can provide an adaptive region partitioning mechanism for multi robot systems. When the number or position of robots changes, the Voronoi cells can be dynamically reconstructed based on the new generating point positions, ensuring that each robot always has an independent and clear control area [25] [26]. This dynamic update feature enables multi-robot systems to maintain efficient collaboration in complex and constantly changing task environments, with strong environmental adaptability and system robustness.

During task execution, Voronoi tessellations can effectively reduce conflicts and resource competition between robots. Due to the natural non-intersection between the Voronoi cells of different robots, the robots can act independently within their designated areas, avoiding task interference and collision risks caused by overlapping areas [27] [28]. Especially in high-density deployment or large-scale collaborative tasks, the task allocation mechanism based on Voronoi tessellations can significantly improve the stability and execution efficiency of the system.

At the same time, Voronoi tessellations has excellent real-time dynamic adjustment capabilities, making it particularly suitable for scenarios that require continuous coverage and ongoing task execution. As robots continue to move and environmental information changes in real-time, the system can maintain a reasonable distribution of each robot in the task space by continuously updating the Voronoi tessellations [29] [30]. This characteristic enables multi-robot systems to have natural adaptability and persistent operational capabilities in tasks such as exploring unknown areas and monitoring dynamic environments.

Based on the above analysis, it can be seen that the Voronoi tessellations not only provides a reasonable way of spatial task partitioning but also provides theoretical support for dynamic optimization control of multi-robot systems. On this basis, by guiding the robots to move toward the centroid of their respective Voronoi cells, the overall performance of the system can be further improved, achieving more efficient and stable collaborative formation control. The next section will provide a detailed introduction to the multi-robot formation algorithm based on centroid Voronoi tessellations (CVT) and its specific implementation method.

## 3.3 Multi-robot Formation Based on Centroidal Voronoi Tessellations

### 3.3.1 The Basic Theory of Centroidal Voronoi Tessellations

Centroid Voronoi tessellations (CVT) are a special form of Voronoi tessellations, in which the generation point of each Voronoi cell is located at the centroid position of its Voronoi cell. Given the task area $Q \in \mathbb{R}^d$, using the density function $\rho(\boldsymbol{q})$ to represent the importance of each point in the task area, the centroid $\boldsymbol{c}_{v_i}$ of the Voronoi cell of robot $i$ can be expressed as:

$$\boldsymbol{c}_{v_i} = \frac{\int_{v_i} \boldsymbol{q}\rho(\boldsymbol{q})d\boldsymbol{q}}{\int_{v_i} \rho(\boldsymbol{q})d\boldsymbol{q}}. \tag{3.2}$$

This definition implies that the robot at pi is located at the center of mass of its Voronoi region, assuming mass distribution given by $\rho(\boldsymbol{q})$. When the position of the robot satisfies $\boldsymbol{p}_i = \boldsymbol{c}_{v_i}$, it constitutes the CVT. Fig. 3.4 shows the difference between Voronoi tessellations and CVT in two-dimensional and three-dimensional environments, where the density function is set to be a constant density function. The robot distribution maps of Voronoi tessellations in Figs. 3.4a and 3.4c show that the robots are not located at the centroid where they generate Voronoi units, and the size of each Voronoi unit is also different. Figs. 3.4b and 3.4d show the distribution of robots divided by CVT. It can be seen that each robot is located at the centroid position of the Voronoi cell, and the size of each Voronoi cell is consistent.

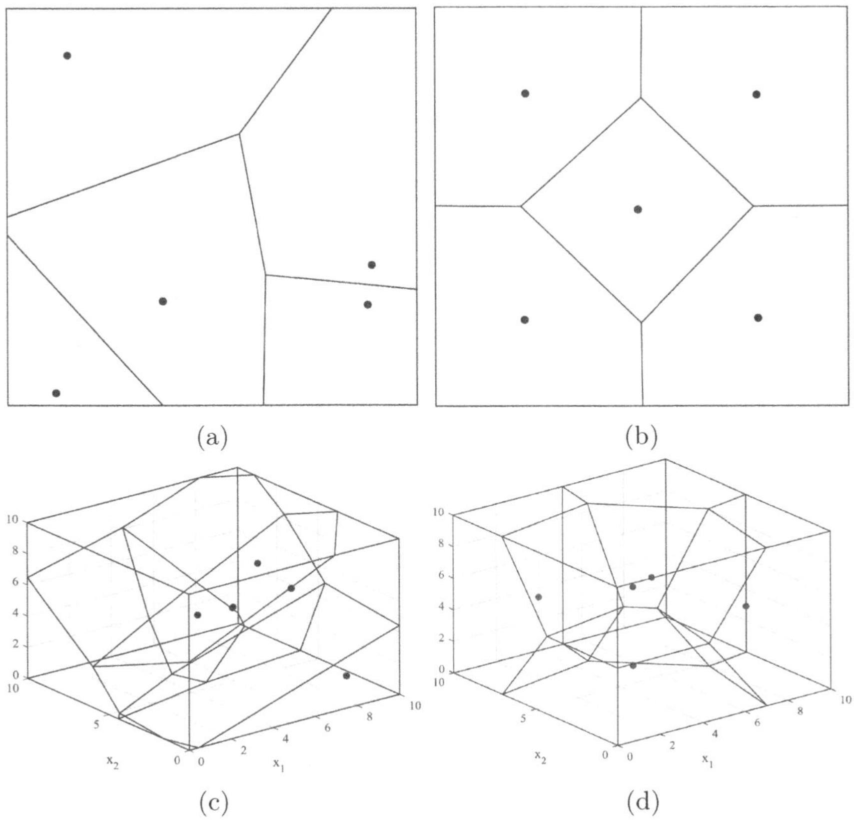


Figure 3.4: Voronoi tessellations and centroid Voronoi tessellations: (a) Two-dimensional Voronoi tessellations; (b) two dimensional centroid Voronoi tessellations; (c) 3D Voronoi tessellations; (d) 3D centroid Voronoi tessellations.

When a multi-robot system performs specific tasks, there is always a certain distance between the position of the robot and the position of the center of mass, which is represented by $f(\boldsymbol{q}) = \|\boldsymbol{q} - \boldsymbol{p}_i\|$. The farther the robot is from the center of mass of the Voronoi unit, the higher the cost of reaching the center of mass. The cost function for centroid Voronoi tessellations of a single robot in the entire task area given a density function can be expressed as:

$$H(\boldsymbol{p}_i) = \sum_{i=1}^{n} \int_{v_i} \|\boldsymbol{q} - \boldsymbol{p}_i\|^2 \rho(\boldsymbol{q}) d\boldsymbol{q}. \tag{3.3}$$

This is often interpreted as a locational optimization functional, where the goal is to minimize the average squared distance between robots and points in their assigned region.

Formula (3.3) aims to reduce the distance between the centroid of the Voronoi cell and the position of the robot, driving the robot to the centroid position of the Voronoi cell while ensuring the minimum cost function. To demonstrate the convergence of the centroid Voronoi tessellations, taking the derivative of (3.3) yields:

$$\frac{\partial H(\boldsymbol{p}_i)}{\partial \boldsymbol{p}_i} = -2\int_{v_i} (\boldsymbol{q} - \boldsymbol{p}_i)\rho(\boldsymbol{q})d\boldsymbol{q} = 2m_{v_i}(\boldsymbol{p}_i - \boldsymbol{c}_{v_i}), \tag{3.4}$$

where $m_{v_i} = \int_{v_i} \rho(\boldsymbol{q})d\boldsymbol{q}$ represents the quality of the Voronoi cell. According to (3.4), when the centroid of the robot and its Voronoi cell coincide, the cost function decreases to 0, indicating that the Voronoi tessellations is convergent.

### 3.3.2 Lloyd's Algorithm and Centroidal Voronoi Tessellations

Lloyd's algorithm, named after Stuart P. Lloyd, is a classic iterative method for computing a Centroidal Voronoi Tessellations. Its core idea is to continuously iterate and optimize the positions of the generated points so that each point is located at the centroid of its Voronoi cell, thereby minimizing a specific energy function (the cost function defined in Formula (3.3)). Each iteration consists of:

1. Constructing the Voronoi diagram for $\{p_i\}$: For the current set of generator points $\{p_i\}_{i=1}^n$, construct the Voronoi diagram over the domain $\Omega$ based on Euclidean distance. This results in a partition of the region $\Omega$ into $n$ Voronoi cells $\{V_i\}_{i=1}^n$, where each cell $V_i$ contains all points that are closer to $p_i$ than to any other generator point $p_j$ (for $j \neq i$).

2. Computing the centroid of each cell $V_i$: For each Voronoi cell $V_i$ obtained in the first step, compute its centroid (center of mass) $c_i$.

3. Moving $p_i$ to the centroid: Move each generator point $p_i$ to the newly computed centroid position $c_i$ of its corresponding Voronoi cell $V_i$.

The above steps 2 and 3 constitute a complete iteration. The algorithm repeats this process until a predetermined termination condition is met (for example, the movement distance of all generated points is less than a certain threshold, or the change in the energy functional is no longer significant). This process converges to a local minimum of $H$ under mild conditions. In a multi-robot setting, this corresponds to robots iteratively adjusting their positions to minimize the energy of the formation.

Lloyd's algorithm is known to converge monotonically to a local minimum of the cost function $H$. Each iteration reduces the value of $H$ because moving a generator to the centroid of its Voronoi cell minimizes the local integral of $\|\boldsymbol{q} - \boldsymbol{p}_i\|^2$ for that cell. The resulting configuration, where $\boldsymbol{p}_i = \boldsymbol{c}_i$ for all $i$, is by definition a Centroidal Voronoi Tessellations. CVTs are characterized by their equilibrium property; the generators are perfectly balanced within their cells, leading to a high-quality, well-distributed partition of the space.

### 3.3.3 Weighted and Anisotropic CVTs

The standard Centroidal Voronoi Tessellation (CVT) assumes that all robots in the system are homogeneous in terms of capabilities and task assignments, and that their sensing and mobility capabilities are isotropic, i.e., symmetric in all directions. However, in realistic multi-robot systems, robots often exhibit differences in load-carrying ability, sensing range, mobility speed, endurance, and assigned task priorities. Moreover, physical constraints such as nonholonomic motion or limited sensor fields of view introduce anisotropy into their coverage behavior. To accommodate these heterogeneities and directional requirements, it is essential to extend the standard CVT model, which leads to the notions of weighted and anisotropic CVTs. Next, in this subsection, we will separately introduce weighted and anisotropic CVTs:

1. Weighted CVTs

Weighted CVTs are designed to model heterogeneous robot capabilities or task-specific priorities. Each robot $i$ is assigned a positive scalar weight $w_i$ that reflects its relative importance, sensing capacity, or computational power. The weighted cost functional is given as:

$$H_w = \sum_{i=1}^{n} w_i \int_{V_i} \|\boldsymbol{q} - \boldsymbol{p}_i\|^2 \rho(\boldsymbol{q}) d\boldsymbol{q}, \tag{3.5}$$

where $\rho(\boldsymbol{q})$ is a spatial density function describing the importance of each location $q \in \Omega$, and $V_i$ is the Voronoi region associated with robot $i$. The weight $w_i$, which encodes the influence or capacity of robot $i$, effectively scales the robot's influence radius: a larger $w_i$ corresponds to a robot capable of covering a wider region, whereas a smaller $w_i$ limits the robot's coverage responsibility.

The minimization of $H_w$ leads to a configuration in which each robot is positioned at the centroid of its weighted Voronoi cell. The weighted centroid $c_i^w$ is defined as:

$$c_i^w = \frac{\int_{V_i} \boldsymbol{q}\, w_i \rho(\boldsymbol{q})\, d\boldsymbol{q}}{\int_{V_i} w_i \rho(\boldsymbol{q})\, d\boldsymbol{q}}. \tag{3.6}$$

When all $w_i$ are equal, Eq. (3.5) naturally reduces to the standard CVT formulation.

2. Anisotropic CVTs.

In many practical applications, such as aerial imaging, radar coverage, and underwater sensing, the coverage efficiency of a robot is direction-dependent. To capture this directional sensitivity, anisotropic CVTs generalize the distance metric used in Voronoi partitioning. Instead of the Euclidean norm, a quadratic form induced by a positive-definite matrix $M_i(\boldsymbol{q})$ is employed:

$$\|\boldsymbol{q} - \boldsymbol{p}_i\|_{M_i}^2 = (\boldsymbol{q} - \boldsymbol{p}_i)^{\mathrm{T}} M_i(\boldsymbol{q})(\boldsymbol{q} - \boldsymbol{p}_i). \tag{3.7}$$

Here, $M_i(\boldsymbol{q})$ defines the local anisotropy at position $\boldsymbol{q}$ for robot $i$. For example, when $M_i$ is elongated along a particular axis, the corresponding Voronoi region will also stretch in that direction, aligning with the robot's preferred sensing or motion orientation.

The anisotropic cost functional can then be expressed as:

$$H_a = \sum_{i=1}^{n} \int_{V_i} \|\boldsymbol{q} - \boldsymbol{p}_i\|^2 \rho(\boldsymbol{q}) \|_{M_i(q)}^2 d\boldsymbol{q}, \tag{3.8}$$

where the matrix field $M_i(\boldsymbol{q})$ allows the tessellation to adapt to the geometry of the environment or task domain. This anisotropy enables the CVT to better conform to obstacles, terrain features, or directionally constrained sensor footprints.

3. Weighted-Anisotropic CVTs.

In practice, both heterogeneity and anisotropy coexist in multi-robot systems. A more general formulation that combines both effects is the weighted-anisotropic CVT, which integrates scalar weights and matrix metrics simultaneously:

$$H_{wa} = \sum_{i=1}^{n} w_i \int_{V_i} \|\boldsymbol{q} - \boldsymbol{p}_i\|^2 \rho(\boldsymbol{q}) \|_{M_i(q)}^2 d\boldsymbol{q}. \tag{3.9}$$

This formulation provides a unified mathematical foundation to model robots that differ not only in strength or sensing capacity but also in directional performance. For instance, aerial drones with forward-facing cameras or wheeled robots with limited turning angles can be effectively modeled within this framework.

4. Applications and Implications

Weighted and anisotropic CVTs offer powerful modeling tools for cooperative coverage control, target tracking, and distributed sensing. Weighted CVTs handle heterogeneity in processing or sensing capabilities through scalar weights, while anisotropic CVTs address directional characteristics in perception and motion through matrix metrics. The two concepts can also be combined to meet complex and diverse application requirements such as adaptive surveillance, resource allocation, or energy-aware formation design in real-world environments.

In summary, these generalizations extend the classical CVT from a purely geometric concept to a versatile analytical framework capable of describing realistic, heterogeneous, and anisotropic multi-robot systems. They form the theoretical basis for the design of distributed, scalable, and adaptive control algorithms in multi-agent coordination tasks.

### 3.3.4 The Proof of Centroidal Voronoi Tessellations

In order to gain a deeper understanding of the multi-robot formation control method based on CVT, it is necessary to formally prove several key mathematical properties of CVT. These properties are not only the foundation of algorithm design and implementation but also directly related to the practical feasibility and security of the system. Next, three theoretical proofs will be presented regarding the safe mobility of robots within the task area, as well as the integrity and consistency of Voronoi tessellations:

(1) The proof of robot safety

For two robots $\boldsymbol{p}_i$, $\boldsymbol{p}_j$, and their Voronoi units $v_i$, $v_j$, use $\dot{\boldsymbol{p}}_i$, $\dot{\boldsymbol{p}}_j$ to represent any point in the Voronoi units of the two robots. The distance $d$ between two arbitrary points $\dot{\boldsymbol{p}}_i$ and $\dot{\boldsymbol{p}}_j$ satisfies $d_{ij} = \left\| \dot{\boldsymbol{p}}_i - \dot{\boldsymbol{p}}_j \right\| \geq 0$.

Proof: For arbitrary point $\dot{\boldsymbol{p}}_i \in v_i$: $\|\boldsymbol{p}_i - \dot{\boldsymbol{p}}_i\|^2 \leq \left\|\boldsymbol{p}_j - \dot{\boldsymbol{p}}_i\right\|^2$, for all $i$, and for arbitrary point $\dot{\boldsymbol{p}}_j \in v_j$: $\left\|\boldsymbol{p}_j - \dot{\boldsymbol{p}}_j\right\|^2 \leq \left\|\boldsymbol{p}_i - \dot{\boldsymbol{p}}_j\right\|^2$, for all $j$. This is equivalent to $\|\boldsymbol{p}_i - \dot{\boldsymbol{p}}_i\|^2 + \left\|\boldsymbol{p}_j - \dot{\boldsymbol{p}}_j\right\|^2 \leq \left\|\boldsymbol{p}_j - \dot{\boldsymbol{p}}_i\right\|^2 + \left\|\boldsymbol{p}_i - \dot{\boldsymbol{p}}_j\right\|^2$, for all $i, j$.

By $\left\|\boldsymbol{p}_i - \boldsymbol{p}_j\right\| \leq \left\|\boldsymbol{p}_j - \boldsymbol{p}_i\right\|$, one can get $\left(\dot{\boldsymbol{p}}_i - \dot{\boldsymbol{p}}_j\right)^T \left(\boldsymbol{p}_i - \boldsymbol{p}_j\right) \geq \left\|\boldsymbol{p}_i - \boldsymbol{p}_j\right\|$, by $\|a\|\|b\| \geq \|a^T b\|$, one can get:
$\frac{\|\dot{\boldsymbol{p}}_i - \dot{\boldsymbol{p}}_j\|\|\boldsymbol{p}_i - \boldsymbol{p}_j\|}{\|\boldsymbol{p}_i - \boldsymbol{p}_j\|} \geq \frac{(\dot{\boldsymbol{p}}_i - \dot{\boldsymbol{p}}_j)^T (\boldsymbol{p}_i - \boldsymbol{p}_j)}{\|\boldsymbol{p}_i - \boldsymbol{p}_j\|} \geq \frac{\|\boldsymbol{p}_i - \boldsymbol{p}_j\|}{\|\boldsymbol{p}_i - \boldsymbol{p}_j\|}$. so, it can be proven that $\|\dot{\boldsymbol{p}}_i - \dot{\boldsymbol{p}}_i\| \geq 0$, for two arbitrary points $\dot{\boldsymbol{p}}_i \in v_i$, $\dot{\boldsymbol{p}}_j \in v_j$, $i \neq j$. The distance between robots is greater than or equal to zero, and both robots in adjacent Voronoi cells are at a safe distance.

(2) The proof of Voronoi cell non-empty property

When the robot is in the center of mass of the Voronoi tessellations configuration, each Voronoi cell is non-empty, that is, for $\boldsymbol{p}_i \in v_i$, $\forall i \in n$, $v_i \neq \varnothing$.

Proof: The formula $v_i = \left\{q \in Q \mid \|q - p_i\|^2 \leq \|q - p_j\|^2\right\}$, when $\boldsymbol{q} = \boldsymbol{p}_i$, can be changed to $v_i = \left\{q \in Q \mid 0 \leq \|q - p_j\|^2\right\}$, therefore, it can be obtained: $0 \leq \|q - p_j\|^2$.

According to the proof of robot safety, $d_{ij} \geq 0$, the minimum distance between two robots is $d_{ij} = 0$, and $d_{ij} = \left\|\boldsymbol{p}_i - \boldsymbol{p}_j\right\|$, and it can be obtained that $0 \leq d_{ij}$, i.e, $0 \leq \left\|\boldsymbol{p}_j - \boldsymbol{p}_i\right\|^2$ can be established, it can be obtained that $\boldsymbol{p}_i \in v_i$, $\forall i \in n$, $v_i \neq \varnothing$, for $d_{ij} \geq 0$, which means that there must be a robot in each Voronoi cell.

(3) The proof of Voronoi cell non-intersection

For robots $\boldsymbol{p}_i$, $\boldsymbol{p}_j$ and their Voronoi cells $v_i$, $v_j$, when $i \neq j$, $v_i \cap v_j = \varnothing$, the Voronoi cells will not intersect with each other.

Proof: For arbitrary point $\dot{\boldsymbol{p}}_i \in v_i$, there exists $\left\|\boldsymbol{p}_j - \dot{\boldsymbol{p}}_j\right\|^2 \leq \|\boldsymbol{p}_i - \dot{\boldsymbol{p}}_i\|^2$. Therefore one can get $\left\|\boldsymbol{p}_j - \dot{\boldsymbol{p}}_i\right\|^2 \leq \left\|\boldsymbol{p}_j - \dot{\boldsymbol{p}}_i\right\|^2$.

Obviously, the result is contradictory, that is, the assumption is not valid, that is, $v_i \cap v_j = \varnothing$, $\forall i \neq j$.

Through the proofs of the above three properties, it can be seen that the CVT not only has clear definitions and boundary tessellations logic from the geometric angles but also has good theoretical guarantees for its practical applications in multi-robot systems. Robots can always remain within their controllable area during operation, avoiding the occurrence of conflicts or overlapping areas. At the same time, each robot is assigned to a specific task area, ensuring the integrity of the multi-robot system in executing tasks. These mathematical characteristics provide a solid theoretical foundation for the subsequent applications of formation algorithms.

### 3.3.5 Formation Control via CVT

The above content demonstrates from the perspectives of geometry and optimization that CVT can generate an ideal spatial arrangement for robots. We now embed CVT into a dynamic control law for multi-robot formation, integrating this static geometric configuration with the dynamic control process of robots, and design a control law that enables the robot system to autonomously and asymptotically converge to the target CVT configuration.

The control strategy we adopt is intuitive and effective: driving each robot to move toward the centroid of its current Voronoi cell. This strategy directly stems from the definition of CVT—the system achieves CVT configuration if and only if all robots are located at the centroids of their respective Voronoi cells. Let $p_i(t)$ be the position of robot $i$ at time $t$. The control input $u_i(t)$ guiding $p_i$ toward the centroid $c_i$ of its current cell is:

$$u_i(t) = -k(p_i(t) - c_i(t)) \tag{3.10}$$

where $k > 0$ is a gain coefficient. Under this law, robots asymptotically converge to the CVT configuration.

One key advantage of this control law resides in its intrinsic decentralized potential. An important property of Voronoi diagrams is that the shape of a Voronoi cell and the calculation of its centroid depend only on the positions of that robot and its Voronoi neighbors. Each robot can compute its control input using only local information—its own position and those of its Voronoi neighbors. Then generate control commands according to the control law. This endows the algorithm with good scalability and robustness against disturbances. The Voronoi diagram can be computed via distributed algorithms such as:

1. Flood-fill neighbor broadcasting: This is a message-passing-based protocol. Robots periodically broadcast their position information to other robots within communication range. These messages are received by neighbors and continue to be forwarded. After several rounds of broadcasting, robots can collect sufficient position information to locally construct a partial Voronoi diagram fragment, which is adequate for computing their own cells. This method is relatively simple to implement, but may involve certain communication delays.

2. Local convex hull construction: Robots first identify all neighbors within their direct communication range. Then, by computing the local convex hull of these neighboring points and incorporating geometric constraints, the approximate positions of Voronoi cell boundaries can be inferred. This method places more emphasis on exploiting local geometric relationships and has certain requirements for the communication range.

3. Power diagram methods for weighted cases: When the system extends to weighted CVT, the corresponding geometric structure is the Power Diagram. Fortunately, the Power Diagram also possesses local properties, and corresponding distributed algorithms exist. Robots need to exchange position and weight information, and the computational principle is similar to that of standard Voronoi diagrams, but employs a weighted distance metric.

Distributed implementation is the key to applying CVT theory to practical multi-robot systems. It leverages the local properties of Voronoi diagrams to decompose the global optimization problem into a series of local decision problems that can be processed in parallel, thereby constructing a highly autonomous, reliable, and scalable cooperative control system.

## 3.4 Simulation and Analysis of Multi-robot Formation

The multi-robot formation control algorithm based on CVT takes the centroid of the Voronoi cell where the robot is located as the tracking target point, that is Euclidean distance between the robot position point and the distributional centroid of the Voronoi cell is regarded as the tracking error. When the tracking error value is less than the convergence threshold of the formation, it indicates that the formation has reached a convergence state. Otherwise, it continues to iterate until the error meets the requirements. The main advantages of applying the CVT to multi-robot formation control algorithms are: first, it can allocate a Voronoi cell to each robotic agent to prevent collisions between agents; second, the design of formation control laws is simple. The multi-agent formation process under the CVT is as follows:

(1) Set $n$ initial position points of the intelligent agents as the Voronoi generation points;

(2) Generate Voronoi cells for each agent based on the initial position point coordinates;

(3) Provide the density function distribution and obtain the centroid position of each Voronoi cell based on the centroid calculation formula;

(4) Drive the robot to the center of mass position and determine whether the distance error between the robot position and the center of mass is less than the preset small value;

(5) If it is less than the preset value, it indicates that the algorithm is converging and the iteration stops; otherwise, use the new position of the robot after movement as the new generating point for the Voronoi cell and return to step (2).

The following subsections will demonstrate the simulation of multi-robot formation control in two-dimensional and three-dimensional environment under different density functions.

### 3.4.1 Formation Simulation in 2D Environments

First, let us simulate the formation control of multi-robot in two-dimensional environment. Fig. 3.5 shows the centroid Voronoi partitioning of multiple robots in a constant density environment, where the red circles represent the positions of the robots in the two-dimensional environment and the red crosses represent the centroid positions of the corresponding Voronoi units. By setting the initial position of the robot to the bottom left corner of the task environment, it can be seen that the robot gradually reduces its distance from the center of mass, ultimately achieving centroid Voronoi partitioning. In a constant density environment, the Voronoi cell area of each robot is consistent, achieving uniform task area division.

Then, a simulation of multi-robot formation control under two-dimensional Gaussian density function is shown in Fig. 3.6, with the peak of the Gaussian density function set in the upper right corner of the task area. It can be seen that after completing the CVT, the robots gather in the upper right corner of the task area. From this, it can be seen that by changing the position of the center of mass, the robot can

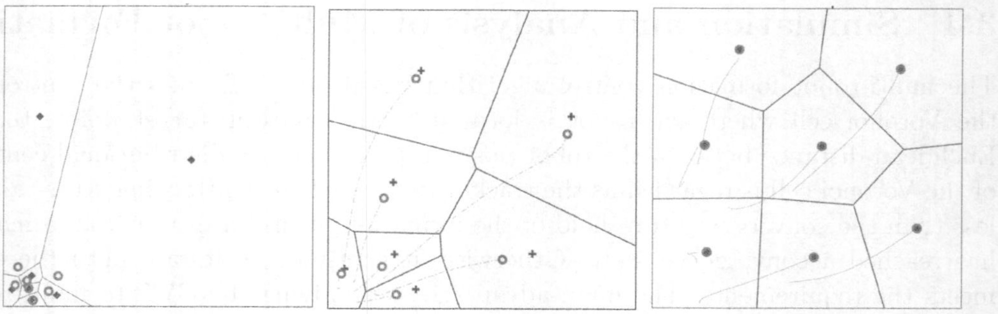

(a) (b) (c)

Figure 3.5: The process of centroid Voronoi tessellations under two-dimensional constant density function: (a) Initial Voronoi tessellations; (b) the robot moves toward the center of mass; (c) centroid Voronoi tessellations.

be driven to the area of interest, thereby improving the accuracy and speed of task completion.

### 3.4.2 Formation Simulation in 3D Environments

Next, we simulate the formation control of multi-robots in a three-dimensional environment. Fig. 3.7 shows the centroid Voronoi tessellations of multi-robots in a constant density environment. The black lines in the figure represent the boundaries of the Voronoi units, and the colored curves represent the motion trajectories of the robots. It can be seen that in a three-dimensional environment, the CVT can still partition robots into different Voronoi units. With the iteration of the algorithm, the size of each Voronoi cell tends to be consistent, achieving uniform task space tessellations.

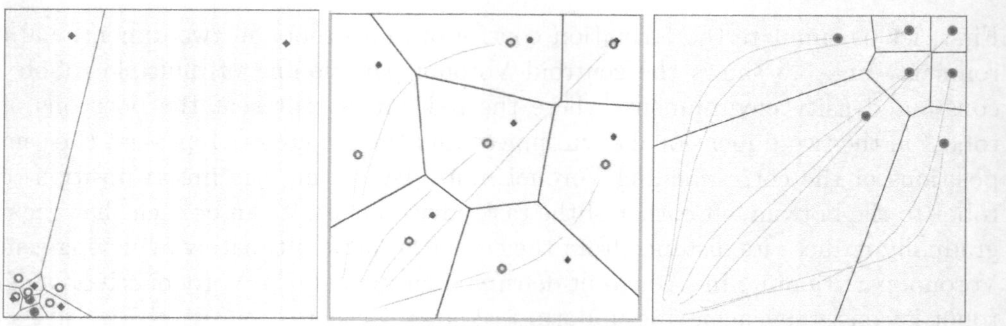

(a) (b) (c)

Figure 3.6: The process of centroid Voronoi tessellations under two-dimensional Gaussian density function centered at the top right corner: (a) Initial Voronoi tessellations; (b) the robot moves toward the center of mass; (c) centroid Voronoi tessellations.

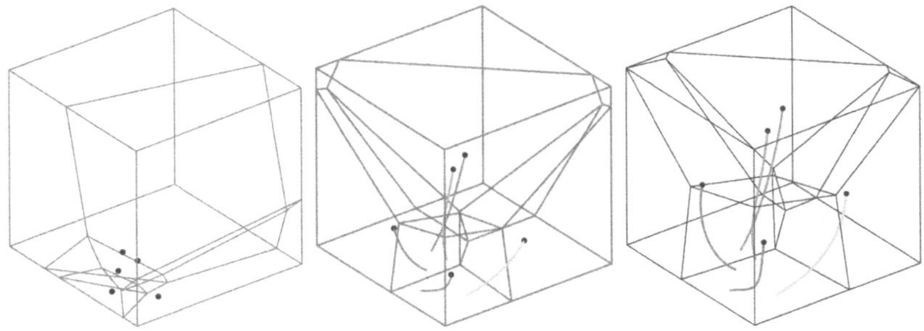


Figure 3.7: The process of centroid Voronoi tessellations under three-dimensional constant density function: (a) Initial Voronoi tessellations; (b) the robot moves toward the center of mass; (c) centroid Voronoi tessellations.

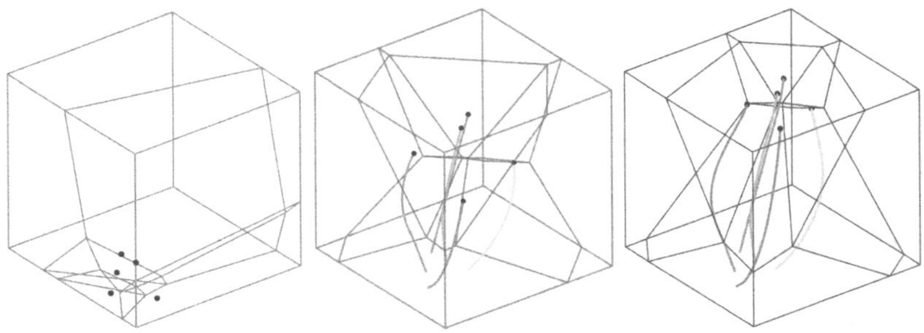


Figure 3.8: The process of centroid Voronoi tessellations under three-dimensional Gaussian density function centered at the top right corner: (a) Initial Voronoi tessellations; (b) the robot moves toward the center of mass; (c) centroid Voronoi tessellations.

The simulation of multi-robot formation control under three-dimensional Gaussian density function is shown in Fig. 3.8, with the peak of the Gaussian density function set in the upper right corner of the three-dimensional task space. It can be seen that after completing the CVT, the robots gather in the upper right corner of the task space. From this example, it can be seen that the CVT still has a good adaptability in three-dimensional space. By changing the density function, the robots can be effectively controlled to complete tasks in three-dimensional space.

## 3.5 Chapter Summary

This chapter focuses on multi-robot collaboration and formation control based on centroidal Voronoi tessellation. It first explains the advantages of applying Voronoi tessellation in task space division and collision avoidance. Then, a CVT-based multi-robot formation control algorithm is proposed, and its safety, completeness, and effectiveness are verified through theoretical derivation. The chapter also introduces centroidal Voronoi tessellation as a powerful method for decentralized and adaptive formation control in multi-robot systems. By minimizing positioning energy, robots naturally self-organize into efficient configurations while maintaining robustness and scalability. In the following sections, this model is extended to dynamic task allocation, formation switching, and obstacle avoidance. Additionally, the chapter presents simulation experiments conducted in both two-dimensional and three-dimensional environments, validating the algorithm's adaptability and convergence across various scenarios.

## Bibliography

[1] Chenyang Huang, Tiantian Xu, and Xinyu Wu. Leader–follower formation control of magnetically actuated millirobots for automatic navigation. *IEEE/ASME Transactions on Mechatronics*, 29(2):1272–1282, 2023.

[2] Hossein Bolandi, Hossein Moradi Pari, and Mohammad Reza Izadi. Multiple spacecraft formation and station-keeping control in presence of static attitude constraint via decentralized virtual structure approach. *Journal of Aerospace Engineering*, 35(1):04021104, 2022.

[3] Jiabao Wen, Jiachen Yang, Yang Li, Jingyi He, Zhengjian Li, and Houbing Song. Behavior-based formation control digital twin for multi-AUG in edge computing. *IEEE Transactions on Network Science and Engineering*, 10(5):2791–2801, 2022.

[4] YangQuan Chen and Zhongmin Wang. Formation control: a review and a new consideration. In *2005 IEEE/RSJ International Conference on Intelligent Robots and Systems*, pages 3181–3186. IEEE, 2005.

[5] Junseok Boo and Dongkyoung Chwa. Fuzzy integral sliding mode observer-based formation control of mobile robots with kinematic disturbance and unknown leader and follower velocities. *IEEE Access*, 10:76926–76938, 2022.

[6] Ruimin Zhou, Wenqian Ji, Qingzheng Xu, and Wenjie Si. Collision avoidance and connectivity preservation for time-varying formation of second-order multi-agent systems with a dynamic leader. *IEEE Access*, 10:31714–31722, 2022.

[7] Jie Mei and Yao Tan. Formation control of mobile robots using bearing-only measurements. *Control Theory & Applications*, 38(7):1043–1050, 2021.

[8] Shen Guofeng, Zhou Mingtuo, Li Jian, Wang Huajun, Li Kai, and Yang Yang. Research and design of fog computing-enabled mobile robots fleet tracking. *Journal of University of Chinese Academy of Sciences*, 38(4):557, 2021.

[9] Lu Liu, Dan Liu, Yuhang Ma, Chenyu Yang, Huiyu Zhang, Bin Yao, Zengxing Zhang, Zhidong Zhang, and Chenyang Xue. Bipartite tracking formation control of nonlinear multi-agent systems using adaptive output–feedback protocols. *IEEE Access*, 10:70699–70711, 2022.

[10] Craig W Reynolds. Flocks, herds and schools: A distributed behavioral model. In *Proceedings of the 14th Annual Conference on Computer Graphics and Interactive Techniques*, pages 25–34, 1987.

[11] Levent Bayindir and Erol Sahin. Modeling self-organized aggregation in swarm robotic systems. In *2009 IEEE Swarm Intelligence Symposium*, pages 88–95. IEEE, 2009.

[12] Tucker Balch and Maria Hybinette. Social potentials for scalable multi-robot formations. In *Proceedings 2000 ICRA. Millennium Conference. IEEE International Conference on Robotics and Automation. Symposia Proceedings (Cat. No. 00CH37065)*, volume 1, pages 73–80. IEEE, 2000.

[13] Wei Ren and Randal W Beard. Consensus seeking in multiagent systems under dynamically changing interaction topologies. *IEEE Transactions on Automatic Control*, 50(5):655–661, 2005.

[14] Ruimin Zhou, Wenqian Ji, Qingzheng Xu, and Wenjie Si. Collision avoidance and connectivity preservation for time-varying formation of second-order multi-agent systems with a dynamic leader. *IEEE Access*, 10:31714–31722, 2022.

[15] Matteo Santilli, Pratik Mukherjee, Ryan K Williams, and Andrea Gasparri. Multi-robot field of view control with adaptive decentralization. *IEEE Transactions on Robotics*, 38(4):2131–2150, 2022.

[16] Enming Wu, Yidong Sun, Jianyu Huang, Chao Zhang, and Zhiheng Li. Multi-UAV cluster control method based on virtual core in improved artificial potential field. *IEEE Access*, 8:131647–131661, 2020.

[17] Luigi Feola and Vito Trianni. Adaptive strategies for team formation in minimalist robot swarms. *IEEE Robotics and Automation Letters*, 7(2):4079–4085, 2022.

[18] M Anthony Lewis and Kar-Han Tan. High precision formation control of mobile robots using virtual structures. *Autonomous Robots*, 4:387–403, 1997.

[19] Jiehong Wu, Liangkai Zou, Liang Zhao, Ahmed Al-Dubai, Lewis Mackenzie, and Geyong Min. A multi-UAV clustering strategy for reducing insecure communication range. *Computer Networks*, 158:132–142, 2019.

[20] Charbel Toumieh and Alain Lambert. Decentralized multi-agent planning using model predictive control and time-aware safe corridors. *IEEE Robotics and Automation Letters*, 7(4):11110–11117, 2022.

[21] Shuai Zhang and Jia Pan. Collecting a flock with multiple sub-groups by using multi-robot system. *IEEE Robotics and Automation Letters*, 7(3):6974–6981, 2022.

[22] He Yin, Shuxiang Guo, and Meng Liu. A virtual linkage-based dual event-triggered formation control strategy for multiple amphibious spherical robots in constrained space with limited communication. *IEEE Sensors Journal*, 22(13):13395–13406, 2022.

[23] Meng Zhou, Jianyu Li, Chang Wang, Jing Wang, and Li Wang. Applications of Voronoi diagrams in multi-robot coverage: A review. *Journal of Marine Science and Engineering*, 12(6):1022, 2024.

[24] Jeongeun Kim and Hyoung Il Son. A Voronoi diagram-based workspace partition for weak cooperation of multi-robot system in orchard. *IEEE Access*, 8:20676–20686, 2020.

[25] Alexandre Eudes, Sylvain Bertrand, Julien Marzat, and Ioannis Sarras. Distributed control for multi-robot interactive swarming using Voronoi partioning. *Drones*, 7(10):598, 2023.

[26] Hao Cheng, Bailing Tian, Xuewei Zhang, and Hongming Shen. A distributed algorithm for multi-robot task allocation via weighted buffered Voronoi partition. *IEEE Transactions on Industrial Electronics*, 2024.

[27] Yiran Ding. Distributed multi-robot collision avoidance using the Voronoi-based method. In *Journal of Physics: Conference Series*, volume 1948, page 012015. IOP Publishing, 2021.

[28] Hai Zhu, Bruno Brito, and Javier Alonso-Mora. Decentralized probabilistic multi-robot collision avoidance using buffered uncertainty-aware Voronoi cells. *Autonomous Robots*, 46(2):401–420, 2022.

[29] Hongyang Zhao, Yanan Guo, Yi Liu, and Jing Jin. Multi-robot unknown environment exploration and obstacle avoidance based on a Voronoi diagram and reinforcement learning. *Expert Systems with Applications*, 264:125900, 2025.

[30] Dong Yan, Yufeng Xiao, Shizun Sheng, Zhenyu Ren, and Shuang Yang. Radioactive sources search method based on multi-robot and Voronoi partition. *Applied Radiation and Isotopes*, 212:111475, 2024.

CHAPTER 4

# Multi-Robot Formation Switching Based on Voronoi Tessellations

## 4.1 Introduction

MULTI robot formation control refers to the coordinated operation of multiple robots to achieve structured collective behaviors through formation algorithms and cooperative control strategies. With the rapid development of autonomous systems, such approaches have found applications across a wide range of domains, including unmanned ground vehicles for transportation and reconnaissance, unmanned aerial vehicles for surveillance and delivery, spacecraft swarms for distributed sensing, and autonomous underwater vehicles for oceanographic exploration. Compared with single-robot systems, multi-robot formations can accomplish tasks that are infeasible for a single robot, such as search and rescue in hazardous environments, cooperative transportation of large payloads, environmental monitoring of phenomena such as gas leaks and oil spills.

Control strategies for multi-robot formations can be categorized into centralized and distributed paradigms. In centralized control, a central decision-making unit computes trajectories and allocates tasks based on global formation information. This paradigm offers advantages in terms of globally optimal decision-making, simplified coordination, and reduced inter-robot communication, since only the central node manages information flow. Representative studies include Guo et al. [1] extended the GPMP2 motion planning framework to multi-robot systems, and Wang et al. [2] combined artificial potential fields with centralized formation control to enable multi-vehicle formation assembly. However, centralized control suffers from two major drawbacks: poor scalability as the robot number grows, and vulnerability to single-point failures that can compromise the entire system. Distributed control eliminates reliance on a central authority by exploiting local information and inter-robot interactions. Robots coordinate through neighbor-to-neighbor communication and consensus, which makes the system inherently more scalable and robust against node

DOI: 10.1201/9781003714569-4

failures. Huang et al. [3] proposed a distributed optimization framework based on relative positions and virtual reference centers, employing the Gradient Projection (GP) algorithm to minimize formation distances and the Hungarian algorithm to optimize task assignments. Feola et al. [4] investigated minimalist swarms and demonstrated how adaptive modulation of random walk persistence can yield self-organized aggregation resembling molecular-scale interactions in active colloids. Rezeck et al. [5] further introduced a chemistry-inspired heuristic based on Gibbs Random Field (GRF), enabling fully decentralized formation control with only local interactions. While distributed control enhances robustness and scalability, it often faces challenges in achieving global optimality, ensuring stability in cluttered environments, and mitigating communication overhead.

In recent years, VT and in particular CVT, have emerged as powerful tools for multi-robot coordination. Voronoi-based methods naturally partition the environment into regions of influence, ensuring equitable spatial coverage and inherent collision avoidance. Zheng et al. [6] achieved heterogeneous swarm formation transitions using weighted Voronoi diagrams, while Zheng et al. [7] integrated uniformity and spatial coherence into a unified objective function using 2-Wasserstein distance-based interpolation to optimize switching strategies. For density function-guided methods, Bai et al. [8] designed repulsive density functions based on obstacle-detection data to enable adaptive obstacle avoidance, while Xu et al. [9] controlled formation contraction via time-varying density functions. Further refinements, include time-varying diffeomorphisms [10], improved mobility in obstacle-rich environments and etc. Applications were extended to dynamic scenarios, such as Guo et al. [11] simulating flood zones for UAV coverage optimization, Xu et al. [12] configuring hemispherical constraint regions for multi-UAV photography, and Teruel et al. [13] developing controllers for time-varying density functions, to name a few. However, two challenges persist: constraint deformation methods struggle with fine-tuning individual robot positions, while density-guided approaches risk increased path lengths and positional errors due to Voronoi cell randomness. In extreme cases, centroids may misalign with high-density regions, leading to formation failure. These limitations highlight the trade-offs between global shape control and local adaptability in CVT-based strategies.

Despite these advances, several open challenges remain. Constraint-deformation methods often lack flexibility in fine-tuning individual robot trajectories, leading to rigid or suboptimal configurations. Density-guided approaches, while adaptive, are prone to Voronoi randomness, which can increase path lengths, introduce positional errors, and even cause centroid misalignment with high-density regions. In extreme cases, formation degradation or outright switching failures can occur. These challenges reveal a fundamental trade-off between global shape regulation and local adaptability, which lies at the core of CVT-based strategies.

To address these challenges, this chapter presents a dynamic density-guided formation switching framework. Within predefined constraint regions, we employ shape interpolation techniques grounded in optimal transport theory to generate smooth transitions between the current formation density function and the desired formation density function. By leveraging the generated dynamic density to guide the formation and exploiting the inherent collision avoidance properties of the CVT, we can achieve smooth multi-robot formation transitions across distinctive configurations. This method ensures that Voronoi cell centroids adaptively align with the evolving high-density regions during transitions, minimizing path deviations and positional errors while maintaining global coherence and local collision avoidance.

## 4.2 Formation Switching Based on Dynamic Density Guidance

### 4.2.1 Traditional Formation Switching

Traditional formation switching relies on the Voronoi partitioning property of CVT and the tendency of robots to converge toward the centroids of their Voronoi cells. By prescribing a density distribution function within the workspace, CVT aligns the centroidal positions of robots with the regions of higher density, thereby achieving a desired spatial arrangement. Different target formations can thus be generated by selecting appropriate density functions. The influence of these density functions is summarized in Table 4.1, where the parameter $\sigma$ represents the standard deviation controlling the spread of the distribution, $r$ defines the radius in ring-shaped functions, and $(x_c, y_c)$ specifies the center of the distribution. Constants $a$, $b$, and $c$ are additional shaping parameters that determine the orientation and offset of the density profile. Through careful tuning of these parameters, a wide variety of canonical formations (line, circle, ellipse, etc.) can be prescribed.

The general switching procedure is as follows: an initial formation is generated using a chosen density function, and then a new density function corresponding to the target formation is imposed on the CVT framework. The robots update their Voronoi regions accordingly and move toward the new centroids, thereby achieving

Table 4.1: Relationship between density function and vino diagram

| $\rho(q)$ | **Density distribution** | **Voronoi diagram** | **Formation** |
|---|---|---|---|
| $e^{-\sigma[ax+by+c]^2}$ | 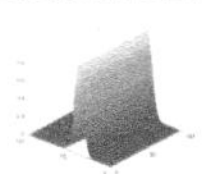 | 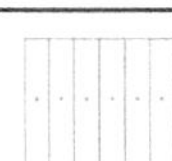 | Straight line |
| $e^{-\sigma[a\|x-x_c\|+b(y-y_c)+c]^2}$ | 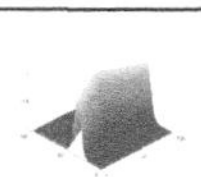 | 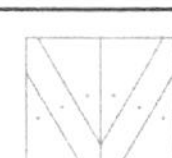 | V-shaped |
| $e^{-\sigma[a(x-x_c)^2+b(y-y_c)^2-r^2]^2}$ | 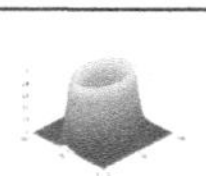 | 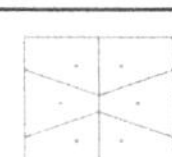 | Circular |

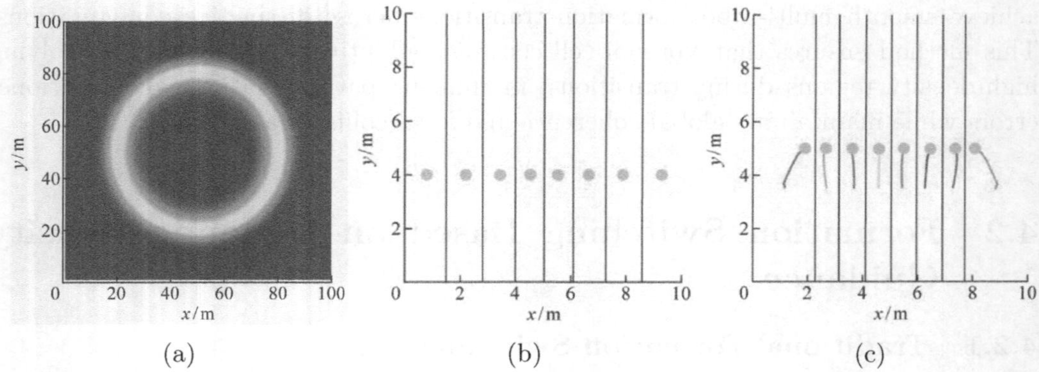


Figure 4.1: Switching ring formation: (a) Circumferential density distribution; (b) initial formation; (c) the formation switch is complete.

formation switching. This method leverages the distributed nature of CVT, ensuring collision-free transitions as each robot remains confined to its Voronoi cell.

However, in certain special configurations of the CVT structure, there are cases where the centroids are not partitioned into high-density regions after re-imposing the density function of the desired formation shape. A typical failure scenario is illustrated in Fig. 4.1. As shown in Fig. 4.1b, a linear formation is first generated using a linear density function. The robots (orange dots) are positioned along a straight line, and their Voronoi partitions are represented by vertical boundaries. When the ring-shaped density function of Fig. 4.1a is introduced into the same workspace, the centroids of the Voronoi cells shift toward their local mass centers. Although each robot follows its centroidal trajectory, the final positions do not align with the desired circular structure, as observed in Fig. 4.1c. Instead, the robots remain trapped near their initial linear configuration, failing to complete the formation transition.

This phenomenon arises because the Voronoi tessellation depends on the current robot distribution: if the initial arrangement does not sufficiently overlap with the high-density regions of the target distribution, the centroids may converge to locally consistent but globally incorrect equilibria. In other words, traditional CVT-based switching is strongly influenced by initial conditions and the geometry of Voronoi partitions, which can lead to mismatches between the prescribed density and the realized formation.

Therefore, while CVT offers flexibility and inherent collision avoidance, its direct application to formation switching is limited when density functions change drastically. This motivates the need for enhanced strategies, such as dynamic density-guided interpolation, which gradually morphs one density into another and ensures that centroid trajectories remain aligned with the evolving target distribution. Such an approach will be introduced in the next section.

### 4.2.2 Density Function Shape Interpolation

To overcome the limitations of direct density switching, this section adopts the shape interpolation method to generate transitional density distributions between two density functions, thereby enabling smooth transformations between formations. Shape interpolation [14] is a computer graphics technique that generates intermediate forms between two or more distinct shapes. Conventional shape interpolation involves selecting control points from the target shapes, deforming and displacing these points to create intermediate forms. This method requires constructing mesh models for interpolated objects, which involves intensive computational workloads and is unsuitable for the concise nature of CVT density functions.

Unlike traditional mesh-based interpolation methods that require heavy computation of intermediate geometries, this chapter employs an optimal transport theory-based shape interpolation approach [15]. Unlike traditional methods, this technique interpolates probability densities directly applicable to CVT. The Wasserstein distance $W_2$ between probability density distributions $\rho_0 \in \Omega_1$ and $\rho_1 \in \Omega_2$ is defined as:

$$W_2(\rho_0, \rho_1) = \sqrt{\inf_{\xi \in \mathcal{E}(\rho_0, \rho_1)} \iint_{\Omega_1 \times \Omega_2} d^2(x, y)\, \mathrm{d}\xi(x, y)}, \tag{4.1}$$

where $\mathcal{E}$ represents the transportation plan from distribution $\alpha$ to $\beta$; $d(x, y)$ is the distance between points $x$ and $y$; $\xi(x, y)$ is the joint distribution function in the two-dimensional Wasserstein distance, representing the maximum probability distribution from probability distribution $\Omega_1$ to probability distribution $\Omega_2$; $\mathrm{d}\xi(x, y)$ is the differential measure of the joint distribution function, describing how the mass at a point changes its position during the calculation of the Wasserstein distance.

A weighted barycentric formulation of interpolation can be written as:

$$\rho(\gamma) = \arg\min_{\rho} \left[ (1-\gamma) W_2^2(\rho_0, \rho) + \gamma W_2^2(\rho, \rho_1) \right], \quad \gamma \in [0, 1], \tag{4.2}$$

where $\rho_0$ and $\rho_1$ are the initial and desired density distributions of the formation. By computing the optimal transport path between the two density distributions, selecting a position $y$ along all paths, and applying barycentric interpolation to derive the transitional density distribution, the transformation between the two density functions is realized. Fig. 4.2 illustrates the process of interpolating the transitional density distribution between a straight-line and a toroidal (circular-ring) density distribution using this method.

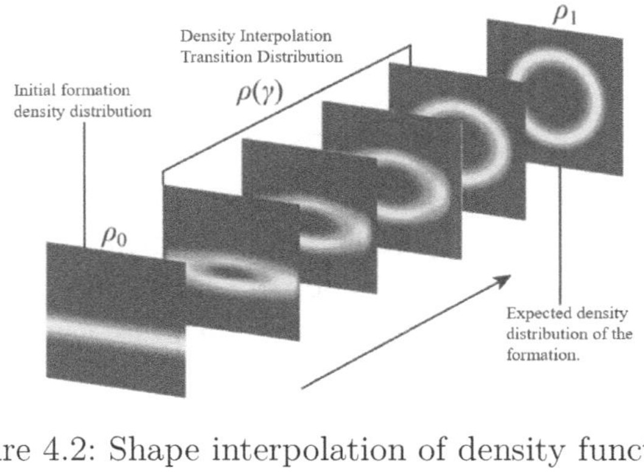


Figure 4.2: Shape interpolation of density function.

Practically, the interpolation yields a sequence of intermediate densities:

$$\{Z_m(q)\}_{m=1}^{\lambda}, \quad Z_1(q) = \rho_0(q),\ Z_\lambda(q) = \rho_1(q). \tag{4.3}$$

where $\lambda$ denotes the number of interpolation steps. These densities are then used sequentially as inputs to the CVT formation control, guiding the robots smoothly from the initial to the target formation.

To ensure successful switching, it is necessary to analyze the cost function minimization of CVT under interpolated densities. Define the cost of robot $i$ located at $p_i$ within Voronoi cell $v_i$ as:

$$E_{v_i}(p_i) = \int_{v_i} Z_m(q)\, \|q - p_i\|^2\, dq. \tag{4.4}$$

Consider a perturbation of the robot's position, denoted by $\varepsilon \in \mathbb{R}^2$. It follows that:

$$\begin{aligned} E_{v_i}(p_i + \varepsilon) - E_{v_i}(p_i) &= \int_{v_i} Z_m(q) \Big( \|q - (p_i + \varepsilon)\|^2 - \|q - p_i\|^2 \Big)\, dq \\ &= \int_{v_i} Z_m(q) \Big( -2\varepsilon^\top (q - p_i) + \|\varepsilon\|^2 \Big)\, dq. \end{aligned} \tag{4.5}$$

Simplifying equation (4.5) yields:

$$\frac{E_{v_i}(\boldsymbol{p}_i + \boldsymbol{\varepsilon}) - E_{v_i}(\boldsymbol{p}_i)}{\epsilon} = \int_{\mathcal{V}_i} Z_m(\boldsymbol{q})\, [\boldsymbol{\varepsilon} + 2(\boldsymbol{p}_i - q)]\, \mathrm{d}\boldsymbol{q}. \tag{4.6}$$

when $\varepsilon$ approaches infinitesimally close to zero, we obtain:

$$\dot{E}_{v_i}(\boldsymbol{p}_i) = 2 \int_{v_i} (\boldsymbol{p}_i - \boldsymbol{q}) Z_m(\boldsymbol{q})\, \mathrm{d}\boldsymbol{q}. \tag{4.7}$$

Expanding the integral gives:

$$\dot{E}_{v_i}(\boldsymbol{p}_i) = 2\boldsymbol{p}_i \int_{v_i} Z_m(\boldsymbol{q})\, \mathrm{d}\boldsymbol{q} - 2 \int_{v_i} \boldsymbol{q} Z_m(\boldsymbol{q})\, \mathrm{d}\boldsymbol{q}. \tag{4.8}$$

If the derivative term equals zero, the minimal solution is:

$$\boldsymbol{p}_i = \frac{\int_{v_i} \boldsymbol{q} Z_m(\boldsymbol{q})\, \mathrm{d}\boldsymbol{q}}{\int_{v_i} Z_m(\boldsymbol{q})\, \mathrm{d}\boldsymbol{q}}. \tag{4.9}$$

At this point, the robot's position coincides with the centroid position of the Voronoi cell. Thus, it can be proven that the CVT can minimize the cost function of the formation, achieving convergence of the formation. In formation tasks, it is necessary to form the desired formation shape. Therefore, the estimated error between the actual distance and the desired distance is used as the cost function:

$$E_v(\boldsymbol{p}) = \sum_{i=1}^{n} \sum_{j \in N_i} \frac{1}{2} \left( \|\boldsymbol{p}_i - \boldsymbol{p}_j\| - d_{ij} \right)^2 = \sum_{i=1}^{n} E_{v_i}(\|\boldsymbol{p}_i - \boldsymbol{p}_j\|), \tag{4.10}$$

where $d_{ij}$ represents the desired distance between two different robots $i$ and $j$; $E_v(\boldsymbol{p})$ denotes the cost function, which is used to measure the distance error between the current robot formation and the desired formation. When this error is zero, it indicates that the desired formation has been achieved. Taking the derivative of $E_{v_i}(\boldsymbol{p}_i)$ yields:

$$\frac{\partial E_{v_i}}{\partial \boldsymbol{p}_i} = \sum_{j \in N_i} \frac{\|\boldsymbol{p}_i - \boldsymbol{p}_j\| - d_{ij}}{\|\boldsymbol{p}_i - \boldsymbol{p}_j\|} (\boldsymbol{p}_i - \boldsymbol{p}_j)^{\mathrm{T}}, \tag{4.11}$$

which describes the task process of a multi-robot system. When the distance difference between two different robots is less than a preset threshold, the weight coefficient is changed to a negative number, thereby continuously controlling the robots to increase the distance between them. Conversely, the distance between them is reduced. When the distance between the two is zero, the control problem is transformed into a consensus problem of the group, that is, the minimization problem of the system cost function. This formalization derivation of optimization problems based on cost functions can enhance the adaptability of the system. By selecting different weights or defining appropriate assumptions, the distribution of the controller can be constrained, thereby ensuring that the system converges within a limited time. Generally, the error threshold is set to 0.02 m. At this time, the error's impact on the system can be ignored, and the iteration is stopped.

The tracking control law for the robot's moving target point is as follows:

$$\boldsymbol{U}_i(t) = K_p \boldsymbol{e}(t) + K_i \int \boldsymbol{e}(t) \, \mathrm{d}t + K_d \frac{\mathrm{d}\boldsymbol{e}(t)}{\mathrm{d}t}, \tag{4.12}$$

where:

$$\boldsymbol{e}(t) = \boldsymbol{p}_i - \boldsymbol{c}_{v_i}, \tag{4.13}$$

and $\boldsymbol{U}_i$ is the control input of the $i$-th robot; $\boldsymbol{c}_{v_i}$ is the centroid of the CVT and represents the robot's moving target point; $\boldsymbol{e}(t)$ is the error between the robot's real-time position and the target position; $K_p$, $K_i$, and $K_d$ are the parameters of the PID controller.

Next, we analyze the stability of the controller and use a Lyapunov function to prove the convergence of the system. First, we separate the integral part of (4.12) :

$$\ddot{\boldsymbol{p}}_i = \boldsymbol{u}_i = -K_p \left( \boldsymbol{p}_i - \boldsymbol{c}_{v_i} \right) - K_d \dot{\boldsymbol{p}}_i. \tag{4.14}$$

The Lyapunov function is defined as:

$$V(\boldsymbol{p}_i) = \frac{1}{2} K_p \left\| \boldsymbol{p}_i - \boldsymbol{c}_{v_i} \right\|^2 + \frac{1}{2} \dot{\boldsymbol{p}}_i^2. \tag{4.15}$$

Taking the derivative of (4.15) yields:

$$\dot{V}(\boldsymbol{p}_i) = K_p(\boldsymbol{p}_i - \boldsymbol{c}_{v_i})\dot{\boldsymbol{p}}_i + \dot{\boldsymbol{p}}_i\ddot{\boldsymbol{p}}_i. \tag{4.16}$$

Substituting the parameter $\ddot{\boldsymbol{p}}_i$ with $\dot{\boldsymbol{p}}_i$ in (4.17):

$$\dot{V}(\boldsymbol{p}_i) = \dot{\boldsymbol{p}}_i\left[K_p(\boldsymbol{p}_i - \boldsymbol{c}_{v_i}) - K_p(\boldsymbol{p}_i - \boldsymbol{c}_{v_i}) - K_d\dot{\boldsymbol{p}}_i\right]. \tag{4.17}$$

Simplifying this expression gives:

$$\dot{V}(\boldsymbol{p}_i) = -K_d\dot{\boldsymbol{p}}_i^2. \tag{4.18}$$

From(4.18), we know that $\dot{V} \leq 0$, it can therefore be concluded that the time derivative of the Lyapunov function is semi-negative definite and hence $V$ is non-increasing. Moreover, since $V$ is positive definite with respect to $(\boldsymbol{p}_i, \dot{\boldsymbol{p}}_i)$, and its derivative is non-positive, the system trajectories are bounded. According to the LaSalle invariance principle, the system will asymptotically converge to the largest invariant set contained in $\{\dot{V} = 0\}$.

Applying LaSalle's invariance principle, define the set:

$$S = \{(\boldsymbol{p}_i, \dot{\boldsymbol{p}}_i) \mid \dot{V} = 0\} = \{(\boldsymbol{p}_i, \dot{\boldsymbol{p}}_i) \mid \dot{\boldsymbol{p}}_i = 0\}. \tag{4.19}$$

To ensure that the system state remains invariant on $S$, it must satisfy $\ddot{p}_i \equiv 0$, which implies that $\boldsymbol{p}_i = \boldsymbol{c}_{v_i}$. Therefore, the largest invariant set can be defined as:

$$M = \{(\boldsymbol{p}_i, \dot{\boldsymbol{p}}_i) \mid \boldsymbol{p}_i = \boldsymbol{c}_{v_i},\ \dot{\boldsymbol{p}}_i = 0\}. \tag{4.20}$$

By LaSalle's invariance principle, all trajectories of the system satisfy:

$$(\boldsymbol{p}_i(t), \dot{\boldsymbol{p}}_i(t)) \to M, \quad t \to \infty. \tag{4.21}$$

At equilibrium, $\boldsymbol{c}_{v_i}$ corresponds to a constant (or equilibrium) value. Hence, the set $M$ contains only one equilibrium point, meaning:

$$(\boldsymbol{p}_i(t), \dot{\boldsymbol{p}}_i(t)) \to (\boldsymbol{c}_{v_i}, 0), \quad t \to \infty. \tag{4.22}$$

Therefore, the system is asymptotically stable at the equilibrium point, the equilibrium $(\boldsymbol{c}_{v_i}, 0)$. This indicates that during the formation switching process, the system can converge to the current target position at each stage, ultimately completing the switch precisely.

### 4.2.3 Formation Switching Algorithm

This section details the implementation of the formation switching algorithm based on dynamic density guidance. The algorithm leverages a smooth transition between density functions to drive the robot swarm through an orderly and continuous transformation.

The process commences by generating a set of dynamic, continuously varying density functions, $Z_m(\boldsymbol{q})$, using the density interpolation technique described previously. This set is derived from the initial formation's density distribution $\rho_0(\boldsymbol{q})$, and the target formation's density distribution $\rho_1(\boldsymbol{q})$. Within this set $Z_1(\boldsymbol{q})$ corresponds to the initial distribution $\rho_0(\boldsymbol{q})$, while $Z_\lambda(\boldsymbol{q})$ represents the final target distribution $\rho_1(\boldsymbol{q})$.

The core of the algorithm is structured as a nested loop. The outer loop iterates through the series of dynamic density functions, from $m = 1$ to $\lambda$. The purpose of the inner loop is to ensure that the formation stably converges under the guidance of the current density function $Z_m(\boldsymbol{q})$. Specifically, the system continuously computes the centroid $\boldsymbol{c}_{v_i}$ of each robot's Voronoi cell, which serves as the robot's next target point. The inner loop terminates only when the position error $\|\boldsymbol{e}\|$ for every robot falls below a predefined threshold $\|\boldsymbol{\varepsilon}\|$, signifying that the formation has stabilized for the current intermediate stage.

Upon the convergence of the inner loop, the outer loop advances to the next density function$Z_{m+1}(\boldsymbol{q})$, and the process is repeated. Through this method of progressive updates and guidance, the positions of the robots in the swarm are adjusted smoothly in accordance with the evolving density distribution. When the guidance process under all $\lambda$ density functions is complete, the entire formation switching procedure concludes, achieving a precise transformation from the initial to the target configuration.

**Algorithm 1** The algorithm flow for CVT formation switching based on dynamic density guidance.

**Input:** Initial positions of robots $\boldsymbol{p}_i$, Boundary of constrained area $\partial Q$, maximum allowable error $\boldsymbol{\varepsilon}$, number of density instances $\lambda$, initial formation density function $\rho_0(\boldsymbol{q})$, desired formation density function $\rho_1(\boldsymbol{q})$
**Output:** Final positions $\boldsymbol{p}_i$ of each robot
Compute the interpolated dynamic density $Z_m(\boldsymbol{q})$ based on densities $\rho_0(\boldsymbol{q})$ and $\rho_1(\boldsymbol{q})$
**for** $m = 1$ to $\lambda$ **do**
    Compute the centroid $\boldsymbol{c}_{v_i}$ of the Voronoi cell using $Z_m(\boldsymbol{q})$ and current robot positions $\boldsymbol{p}_i$
    **while** $[\|\mathrm{e}\| = \|\boldsymbol{p}_i - \mathrm{c}_{v_i}\|] \geq \|\boldsymbol{\varepsilon}\|$ **do**
        Compute the control input $\boldsymbol{U}_i$ for each robot
        Drive robots toward $\boldsymbol{c}_{v_i}$ and update positions $\boldsymbol{p}_i$
        Update the centroid $\boldsymbol{c}_{v_i}$ using $Z_m(\boldsymbol{q})$ and new positions $\boldsymbol{p}_i$
    **end while**
**end for**

## 4.3 Simulation Analysis of Formation Switching

To validate the effectiveness of the dynamic density-guided algorithm proposed in this chapter, we designed and conducted a series of simulations for multi-robot formation switching and reconfiguration. In the simulation environment, the parameters were set as follows: the number of robots in the formation is $i = 8$, their movement range

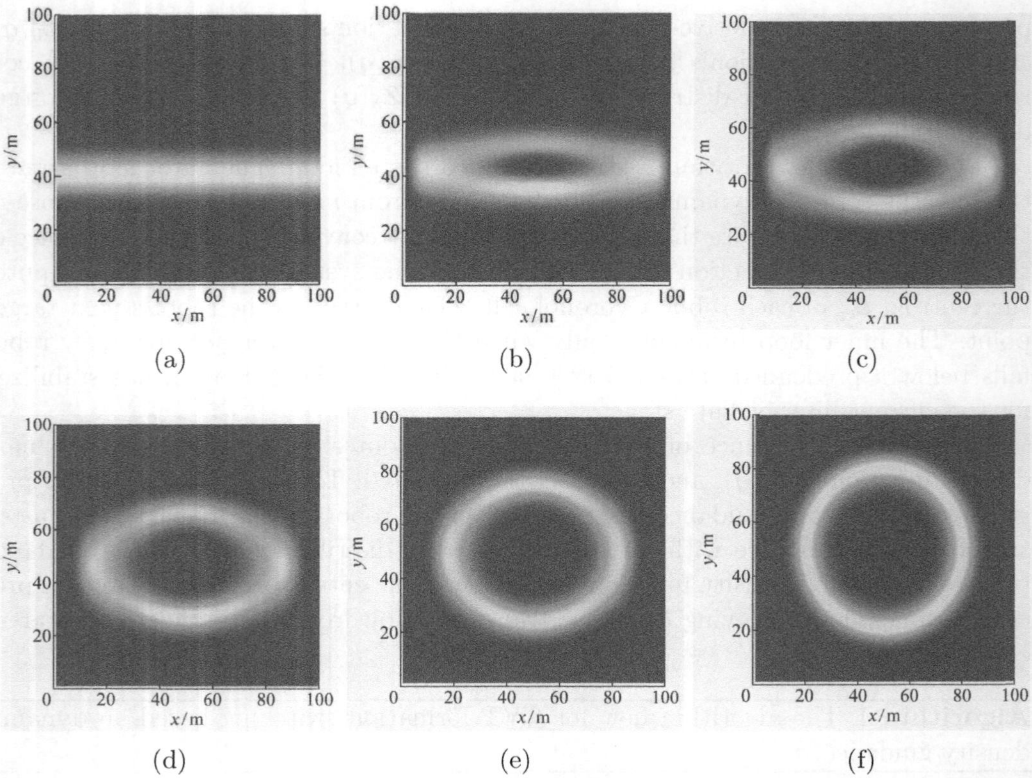


Figure 4.3: Dynamic transition from linear density distribution to annular density distribution: (a) $m = 1$; (b) $m = 4$; (c) $tm = 8$; (d) $m = 12$ (e) $m = 16$; (f) $m = 20$.

is a $10\,\text{m} \times 10\,\text{m}$ square area, and the number of interpolated steps for the dynamic density transition is set to $\lambda = 20$.

### 4.3.1 Formation Switching

(1) Simulation 1

This first simulation is specifically designed to demonstrate how the proposed dynamic density-guided method successfully resolves the formation switching failure that occurs with traditional direct-switching approaches, as highlighted in Fig. 4.3.

Fig. 4.3 visualizes the core mechanism of our method by illustrating the entire evolutionary process of the density function. The process begins with the initial linear distribution of Fig. 4.3a and culminates in the desired circular distribution of Fig. 4.3f. The intermediate subfigures, Fig. 4.3b through Fig. 4.3e, depict the smooth, transitional density fields generated by our algorithm, which create a continuous guidance path for the formation change.

Fig. 4.4 documents the resulting robot trajectories as they are guided by this evolving density function. It clearly shows the robots progressively moving from their initial straight-line arrangement in Fig. 4.4a to a stable circular formation in Fig. 4.4f. This successful outcome provides a stark contrast to the failure case from Fig. 4.1,

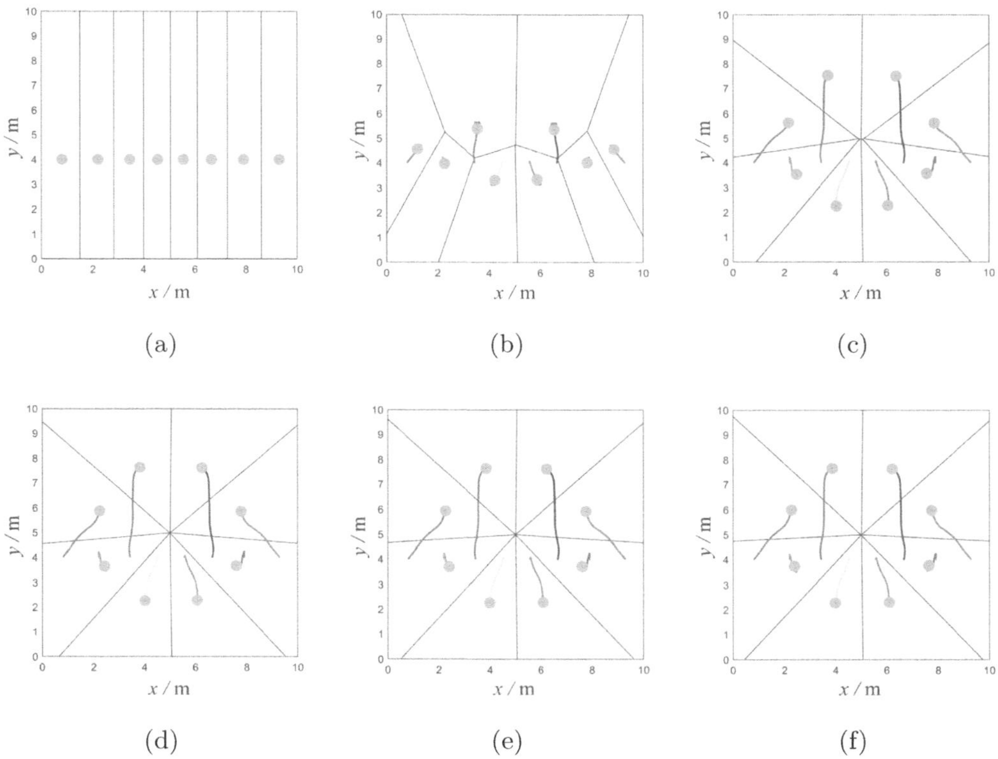


Figure 4.4: Dynamic density guided formation switching process of simulation 1: (a) $k = 1$; (b) $k = 14$; (c) $k = 33$; (d) $k = 47$; (e) $k = 56$; (f) $k = 63$.

thereby validating that our method effectively overcomes the limitations of direct density switching in challenging CVT configurations.

(2) Simulation 2

Formation switching from a circular formation to a linear formation. In this simulation, dynamic density-guided formation switching is compared with direct desired density-guided switching in terms of position error and cost function. First, the direct expectation density switching method is established as a performance baseline. Fig. 4.5c illustrates the robot trajectories as they are guided abruptly by the final linear density function shown in Fig. 4.5a, starting from the initial circular formation depicted in Fig. 4.5b. While this direct approach successfully achieves the desired formation, its transitional performance is suboptimal.

In contrast, the density function interpolation method proposed in this chapter employs a smooth transition. Fig. 4.6 fully documents this process, showing the density function evolving from its initial circular form in Fig. 4.6a to the target linear form in Fig. 4.6f through a series of calculated intermediate distributions illustrated in Fig. 4.6b to Fig. 4.6e. This smooth guidance results in the highly controlled robot movements documented in Fig. 4.7, successfully leading the formation to the desired shape.

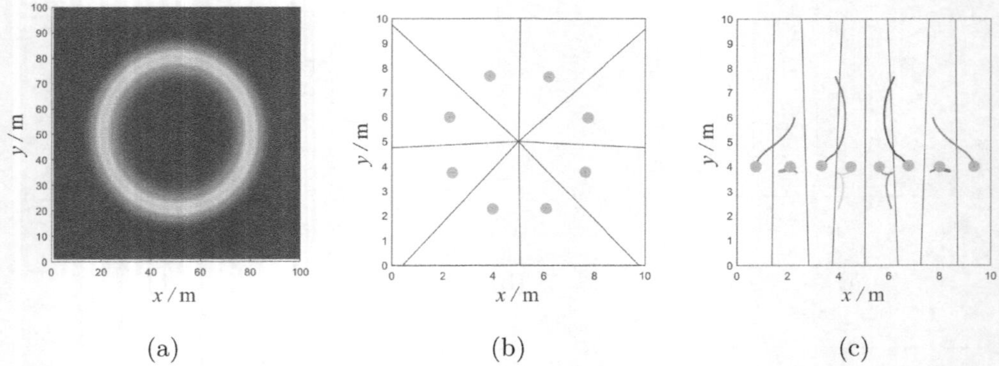


Figure 4.5: Formation switching directly guided by linear density function: (a) Desired formation density; (b) initial formation; (c) formation switching.

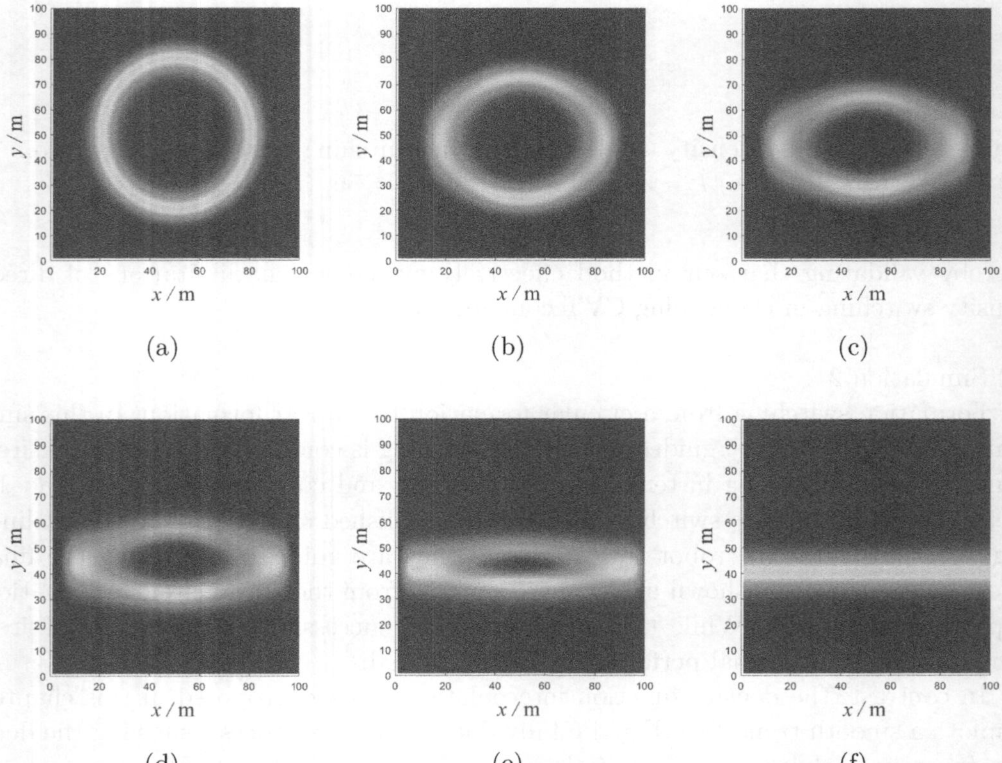


Figure 4.6: Dynamic transition from annular density distribution to linear density distribution; (a) $m = 1$; (b) $m = 4$; (c) $m = 8$; (d) $m = 12$; (e) $m = 16$; (f) $m = 20$.

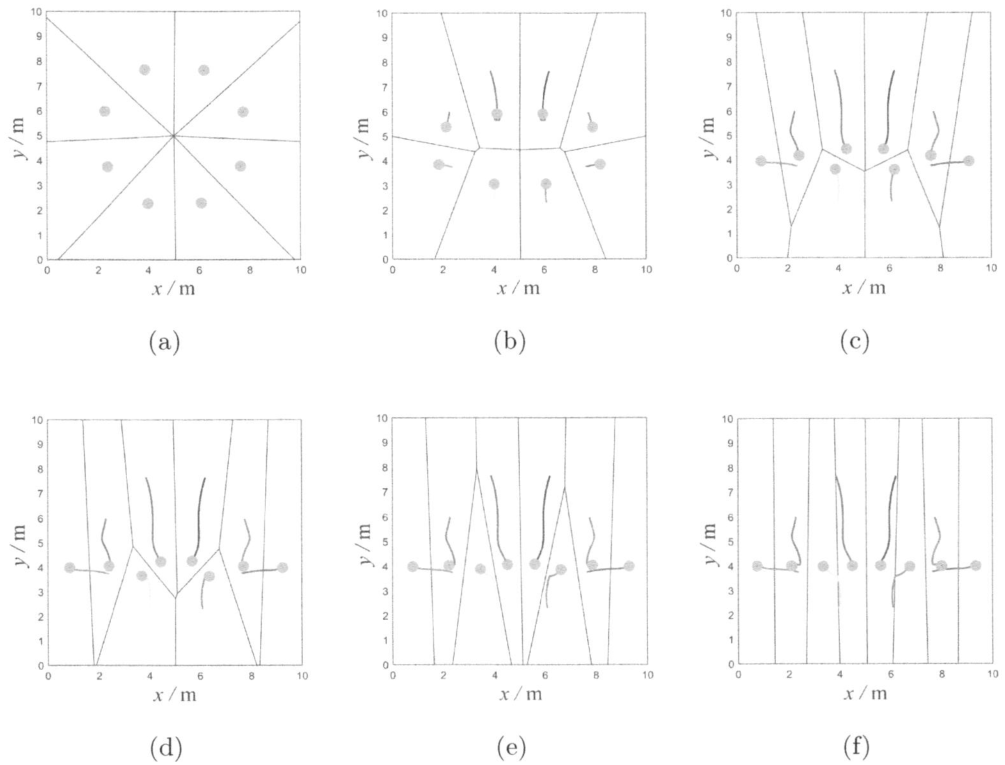


Figure 4.7: Dynamic density guided formation switching process of Simulation 2: (a) $k = 1$; (b) $k = 15$; (c) $k = 29$; (d) $k = 42$; (e) $k = 56$; (f) $k = 69$.

The core findings of this simulation are revealed in the performance metrics. Fig. 4.8 directly compares the position error $\|\boldsymbol{e}\|$of the two methods, revealing that our dynamic guidance maintains a consistently lower and more stable error profile throughout the transition. This superiority is further corroborated by the formation cost $E$ analysis in Fig. 4.9, which shows our method avoids the significant cost spike observed in the direct-switching approach. Collectively, these data provide clear quantitative evidence that the proposed method significantly optimizes the entire switching process, ensuring greater stability and efficiency.

### 4.3.2 Formation Reconfiguration

(1) Simulation 3

Formation splitting. The formation is directly guided by the desired density, as shown in Fig. 4.10. This simulation evaluates the algorithm's robustness on a topologically challenging reconfiguration task: splitting a single, connected formation into two distinct, spatially separated groups. In this simulation, the robots switch from a linear formation to two semi-surrounding formations. Fig. 4.10a shows the desired density distribution for the split formation. Fig. 4.10b shows the linear formation generated in Simulation 2, which serves as the initial formation for this simulation.

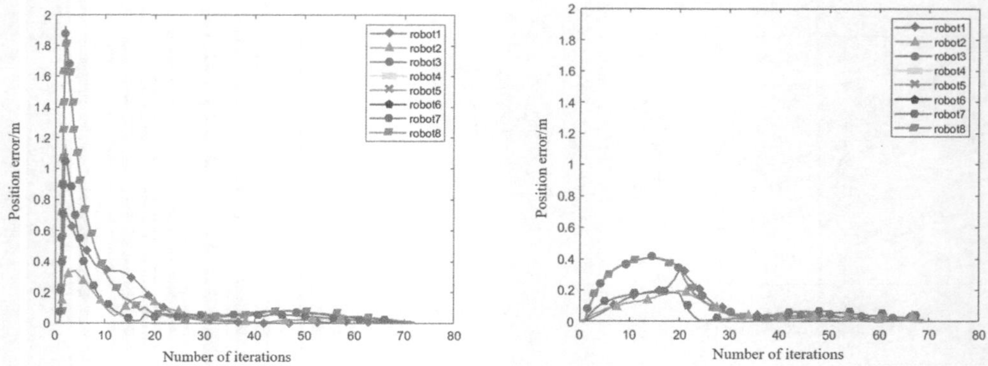


(a) (b)

Figure 4.8: Robot position error information switching process of Simulation 2: (a) The expectation density directly guides the transition; (b) dynamic density guided switching.

Fig. 4.10c illustrates the process of directly using the desired density to split the formation. It can be observed that the centroids of each Voronoi cell do not fall into the high-density areas, and the formation fails to split into the designated two groups.

The dynamic density $Z_m(\boldsymbol{q})$ for formation splitting, generated by the density function interpolation method used in this chapter, is shown in Fig. 4.11. The initial linear formation density function is shown in Fig. 4.11a, and the desired formation density distribution is shown in Fig. 4.11b. The interpolated density distributions are shown in Fig. 4.11b to Fig. 4.11e, respectively. As shown in Fig. 4.12, the dynamic density guides the robot formation to split from a linear formation into two semi-surrounding formations. The simulation results demonstrate that the proposed

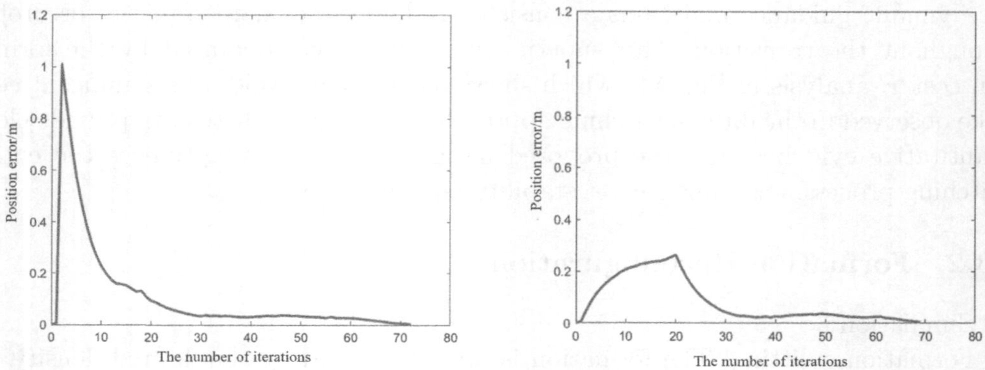


(a) (b)

Figure 4.9: Robot cost function of formation switching process in Simulation 2: (a) The expectation density directly guides the transition; (b) dynamic density guided switching.

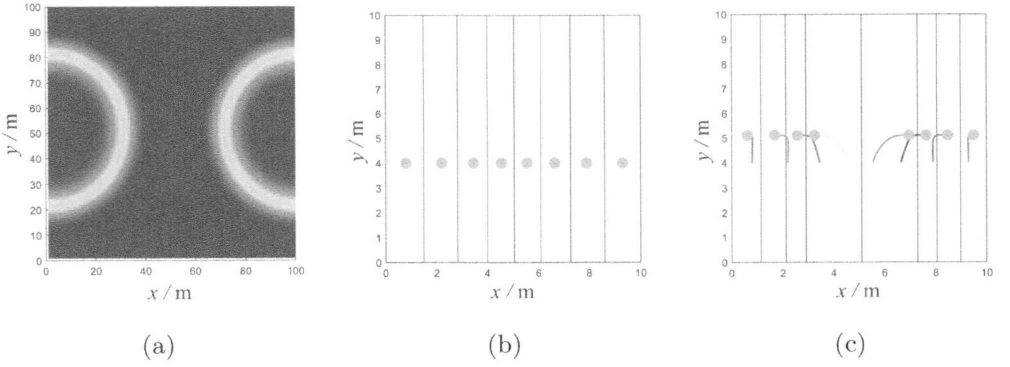


Figure 4.10: Desired density directly guided formation splitting: (a) Formation density of the target; (b) initial formation; (c) formation switching process.

method enables the robot formation to successfully complete the formation switch and achieve the desired formation split.

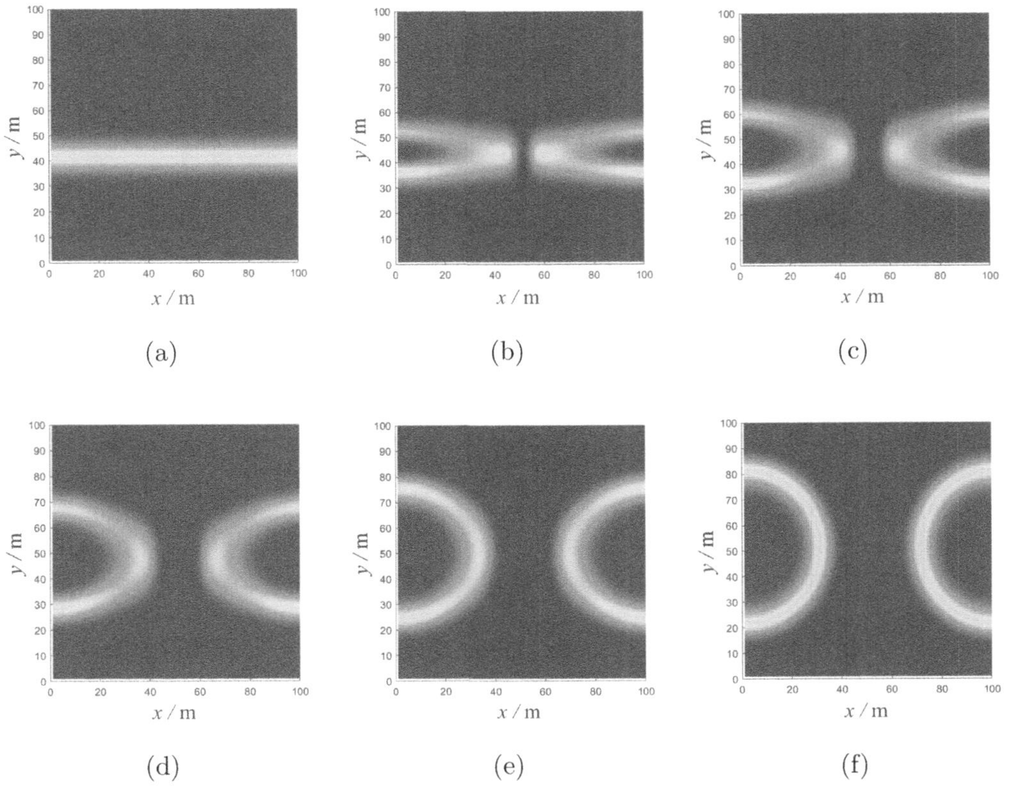


Figure 4.11: Dynamic transition from linear density distribution to expected density distribution: (a) $m = 1$; (b) $m = 4$; (c) $m = 8$; (d) $m = 12$; (e) $m = 16$; (f) $m = 20$.

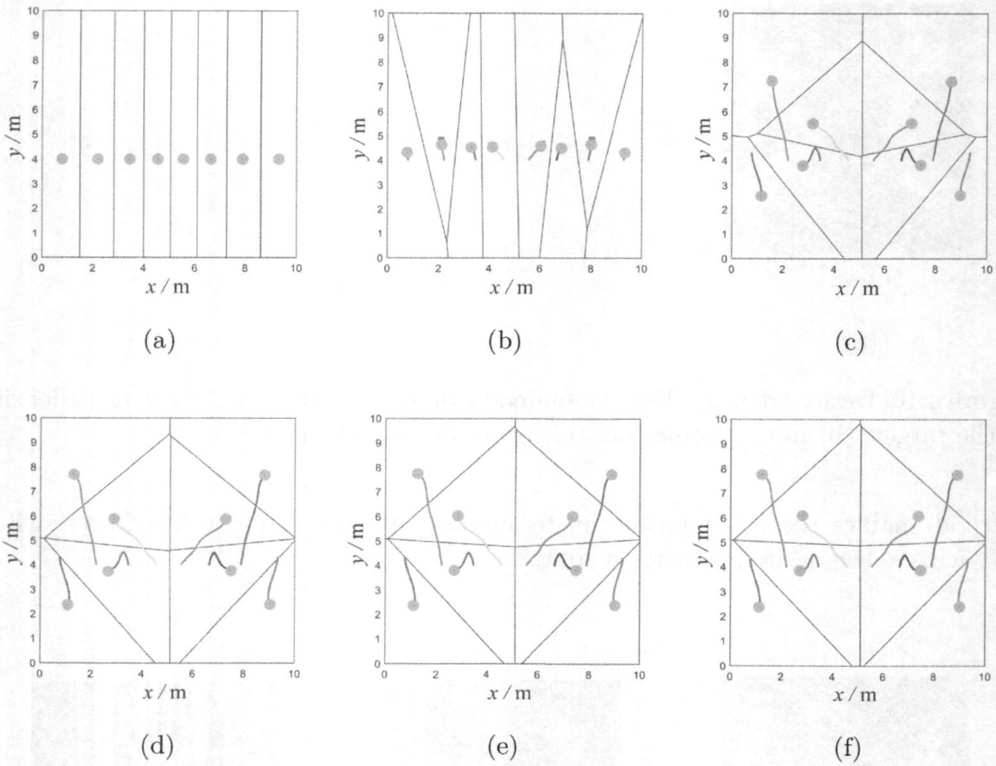


Figure 4.12: Dynamic density guided formation splitting process: (a) $k = 1$; (b) $k = 13$; (c) $k = 25$; (d) $k = 37$; (e) $k = 49$; (f) $k = 60$.

(2) Simulation 4

Formation merging. The formation is directly guided by the circular density function to merge, as shown in Fig. 4.13. This final simulation investigates the inverse topological challenge of Simulation 3: merging two spatially distinct formations into a single, cohesive group. In this simulation, the robots switch from two semi-surrounding formations to a circular formation. Fig. 4.13a shows the desired density distribution for the merged formation. Fig. 4.13b shows the two semi-surrounding formations generated in Simulation 3, which serves as the initial formation for this simulation. Fig. 4.13c illustrates the process of directly using the circular desired density to merge the formation. It can be observed that the robot formation successfully merges from two semi-surrounding formations into a circular formation.

The dynamic density $Z_m(\boldsymbol{q})$ for formation merging, generated by the density function interpolation method used in this chapter, is shown in Fig. 4.14. The initial formation's circular density function is shown in Fig. 4.14a and the desired linear formation density function is shown in Fig. 4.14f The interpolated density distributions are shown in Fig. 4.14b to Fig. 4.14e. As shown in Fig. 4.15, guided by the dynamic

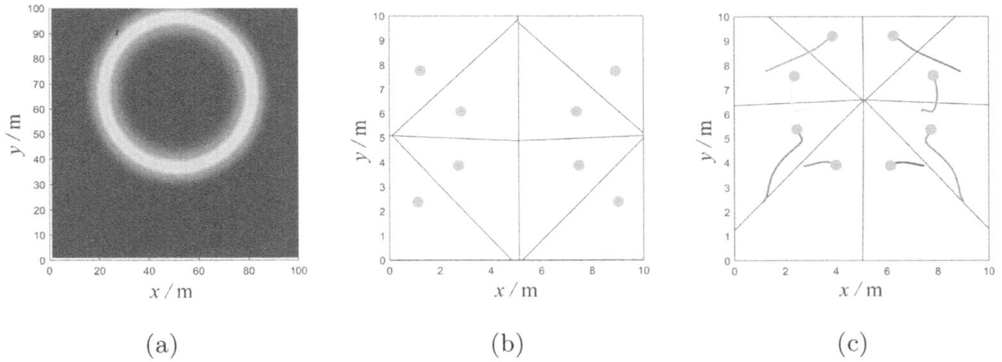


Figure 4.13: Direct formation merging guided by annular density function: (a) Formation density of the target; (b) initial formation; (c) formation switching process.

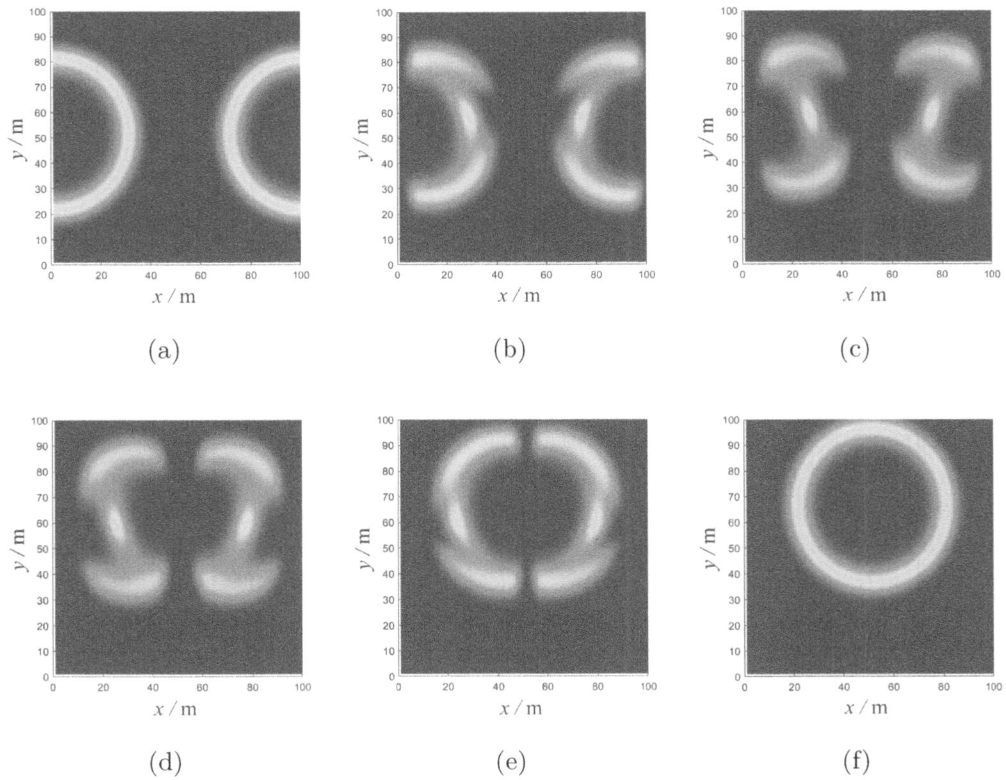


Figure 4.14: Dynamic transition from initial formation density distribution to annular density distribution: (a) $m = 1$; (b) $m = 4$; (c) $tm = 8$; (d) $m = 12$; (e) $m = 16$; (f) $m = 20$.

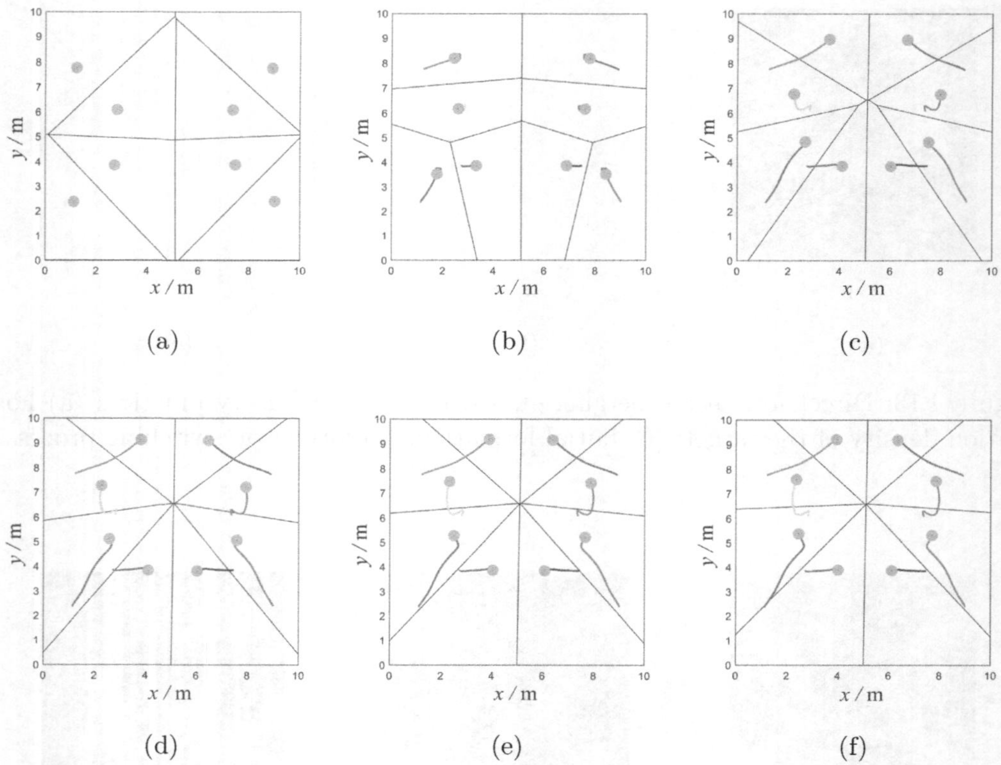


Figure 4.15: Dynamic density guided formation merging process: (a) $k = 1$; (b) $k = 15$; (c) $k = 29$; (d) $k = 44$; (e) $k = 58$; (f) $k = 73$.

density, the formation successfully switches from two semi-surrounding formations to a circular formation, thus completing the desired formation merge.

As shown in Fig. 4.16, it illustrates the changes in the error $\|\boldsymbol{e}\|$ between the robot position and the centroid position during the formation reconstruction process using two methods. The changes in the cost $E$ during the formation switching process using the two methods are shown in Fig. 4.17.It can be observed that the method used in this chapter can reduce the position error of the robot tracking the centroid. Meanwhile, the overall cost of the formation during the merging process also remains at a low level. As shown in Table 4.2, it summarizes the average position error and iteration times of the formation for all simulations. It can be seen that the proposed method prevents a significant increase in the number of iterations during the formation switching process. Moreover, based on the average position error calculation of the formation in Simulations 2 and 4, the overall position error is reduced by 5.4%.

## 4.4 Chapter Summary

This chapter proposes a dynamic density-guided multi-robot formation switching method based on the CVT formation control algorithm. This method uses shape

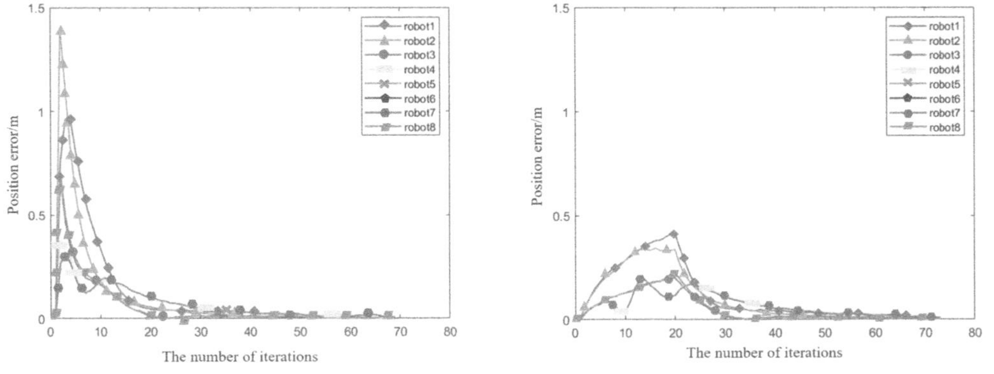


(a) (b)

Figure 4.16: Position error of robot in formation merging process of Simulation 4: (a) The expectation density directly guides the transition; (b) dynamic density guided switching.

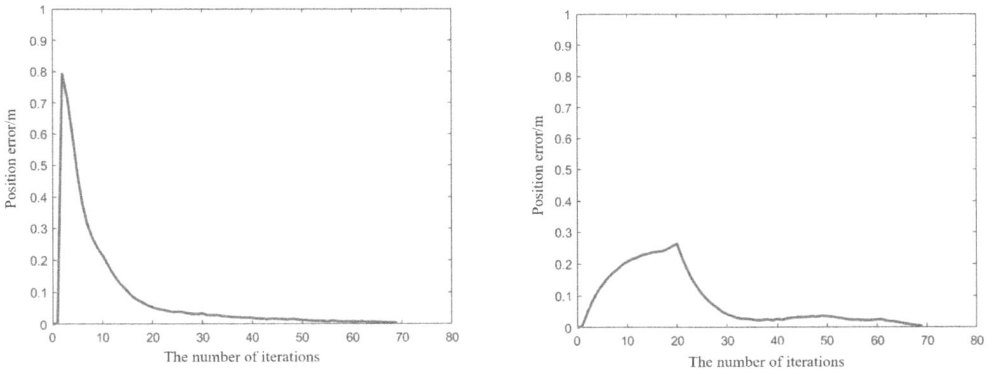


(a) (b)

Figure 4.17: Robot cost function of formation merging process of Simulation 4: (a) Expectation density guides transition; (b) dynamic density guided switching.

interpolation techniques to create a dynamic density transition from the initial formation density function to the desired density function. It enables formation switching and reconstruction tasks for CVT formations even when the density function model changes. Simulation results show that, compared to directly using the desired density function for switching, this method not only solves some formation switching failure problems but also does not significantly increase the number of iterations. Additionally, during the formation switching process, the average position error of the formation is reduced by 5.4%.

Table 4.2: Average position error and iteration times of formation switching process

| Simulation | Switch mode | Formation average position error | Number of iterations |
|---|---|---|---|
| 1 | Directly switch | - | - |
| | Method of this paper | 0.038 | 63 |
| 2 | Directly switch | 0.037 | 72 |
| | Method of this paper | 0.035 | 69 |
| 3 | Directly switch | - | - |
| | Method of this paper | 0.045 | 60 |
| 4 | Directly switch | 0.037 | 69 |
| | Method of this paper | 0.035 | 73 |

# Bibliography

[1] Shuang Guo, Bo Liu, Shen Zhang, Jifeng Guo, and Changhong Wang. Continuous-time Gaussian process trajectory generation for multi-robot formation via probabilistic inference. In *IEEE/RSJ International Conference on Intelligent Robots and Systems*, pages 9247–9253. IEEE, 2021.

[2] Shufeng Wang, Junxin Zhang, and Junyou Zhang. Intelligent vehicles formation control based on artificial potential field and virtual leader. *Journal of Shanghai Jiao Tong University*, 54(3):305–311, 2020.

[3] Jingyi Huang, Shuaiyu Zhou, Hua Tu, Yuhong Yao, and Qingshan Liu. Distributed optimization algorithm for multi-robot formation with virtual reference center. *IEEE/CAA Journal of Automatica Sinica*, 9(4):732–734, 2022.

[4] Luigi Feola and Vito Trianni. Adaptive strategies for team formation in minimalist robot swarms. *IEEE Robotics and Automation Letters*, 7(2):4079–4085, 2022.

[5] Paulo Rezeck and Luiz Chaimowicz. Chemistry-inspired pattern formation with robotic swarms. *IEEE Robotics and Automation Letters*, 7(4):9137–9144, 2022.

[6] Liping Zheng, Yajun Cheng, Chenglong Zhou, Ting Jiang, and Xiaoping Liu. Research on smooth formation control of heterogeneous crowds. *Journal of Computer-Aided Design & Computer Graphics*, 27(10):1963–1970, 2015.

[7] Xinyu Zheng, Chen Zong, Jingliang Cheng, Jian Xu, Shiqing Xin, Changhe Tu, Shuangmin Chen, and Wenping Wang. Visually smooth multi-UAV formation transformation. *Graphical Models*, 116:101111, 2021.

[8] Yang Bai, Yujie Wang, Xiaogang Xiong, Mikhail Svinin, and Evgeni Magid. Adaptive multi-agent control with dynamic obstacle avoidance in a limited region. In *American Control Conference*, pages 4695–4700, Atlanta, USA, 2022. IEEE.

[9] Xiaotian Xu and Yancy Diaz-Mercado. Multi-robot control using coverage over time-varying domains. In *International Symposium on Multi-Robot and Multi-Agent Systems*, pages 179–181, New Brunswick, USA, 2019. IEEE.

[10] Xiaotian Xu and Yancy Diaz-Mercado. Multi-robot control using coverage over time-varying non-convex domains. In *IEEE International Conference on Robotics and Automation*, pages 4536–4542, Paris, France, 2020. IEEE.

[11] Dongfeng Guo, Yang Bai, Mikhail Svinin, and Evgeni Magid. Robust adaptive multi-agent coverage control for flood monitoring. In *International Siberian Conference on Control and Communications*, pages 1–5, Kazan, Russia, 2021. IEEE.

[12] Xiaotian Xu, Guangyao Shi, Pratap Tokekar, and Yancy Diaz-Mercado. Interactive multi-robot aerial cinematography through hemispherical manifold coverage. In *IEEE/RSJ International Conference on Intelligent Robots and Systems*, pages 11528–11534, Kyoto, Japan, 2022. IEEE.

[13] Enrique Teruel, Rosario Aragues, and Gonzalo López-Nicolás. A practical method to cover evenly a dynamic region with a swarm. *IEEE Robotics and Automation Letters*, 6(2):1359–1366, 2021.

[14] Marvin Eisenberger, David Novotny, Gael Kerchenbaum, Patrick Labatut, Natalia Neverova, Daniel Cremers, and Andrea Vedaldi. Neuromorph: Unsupervised shape interpolation and correspondence in one go. *IEEE/CVF Conference on Computer Vision and Pattern Recognition*, pages 7473–7483, 2021.

[15] Gabriel Peyré, Marco Cuturi, et al. Computational optimal transport: With applications to data science. *Foundations and Trends in Machine Learning*, 11(5-6):355–607, 2019.

CHAPTER 5

# Multi-Robot Safety and Collision Avoidance Based on Voronoi Tessellations

## 5.1 Introduction

In multi-robot systems, ensuring the safety and collision avoidance capabilities of each robot during collaborative task execution is a key issue in achieving stable system operation and efficient task completion [1]. As the complexity of task space increases and the number of robots increases, the risk of collisions between robots and between robots and the environment also significantly increases [2]. Therefore, designing obstacle avoidance and region partitioning strategies with high robustness and adaptability is crucial for improving the overall performance of the system [3] [4].

In the research of collision avoidance among multi-robot, the academic community has proposed various distributed and centralized control strategies to address security challenges in complex collaborative scenarios. The repulsive control method based on potential field is a common early approach, which achieves mutual avoidance by introducing a repulsive potential field between robots [5] [6]. Subsequently, consistency control gradually became a mainstream method, especially in the graph theory framework, where the interaction between robots was characterized by the adjacency matrix and Laplacian matrix, and collision detection conditions were combined to design collision avoidance terms, effectively improving the system's coordination ability [7] [8]. In addition, the minimum distance preservation method is widely used, which ensures that the minimum distance between robots always meets the requirements by setting a safe distance threshold [9]. While effective for inter-robot negotiation, many of these methods primarily focus on maintaining separation distances and may not inherently provide a framework for efficient spatial task allocation, which is critical in cluttered environments.

Researchers have proposed various path planning and area adjustment mechanisms to enhance the system's environmental adaptability for obstacle avoidance between robots and obstacles. The artificial potential field method is one of the most

DOI: 10.1201/9781003714569-5

representative local obstacle avoidance methods, which guides the robots to move around obstacles by constructing a potential field that attracts target points and repels obstacles [10] [11]. However, traditional APF (Artificial Potential Fields) is prone to getting stuck in local optima and unreachable objectives. To overcome these shortcomings, various improved potential field methods have been introduced in the research to enhance global performance [12] [13]. In addition, graph search based algorithms such as A *, RRT [14] [15], and their various variants have shown superior performance in global path planning [16] [17]. In recent years, learning based obstacle avoidance methods have shown strong generalization ability in unstructured environments and have gradually become a research hotspot [18] [19]. However, global planning methods often struggle with dynamic obstacles and high computational costs, while local reactive methods lack guarantees of completeness, highlighting the need for a structure that balances local reactivity with organized, distributed coordination.

To address these challenges, Voronoi tessellations as a classic spatial partitioning method, is widely used in coverage control, path planning, and formation tasks of multi robot systems due to its clear geometric structure and ease of distributed implementation. Especially, the CVT have shown strong potential in distributed optimization. However, traditional Voronoi tessellations are often based on ideal assumptions: robots are viewed as particles, their positioning accuracy is infinitely close to ideal, the environment is unobstructed, and their shape is regular. These assumptions do not hold true in practical applications, especially when faced with factors such as robot entity size, positioning error, and non-convexity of the environment, which can easily lead to problems such as overlapping regions, path conflicts, or unreachable targets.

This chapter conducts a systematic exploration of the above issues, focusing on the research of multi-robot obstacle avoidance and safety strategies based on Voronoi tessellations. First, a buffer uncertainty Voronoi unit is introduced to construct a safety boundary by incorporating robot size and positioning error weights on the basis of traditional Voronoi boundaries, effectively reducing the probability of collisions between robots. Furthermore, a dynamic buffer Voronoi tessellations method is proposed to estimate the robot's position uncertainty based on particle filtering and confidence ellipse, achieving adaptive boundary adjustment and improving collision avoidance robustness.

In terms of obstacle avoidance between robots and environmental obstacles, this chapter proposes an improved approach that combines obstacle constraints, separating hyperplanes, artificial potential field method (APF), and dynamic window method (DWA) with two directions: Voronoi boundary reconstruction and obstacle avoidance algorithm fusion. These methods not only effectively avoid the problems of local optima and unreachable targets caused by obstacles during robot motion but also achieve stable convergence and efficient coverage of the algorithm in various non-convex environments.

Through in-depth theoretical analysis, model derivation, and simulation experiments, this chapter demonstrates the potential of Voronoi tessellations to achieve high security collaborative control in multi robot systems, and provides a feasible and efficient solution for practical multi-agent scenarios.

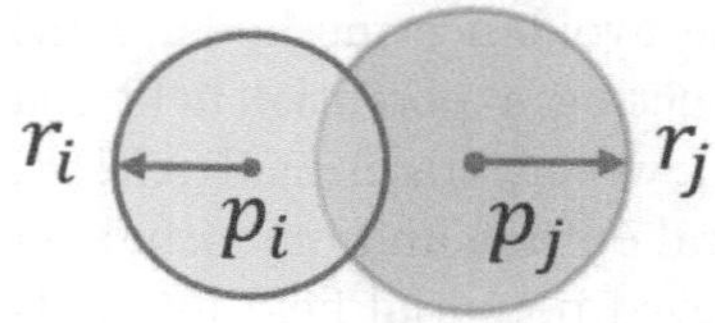


Figure 5.1: Position error range.

## 5.2 Formation switching based on dynamic density guidance

The foundational assumption of traditional CVT—that robots are dimensionless points—is a primary source of failure in practical applications. To establish a robust safety framework, this section addresses the critical challenge of inter-robot collision avoidance by augmenting the geometric structure of Voronoi tessellations. We begin by introducing a static safety margin that accounts for the physical size and a fixed uncertainty level of the robots. Subsequently, we develop a dynamic approach that adapts these safety margins in real-time by quantifying and responding to the continuously changing localization uncertainty of each robot.

### 5.2.1 Buffered Uncertainty Voronoi Cell

For a distributed multi-robot system operating in a task environment, a collision avoidance control strategy should be designed, especially for robots that cannot be presented as points. Although the CVT provides an obstacle avoidance attribute for robots, in practicality, they have real sizes. When the robots are simultaneously located at the edge of a Voronoi cell, a collision may occur, even if the robots' centers are located in the cell. As shown in Fig. 5.1, the blue and green circles are the upper bounds of the position errors of robots $p_i$ and $p_j$. The robots collide with each other in the part where the position error circles intersect.

To solve the above mentioned problems, this study introduces the concept of a buffer zone and appropriately shrinks the edge of the original Voronoi cell so that the robots' centers are located at the edge of the Voronoi cell and the robots do not collide with each other.

The traditional Voronoi cell boundary renders the robots $i$ and $j$ equidistant from their shared Voronoi edges. The buffer zone in this study adds weights based on the original equation, as follows:

$$v_i^b = \left\{ q \in Q \mid \|p_i - q\|^2 \leq \|p_j - q\|^2 - \omega_{ij} \right\}, \tag{5.1}$$

$$\omega_{ij} = (r_i + r_j) \|p_i - p_j\|, \tag{5.2}$$

where $\omega_{ij}$ is the cell weight value between robot $i$ and its neighbor $j$, and $r$ is the radius of the robot. (5.1) and (5.2) create a gap between the boundaries of the Voronoi cells as a safe buffer zone to ensure that robots that are close to the edge of the cell do not collide.

Notably, each robot can measure its position through a positioning technology or communicate with its neighboring robots. However, owing to errors in the positioning system, the position of each robot cannot be accurately obtained. Therefore, in general, each robot has different positioning uncertainties. Assuming that the upper bound of the positioning uncertainty of each robot $R_i^u$ is known, to describe the positioning uncertainty, the positioning uncertainty factor $\lambda_i \in [0, 1]$ is introduced to measure the positioning uncertainty (positioning error) of the robot. If $\lambda_i$ is close to 1, the radius of the robot's positioning uncertainty is closer to the upper bound (that is, the positioning error is significant). On the contrary, the lower the robot's positioning uncertainty, the smaller the positioning error.

$$\omega_{ij} = \left[1 + \frac{1}{2}(\lambda_i - \lambda_j)\right](r_i + r_j)\|p_i - p_j\| - \frac{1}{2}(\lambda_i - \lambda_j)\|p_i - p_j\|^2, \tag{5.3}$$

$$v_i^{b,u} = \left\{q \in Q \mid \|p_i - q\|^2 \leq \|p_j - q\|^2 - \omega_{ij}\right\}. \tag{5.4}$$

Equation (5.3) defines an adaptive weighting term $\omega_{ij}$ that adjusts the Voronoi boundary based on the relative uncertainty between two robots. The core principle is that a robot with higher uncertainty requires a larger safety region. Specifically, if robot $i$ has greater uncertainty than robot $j$ ( $\lambda_i > \lambda_j$), the value of $\omega_{ij}$ increases, effectively pushing the boundary of cell $v_i^{b,u}$ further into the space of robot $j$. Conversely, if robot $i$ is more certain of its position ($\lambda_i < \lambda_j$), its cell boundary contracts. This mechanism ensures that the burden of maintaining safety is shared proportionally to each robot's localization quality.

$$\partial v_i^{b,u} = \left\{q \in Q \mid \|p_i - q\|^2 - \|p_j - q\|^2 = \omega_{ij}\right\}. \tag{5.5}$$

According to hyperbolic theories, (5.5) represents a single branch equation of a hyperbolic equation. The positions of the two robots are located at the focus points of the long axis. The parameter $v_i^{b,u}$ may either be convex or non-convex because $\omega_{ij}$ may either be positive or negative.

(a) When a buffer zone of a Voronoi cell and the positioning uncertainty of a robot are not considered, $r = 0, \lambda_i = 0$, and $\omega_{ij} = 0$. The distribution of the robots is governed by CVT, as shown in Fig. 5.2a. The cell edge between the robots is located on the vertical line of the robot and is symmetrically distributed.

(b) When the buffer zone of the Voronoi cell is introduced, $r_i \neq 0, \lambda_i = 0$, and $\omega_{ij} = (r_i + r_j)\|p_i - p_j\|$. The distribution of the robots is shown in Fig. 5.2b, wherein the black discs represent the actual robot sizes, and the buffer zone between the robots is located on the vertical bisector between the robots and is symmetrically distributed.

(c) When the buffer zone and the robots' positioning uncertainties are introduced, $r_i \neq 0$ and $\lambda_i \neq 0$, Equation (5.4) is obtained. The distribution of the BUVCs(Buffer Uncertain Voronoi Cells) of multiple robots is shown in Fig. 5.2c. The *red* circles represent the positioning uncertainties of the robots. The buffer zone between the

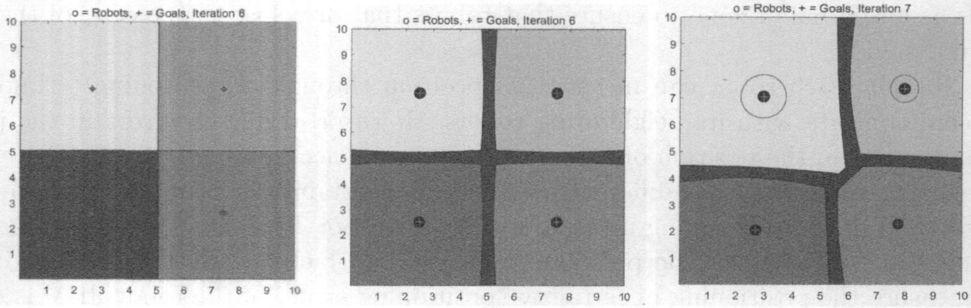


(a) (b) (c)

Figure 5.2: Multi-robot distribution before and after introducing uncertainty and robot size weight: (a) CVT; (b) CVT with buffer zone; (c) BUVC.

robots tends to move toward the small side, demonstrating an asymmetric distribution.

### 5.2.2 Dynamic Buffered Uncertainty Voronoi Cell

While the BUVC framework provides a crucial safety layer, it relies on a predefined, static uncertainty factor $\lambda_i$. In realistic scenarios, a robot's localization quality is not constant; it degrades or improves based on sensor readings, environmental features, and odometry drift. To create a truly robust and adaptive system, it is necessary to estimate this uncertainty dynamically. This section proposes a method to compute the uncertainty weight $\lambda_i$ in real-time by tracking the probability distribution of each robot's position. The above study improves the traditional CVT by adding a buffer for introducing weights, which appropriately shrinks the original Voronoi cell while reducing the possibility of inter-robot collision and enhancing inter-robot safety. However, in addition to having actual physical dimensions [20], the robot the localization range may have uncertainty due to odometry errors and the robot's own perception errors [21], leading to the risk of mutual collision between robots. In this section, the effect of the localization uncertainty of the robots is considered, first the particle filtering algorithm is used to estimate the possible positions of the robots, second the confidence ellipse is used to constrain them and the uncertainty weight of each robot is obtained based on the normalized ellipse area, and finally, buffer Voronoi cells are constructed for the boundary contraction, which improves the performance of the collision avoidance of the multiple robots in formation and the robustness of the algorithms.

(1) Particle filtering algorithm

The particle filtering algorithm [66] is a sequential Monte Carlo method based on a Bayesian framework that estimates the state $x_k$ of the target by updating a set of particles $\{x^i_{0:k}, w^i_k\}_{i=1}^N$ with a stateful quantity $\{X^i_{0:k}, i = 1, \ldots, N\}$ and weight information $\{w^i_k, i = 1, \ldots, N\}$. The weights are updated over time iterations.

When the number of sample points is sufficiently large, the estimate of the unknown state will be infinitely close to the posterior probability distribution $p$ of the

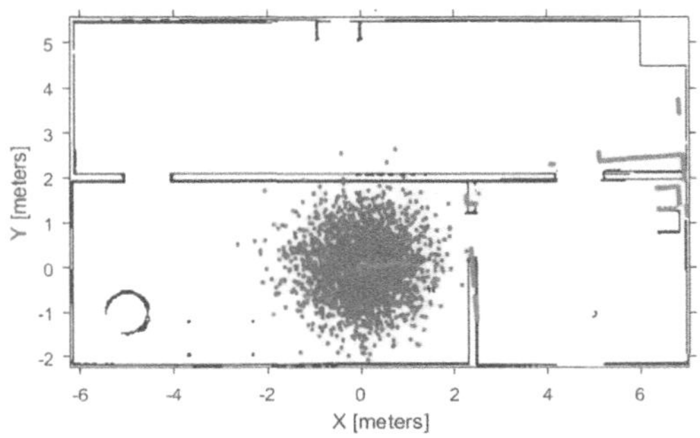


Figure 5.3: Robot position points based on particle filter estimation.

unknown distribution $X_{1:k}$. Consider the sequence of states $\{X_{0:k}, i = 1, \ldots, N\}$ of the target, starting to obey the state transfer equation:

$$x_k = f_k\left(x_{k-1}, v_{k-1}\right), \tag{5.6}$$

where $f_k$ denotes the state transfer function of the system, $x_{k-1}$ denotes the system state at the previous moment, $v_{k-1}$ is a sequence of independently and uniformly distributed noise, and $k$ denotes the maximum number of particles. The blue scatter in Fig. 5.3 represents the particle points estimated by the particle filtering algorithm. The estimated observation equation for the tracked target can be expressed as follows:

$$z_k = h_k\left(x_k, e_k\right), \tag{5.7}$$

where $h_k$ denotes the observation function of the system, $e_k$ denotes an independent uniformly distributed noise sequence.

After obtaining a set of a priori observations $z_{1:k}$ and an initial probability density function $p\left(x_0\right)$, the posterior probability density function $p\left(x_0 \mid z_{1:k}\right)$ at that moment is obtained through a Bayesian framework, so as to derive the filtered value and probabilistic credibility of the target's current state $x_k$.

First making the assumption that the density distribution $p\left(x_{k-1} \mid z_{1:k-1}\right)$ at the moment $k-1$ is known and the information at the next moment can be predicted by this probability density function, the state prediction equation can be expressed as:

$$p\left(x_k \mid z_{1:k}\right) = \int p\left(x_k, x_{k-1} \mid z_{1:k-1}\right) dx_{k-1}. \tag{5.8}$$

The system modeling assumptions are based on the Markov process, combining state and noise information, which can be obtained through Bayes' theorem:

$$p\left(x_k \mid z_{1:k}\right) = \int p\left(x_k \mid x_{k-1}\right) p\left(x_{k-1} \mid z_{1:k-1}\right) dx_{k-1}. \tag{5.9}$$

The posterior density probability function can be updated as:

$$p\left(x_k \mid z_{1:k}\right) = \frac{p\left(z_k \mid x_k\right) p\left(x_k \mid z_{1:k-1}\right)}{p\left(z_k \mid z_{1:k-1}\right)}, \tag{5.10}$$

In the above process, the measure of the state $z_k$ is used to adjust the a priori probability density function. Due to the integral link in $p(x_k \mid z_{1:k})$, it is difficult to find a specific analytical solution for some nonlinear and non-Gaussian systems. Therefore, Monte Carlo methods are introduced for approximate operations, i.e., $x_k^i \sim q\left(x_k^i \mid x_{0:k-1}^i, z_{1:k}\right)$.

Specifically, the approximate integration operation is by randomly sampling a set of randomly weighted sample points $\{X_{0:k}^i, i = 1, \ldots, N\}$ in a known state space and performing a summation and by averaging operation on them, and finally obtaining the approximate probability density function as follows:

$$p\left(x_{0:k} \mid z_{1:k}\right) \approx \sum_{i=1}^{N_s} w_k^j \delta\left(x_{0:k} - x_{0:k}^i\right), \tag{5.11}$$

where $\delta$ denotes the Dirichlet function. The weight update formula is as follows:

$$w_k^i \propto w_{k-1}^i \frac{p\left(z_k \mid x_k^i\right) p\left(x_k^i \mid x_{k-1}^i\right)}{q\left(x_k^i \mid x_{0:k-1}^i, z_{1:k}\right)}, \tag{5.12}$$

where $q$ is the importance probability density function that initializes the weights to $1/N$.

In the robotic observation system, where the importance probability density function generally relies only on the state values at the previous moment and the measurements at the current moment, i.e., $q\left(x_k \mid x_{0:k-1}, z_{1-k}\right) = q\left(x_k \mid x_{k-1}, z_k\right)$, the weighting formula can be updated as:

$$w_k^i \propto w_{k-1}^i \frac{p\left(z_k \mid x_k^i\right) p\left(x_k^i \mid x_{k-1}^i\right)}{q\left(x_k^i \mid x_{k-1}^i, z_{1:k}\right)}, \tag{5.13}$$

where $p\left(x_k^i \mid x_{k-1}^i\right)$ denotes the prior information of the state, $q\left(x_k^i \mid x_{k-1}^i, z_{1:k}\right)$ is the importance function, and $p\left(z_k \mid x_k^i\right)$ is the probability of state transfer. The posterior probability density function is:

$$p\left(x_k \mid z_{1:k}\right) \approx \sum_{i=1}^{N_i} w_k^i \delta\left(x_k - x_k^i\right). \tag{5.14}$$

(2) Quantification of uncertainty

In order to characterize the uncertainty, existing methods [67-69] generally represent the robot error range through uncertainty circles. Although they are able to model uncertainty to a certain extent, the drawback is that the uncertainty circle may overexpands the uncertainty range. As shown in Fig. 5.4a, the circle is the upper boundary of the robot $p_i$ position error, and it is possible for the robot to be located at any point within the circle, whereas Fig. 5.4b constructs a smaller boundary of the uncertainty region, which describes the localization error more accurately.

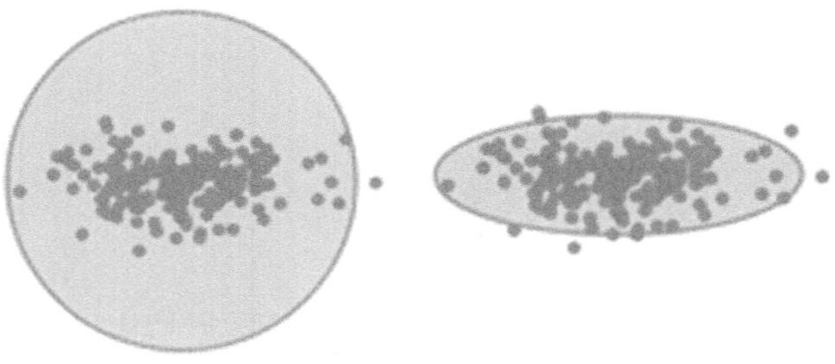

(a) (b)

Figure 5.4: Two different definitions of uncertainty: (a) Uncertainty circle; (b) uncertainty ellipse.

After modeling and analyzing the motion process of the robot, the normally distributed mean $\mu$ and variance $M$ of the state points at that moment are obtained through each iteration:

$$\mu = \begin{bmatrix} \mu_1 \\ \mu_2 \end{bmatrix}^T, \quad M = \begin{bmatrix} \sigma_{11}^2 & \sigma_{12}^2 \\ \sigma_{21}^2 & \sigma_{22}^2 \end{bmatrix} \tag{5.15}$$

where $(\mu_1, \sigma_{11})$ and $(\mu_2, \sigma_{22})$ denote the mean and standard deviation in the $x$ and $y$ axes, respectively. The uncertainty of the location points can be approximated as a probability density function:

$$p(x, \mu, M) = \frac{1}{2\pi |M|^{-\frac{1}{2}}} \exp\left[-\frac{1}{2}(x-\mu)^T M^{-1}(x-\mu)\right]. \tag{5.16}$$

In order to represent the distribution of sample points, confidence ellipse constraints are introduced at the resampling stage. Given a significance level $\beta$, for a set of sample points in the interval $(s_1, s_2)$, the ability to satisfy $(x_p, y_p)$

$$p\,(s_1 < s < s_2) = 1 - \beta, \tag{5.17}$$

where $\beta$ denotes the significance level and $1 - \beta$ denotes the confidence probability. The interval is called the confidence interval and can be visualized as a confidence ellipse $\mathcal{A}$. The values $s_1$ and $s_2$ located at the endpoints of the interval are called the lower and upper confidence limits. The error ellipses under different confidence intervals are shown in Fig. 5.5, respectively.

The standard equation for defining a confidence ellipse is:

$$\left(\frac{x}{\sigma_x}\right)^2 + \left(\frac{y}{\sigma_y}\right)^2 = s, \tag{5.18}$$

where the center coordinates of the confidence ellipse are $(\mu_1, \mu_2)$, and $\sigma_x$ and $\sigma_y$ represent the variances on the $x$ axis and $y$ axis.

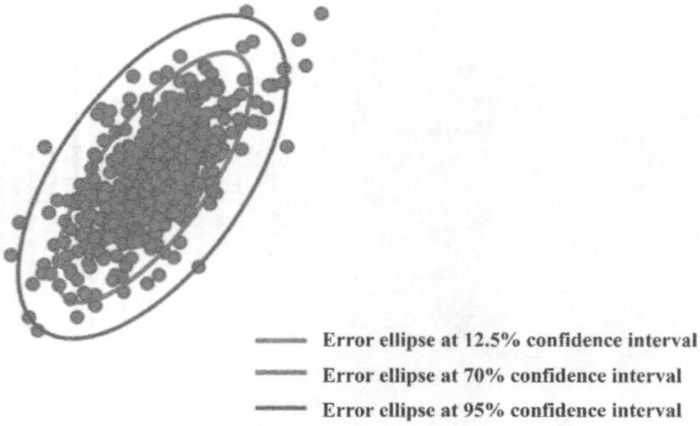


Figure 5.5: Confidence ellipse.

The covariance at time $t$ is obtained by the following formula:

$$\operatorname{cov}(x, y) = E[(x - E(x))(y - E(y))]. \tag{5.19}$$

For the confidence ellipse with the center of the sample point not at the origin, first calculate the covariance matrix of the random variables $x$ and $y$, and then calculate the eigenvectors and eigenvalues of the matrix. The confidence ellipse equation is obtained as follows:

$$\frac{(x - x_p)^2}{\gamma_1} + \frac{(y - y_p)^2}{\gamma_2} = s, \tag{5.20}$$

where $\gamma_1$ and $\gamma_2$ respectively represent the eigenvalues of the long and short axes of the ellipse, and $(x_p, y_p)$ represent the center of the confidence ellipse.

When the confidence ellipse has a certain rotation relative to the coordinate axis, the angle between the $y$ axis and $x$ axis of the ellipse in the positive direction is:

$$\varphi = \arctan\left(\frac{\gamma_1(y)}{\gamma_2(x)}\right). \tag{5.21}$$

Meanwhile, if the rotation coordinates are represented as $(x', y')$, then the coordinates $(x, y)$ can be expressed as:

$$\begin{cases} x = x' \cos\varphi + y' \sin\varphi \\ y = -x' \sin\varphi + y' \cos\varphi. \end{cases} \tag{5.22}$$

Combining (5.20) and (5.22), and through some simplification, the final confidence ellipse constraint is obtained as follows:

$$\begin{aligned} &\frac{((x' - x_p \cos\varphi)\cos\varphi + (y' - x_p \sin\varphi)\sin\varphi)^2}{\gamma_1} \\ &+ \frac{(-(x' - y_p \sin\varphi)\sin\varphi + (y' - y_p \cos\varphi)\cos\varphi)^2}{\gamma_2} \leq s. \end{aligned} \tag{5.23}$$

In the multi-robot system, task areas are assigned to each robot based on the CVT. To describe the position uncertainty of the robot, a particle filtering algorithm is introduced to establish a probability model, and the uncertainty is quantified through a confidence ellipse. After obtaining the constraint formula for the confidence ellipse, the area of the robot's confidence ellipse within each Voronoi cell can be obtained:

$$s_i = \pi \gamma_{i1} \gamma_{i2}, \quad i = 1, 2, \ldots, n, \tag{5.24}$$

where $\gamma_{i1}$ and $\gamma_{i2}$ respectively represent the long and short axes of the $i$ confidence ellipse, and $n$ represents the maximum number of robots.

In order to establish a unified evaluation standard, the area of each robot is normalized to obtain the uncertainty weights of different robots:

$$\lambda_i = \frac{S_i}{\sum_{i=1}^{n} S_i}, i = 1, 2, \ldots, n. \tag{5.25}$$

(3) Dynamic buffer centroid Voronoi tessellations

Consider $N$ robots moving in the spatial environment $\mathbb{R}^2$. Each robot $p_i$, $i \in \{1, 2, \ldots, N\}$, has a boundary shape that is approximately a circle with a radius of $r_i$. For a group of $N$ robots in the two-dimensional plane $\mathbb{R}^2$, the Voronoi cell of robot $p_i$ is generally defined as:

$$\mathcal{V}_i = \left\{ p \in \mathbb{R}^2 \middle| \left( p - \frac{p_i + p_j}{2} \right)^T p_{ij} \leq 0 \, , \forall j \neq i \right\}, \tag{5.26}$$

where $p_{ij} = p_j - p_i$.

Considering the actual size $r_s$ of the robot, a buffer zone is introduced to constrain the Voronoi cells. The calculation formula for the buffer Centroid Voronoi tessellations is defined as follows:

$$\mathcal{V}_i = \left\{ p \in \mathbb{R}^2 \middle| \left( p - \frac{p_i + p_j}{2} \right)^T + r_s \left\| p_{ij} \right\| \leq 0 \, , \forall j \neq i \right\}. \tag{5.27}$$

The above formula only considers the current positions of two robots, and the position information is often under ideal conditions. In reality, due to the motion and estimation errors of the robots themselves, the position information is with uncertainty. Therefore, robots performing tasks may collide with each other, leading to task failure.

Therefore, using particle filtering to estimate the two-dimensional Gaussian distribution of the robot's position, a confidence ellipse constraint is proposed to construct an error probability region. At the same time, the CVT is extended by introducing a dynamic buffer zone with uncertain information to constrain the Voronoi cells within a safe radius and avoid collisions with other agents, as shown in Fig. 5.6. The calculation formula for introducing positional uncertainty in Voronoi cells is expressed as:

$$\mathcal{V}_{b,i} = \left\{ p \in \mathbb{R}^2 \middle| \left( p - \frac{p_i + p_j}{2} \right)^T p_{ij} + (1 + \lambda_{ij}) \, r_{ij} \left\| p_{ij} \right\| \leq 0 \, , \forall j \neq i \right\}, \tag{5.28}$$

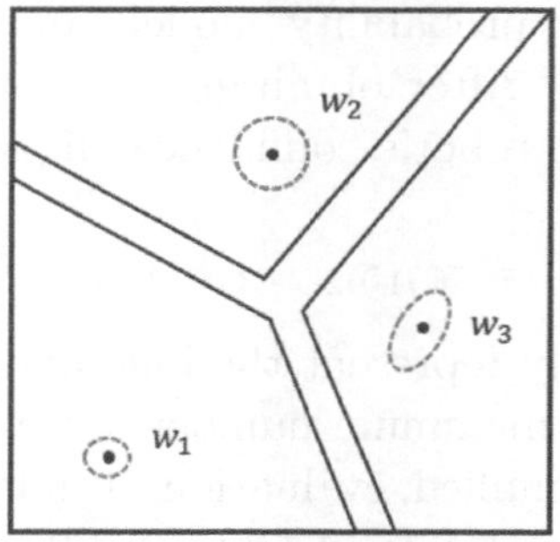


Figure 5.6: Dynamic buffer centroid Voronoi tessellations.

where $\lambda_{ij}$ represents the uncertainty difference between robots $p_i$ and $p_j$, and $r_{ij}$ is the safety radius between robots. Therefore, the edges of each robot's Voronoi cells are contracted to within a safe radius, i.e. $p_i \in \mathcal{V}_{b,i}, p_j \in \mathcal{V}_{b,j}$.

For a given multi-robot configuration, a buffer zone is constructed between the robot $i$ and other agents. When the robot reaches the center of the centroid of the Voronoi cell, the cost function of the system is minimized, and its cost function is as follows:

$$H(p, \eta) = \sum_{i=1}^{n} \int_{\mathcal{V}_{b,i}} \left( \|q - p_i\|^2 - \eta \right) \rho(q) dq, \tag{5.29}$$

where $\rho(q)$ is the known environmental field density information.

In order to prove that the algorithm with the added buffer factor can still ensure convergence, the partial derivative of the cost function formula is obtained as follows:

$$\frac{\partial H(p, \eta)}{\partial p_i} = m_{V_{b,i}} \left( p_i - c_{V_{b,i}} \right), \tag{5.30}$$

where $m_{\mathcal{V}_{b,i}}$ and $c_{\mathcal{V}_{b,i}}$ are the masses and centroids of the corresponding Voronoi areas $\mathcal{V}_{b,i}$, respectively. The above equation indicates that the proposed algorithm can converge when the cost function is minimized.

(4) Simulation result and analysis

Based on the above theoretical foundation, the proposed improved Centroid Voronoi tessellations is simulated and verified. The experiment is mainly divided into two-dimensional and three-dimensional environments, and a circular density function is set in the two-dimensional environment to compare the simulation results before and after introducing the algorithm. In order to demonstrate the adaptability of the improved algorithm in a three-dimensional environment, formation simulations were conducted under constant density and Gaussian density functions, and the performance of the algorithm was verified through simulation results.

a. Voronoi tessellations under different weights

To highlight the extension and improvement of traditional Voronoi tessellations, the differences in VD obtained from Voronoi tessellations, buffered Voronoi tessellations, and dynamic buffered Voronoi tessellations were compared, as shown in Fig. 5.7. The dynamic buffering CVT algorithm considers the impact of non-differentiable

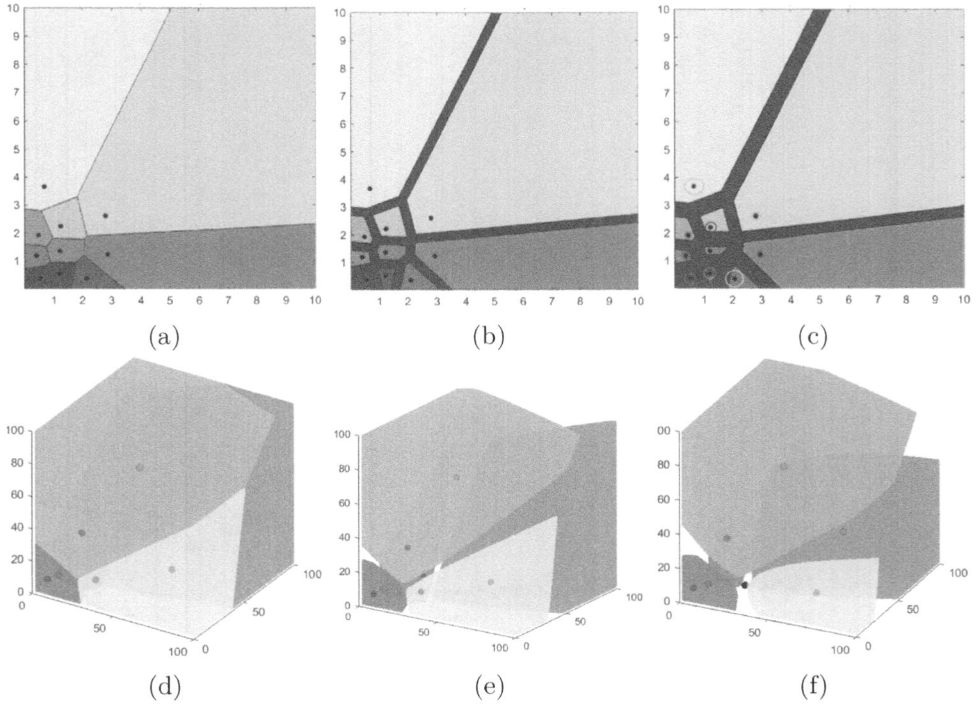

Figure 5.7: Introducing Voronoi tessellations with different weights: (a) 2D-Voronoi; (b) 2D-buffer Voronoi; (c) 2D-dynamic buffer Voronoi; (d) 3D-Voronoi; (e) 3D-buffer Voronoi; (f) 3D-dynamic buffer Voronoi.

agent size and positioning uncertainty on the Voronoi partition, introducing buffer zones between each Voronoi cell to enhance the speed and robustness of formation control algorithms in multi-agent scenarios.

Taking a two-dimensional environment as an example, Fig. 5.7a shows the standard Voronoi tessellations, while Fig. 5.7b only considers the safety radius between two agents as the safety distance, and the width of the tessellations buffer zone is relatively fixed. When considering both the safety radius and positional uncertainty information of the robot, the distribution of Voronoi cells is shown in Fig. 5.7c. At the edge of each Voronoi area, the intelligent agent constructs a variable dynamic buffer zone based on its own uncertainty, thereby improving the obstacle avoidance performance between robots.

b. Simulation in a two-dimensional environment

Set the density function to circular and simulate using 10 intelligent robots. The curves of different colors represent the motion trajectory of the robot, the black dots represent the actual position of the intelligent robot, and the block shaped areas of different colors represent the Voronoi units formed by different agents. Set the size radius of the intelligent robot to $r = 0.25m$, which means the safe distance between robots is $0.5m$. When the distance between two different robots is less than $0.5m$, it indicates a collision. Two sets of simulation comparative experiments were mainly

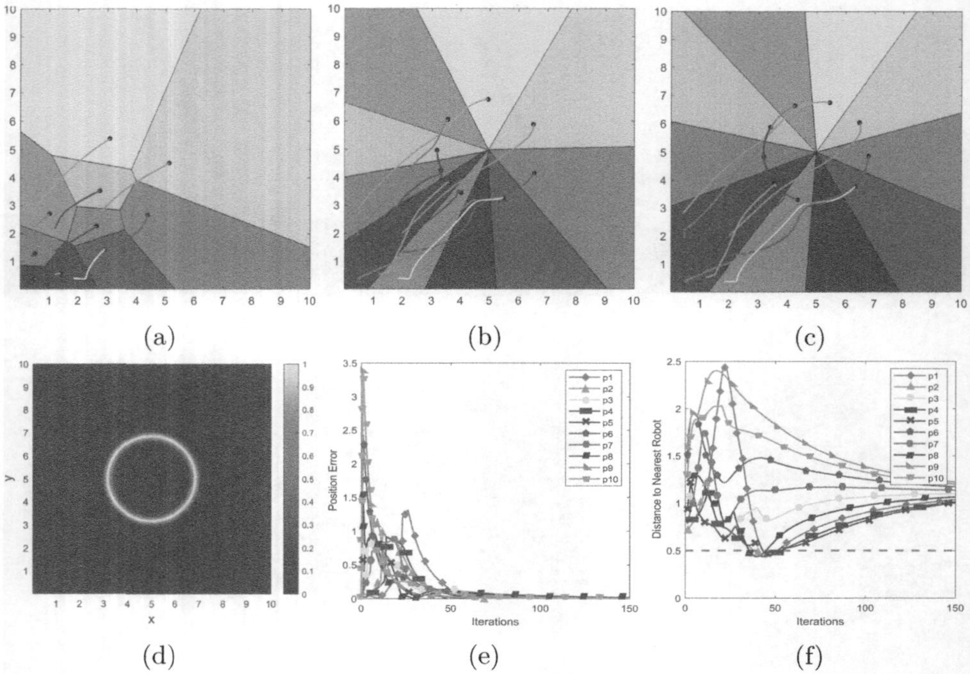


Figure 5.8: Simulation of circular formation under traditional CVT: (a) $k = 10$; (b) $k = 60$; (c) $k = 150$; (d) density function; (e) position error; (f) minimum distance between agents.

conducted in a two-dimensional environment, namely the comparison of traditional CVT and dynamic buffer CVT to form circular formation simulation.

Fig. 5.8 shows the circular formation process under traditional CVT, where Fig. 5.8a to 5.8c are the iterative process of the formation, and Fig. 5.8d is the density function in a two-dimensional environment. Combining the iterative process and density function diagram, it can be seen that multi-robots can continuously track the centroid located in the high-density area according to the characteristics of the density function, and finally reach the final configuration shown in Figs. 5.8c, 5.8e shows the error between the position of the robot and the centroid position. From Fig. 5.8f, it can be seen that in the process of multi-robot iteration, the minimum distance between robots may be less than the safe distance, indicating that traditional algorithms may cause collisions among robots.

Fig. 5.9 shows the circular formation process under dynamic buffering CVT, where Fig. 5.9a to Fig. 5.9c are the iterative process of the formation, and the different colored ellipses represent the confidence ellipses of the robot's estimated position. It can be seen that the improved algorithm can dynamically allocate buffer areas for the Voronoi units according to the size of the confidence ellipse. Fig. 5.9d shows the circular density function. According to the position error plot of the system in Fig. 5.9e, it can be seen that at the 150th iteration, the cost function of the system satisfies the convergence condition. From the red dashed line in Fig. 5.9f representing the safe

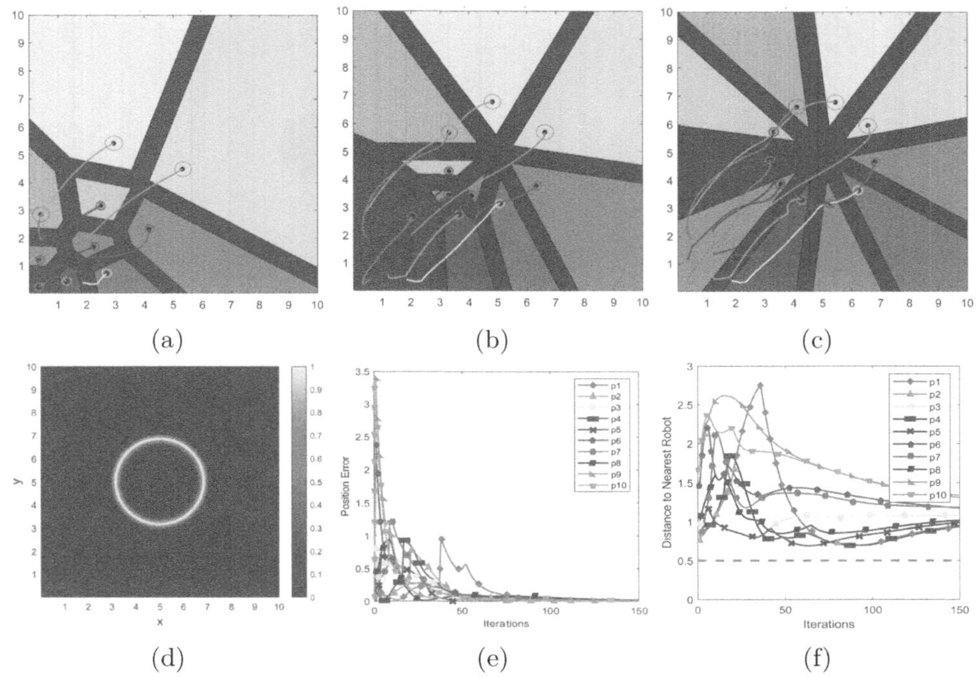


Figure 5.9: Simulation of circular formation under dynamic buffer CVT: (a) $k = 10$; (b) $k = 60$; (c) $k = 150$; (d) density function; (e) position error; (f) minimum distance between agents.

distance, it can be seen that in the process of multi-agent iteration, the minimum distance between robots is always greater than the safe distance, which can avoid collisions between robots and ensure the safety of the formation process.

The comparison between Figs. 5.8 and 5.9 provides definitive evidence of the proposed method's efficacy. In the traditional CVT simulation, Fig. 5.8f clearly shows that the minimum inter-robot distance repeatedly violates the 0.5m safety threshold, confirming that collisions would occur in a physical implementation. In stark contrast, Fig. 5.9f demonstrates that under the Dynamic BUVC framework, the minimum distance is strictly maintained above the safety threshold throughout the entire maneuver. This critically validates that our dynamic uncertainty-aware buffering mechanism successfully translates the theoretical safety concept into practical, collision-free performance, overcoming a fundamental limitation of the standard CVT algorithm.

## 5.3 Safety Between Robots and the Environment

In the previous section, we focused on the collision avoidance problem between robots. However, with the continuous deepening of research, obstacle avoidance between robots and obstacles in complex non-convex environments has gradually become an important challenge for multi-robot systems to achieve safe operation and efficient task execution. In such environments, traditional CVT often struggle to ensure the

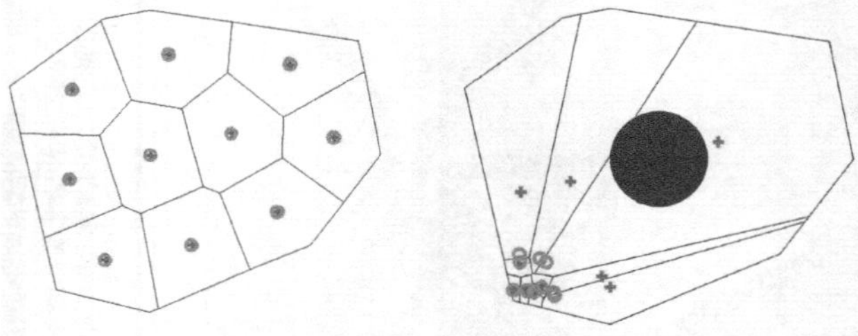

(a) (b)

Figure 5.10: Two different definitions of uncertainty: (a) Convex area; (b) non-convex area.

rationality of spatial partitioning and the reliability of obstacle avoidance due to their inability to effectively handle non-convex regions. Figs. 5.10a and 5.10b respectively show the CVT effects in convex and non-convex environments, where the orange circle represents the current position of the robot and the blue cross represents the centroid position of the corresponding Voronoi cell. It can be seen that in the convex environment shown in Fig. 5.10a, the CVT can achieve good convergence, and the robot eventually tends to coincide with the centroid of its Voronoi cell. In the non-convex environment shown in Fig. 5.10b, there is a black obstacle between the robot and its target centroid. If the corresponding obstacle avoidance improvement is not made to the CVT, the robot will inevitably collide during the process of approaching the centroid. This issue stems from a fundamental conflict: the straight-line path from a robot to its Voronoi centroid, which is the path dictated by the CVT control law, is no longer guaranteed to be collision-free.

To address the above issues, this section will focus on two types of obstacle avoidance methods based on Voronoi tessellations: one is to modify the boundaries of Voronoi units to actively avoid obstacle areas; another type is to integrate obstacle avoidance strategies into centroid Voronoi tessellations, enabling robots to perceive and avoid complex obstacles in real-time during dynamic collaboration. By systematically elaborating and comparing these two methods, the aim is to provide effective solutions for safe obstacle avoidance in non-convex environments for multi-robot systems.

### 5.3.1 Obstacle Avoidance Method Based on Voronoi Cell Boundary Adjustment

One of the primary approaches to ensure safety in environments with obstacles is to modify the geometry of the Voronoi cells themselves, such that each robot's assigned region is inherently collision-free. This section explores two distinct strategies to achieve this. The first strategy, based on spatial constraints, directly subtracts obstacle regions from the standard Voronoi cells. The second strategy employs separating hyperplanes to geometrically construct new cell boundaries that safely isolate robots from both obstacles and each other. We will introduce and simulate these two methods in the following subsections.

**Voronoi Tessellations Based on Obstacle Constraints**

(1) Introduction to the Voronoi tessellations based on obstacle constraints

Assuming that the Voronoi area corresponding to each robot in the swarm is $v_i, i = 1, 2, \ldots n$, and the obstacle area in the environment is $w_i, i = 1, 2, \ldots, n$, to ensure that the Voronoi area does not contain obstacles, the corresponding Voronoi area of the robot should be:

$$V_i = v_i - w_i, i = 1, 2 \ldots n. \tag{5.31}$$

If there are no obstacles or convex areas in the environment, then and the Voronoi area will not change. When there is an obstacle in the Voronoi area, the original Voronoi area will regard the boundary of the obstacle area as the inner boundary to redefine the Voronoi area. After ensuring that each Voronoi cell is within the region, the resulting centroid position may appear outside the region, considering that the region is non-convex. Hence, the centroid position needs to be restricted. This is accomplished by strictly constraining the centroid of the Voronoi area within the Voronoi area, as shown in (5.32).

$$p_i^* \subset V_i, \quad i = 1, 2 \ldots n. \tag{5.32}$$

If the constraint solution is not obtained, it indicates that the position of the centroid is not in the Voronoi cell. Then, the center of mass of the Voronoi area will not be updated until the other Voronoi cells change, which again affects the updating of the center of mass. The target position and robot position relative to the obstacles are divided into the four cases shown in Fig. 5.11.

The point represents the current position of the robots, the cross represents the target position of the robots, the blue shaded area represents the obstacle, and the

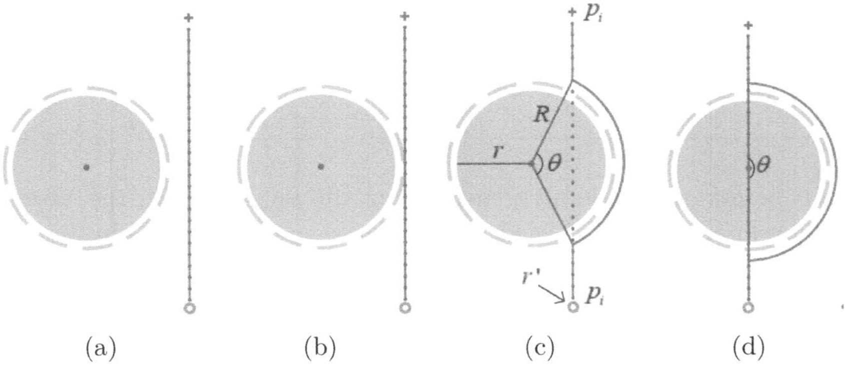


Figure 5.11: Relationship between obstacles and paths: (a) Path does not pass through the obstacle; (b) path is tangential to the obstacle; (c) path is blocked by the obstacle; (d) path passes through the center of the obstacle.

dotted line represents the safe path of the robots relative to the obstacle. The trajectory of the robot is marked with a red line. $r$ is the radius of the obstacle, and $r'$ is the radius of the robot. $R$ is the distance between the center of the robot and the center of the obstacle, where $R = r + r'$, and $\theta$ is the central angle of the arc path.

In Fig. 5.11a, the path between the robot and the target point does not pass through the obstacle, and the robot can directly track its target. At this time, $\theta = 0$. In Fig. 5.11b, the path between the robot and the target point is tangential to the obstacle; hence, the robot can still track the target directly, and $\theta = 0$. In Fig. 5.11c, the straight path between the robot and the target point is blocked by the obstacle. The robot moves along the arc with radius $R$ to avoid collision with the obstacle, and $0 < \theta < \pi$. In Fig. 5.11d, the line between the robot and the target point passes through the center of the obstacle, that is, $\theta = \pi$. The algorithm steps for this case are shown in Algorithm 2.

**Algorithm 2** Constrained CVT algorithm.

```
Randomly generate n points
for (i ≤ n)
Calculate the Voronoi cell of the i point
if there are no obstacles in the Voronoi cell
integrate the Voronoi cell to find the centroid
else replace the inner boundary of the Voronoi cell with the boundary of the obstacle
calculate the centroid
End
if the straight line connecting the robot and the target point is blocked by obstacles
plan a new path
drive the robot to the centroid
else drive the robot to the centroid
end
if the algorithm convergence,
Break
else Jump to step 2, calculate the Voronoi cell again,
End
End
```

Total $n$ points are randomly generated in the region. First, the Voronoi cells of each generated point are calculated, and the presence of obstacles in the Voronoi cells is examined. If there is no obstacle, the centroid of the cell is calculated. Otherwise, the boundary of the obstacle is replaced by the inner boundary of the Voronoi cell. Next, the presence of obstacles on the straight line connecting the robot and the target point is examined. If the path is blocked by an obstacle, plan a new path and drive the robot to the centroid. If not, drive the robot to the centroid directly. The last step is to judge whether the algorithm convergence meets the termination

Table 5.1: Different size of obstacles

| Radius | No. of robots | Center | Iterations | Time consumed(s) |
|---|---|---|---|---|
| $r = 0$ | 10 | (5,5) | 124 | 43.48 |
| $r = 1$ | 10 | (5,5) | 113 | 37.27 |
| $r = 2$ | 10 | (5,5) | 82 | 28.97 |
| $r = 3$ | 10 | (5,5) | 102 | 36.80 |

condition, and then interrupt or jump to the second step to calculate the Voronoi cells again.

(2) Simulation verification of Voronoi tessellations based on obstacle constraints

To evaluate the adaptability and effectiveness of the obstacle-constrained Voronoi tessellation method in various non-convex environments, three sets of systematic simulation experiments were designed. By varying the size, number, and spatial distribution of obstacles, representative non-convex scenarios were constructed to assess the algorithm's performance and robustness under different conditions.

(a) Radius changes of obstacle

The sizes of the obstacles were changed to verify the performance of the algorithm proposed in this section under obstacles of different sizes. The $10m \times 10m$ environment was again used in this experiment. Fig 5.12 shows the final distributions of multi-robot for obstacles of different sizes when the obstacle center is at (5,5) and the radius $r$ is 0, 1, 2, 3. Different areas where robots are distributed are distinguished using different colors, such as light cyan, sky blue, light purple, and so on. The corresponding cost function is shown in Fig 5.13. The minimum values were obtained when the number of iterations was 124, 113, 82, and 102, respectively. It is considered that the swarm reached the stable state. The experimental data are shown in Table 5.1.

From Fig. 5.12 and Table 5.2, it can be seen that the size of the obstacle affects the degree of non-convexity of the environmental area. The algorithm can ensure a reasonable partition of the environment for obstacles of different sizes. However, a greater number of obstacles will increase the computation effort required for the algorithm. When the obstacle radius gradually increases, the area that the robot can effectively divide becomes increasingly smaller, which decreases the number of iterations and time to the steady state. When $r = 2$, the time required is the smallest. Because the area is fixed in the simulation, the obstacle area in the case of $r = 3$ is

Table 5.2: Different locations of obstacle

| Center | Radius | No. of robots | Iterations | Time consumed(s) |
|---|---|---|---|---|
| (3.3) | $r = 2$ | 10 | 107 | 39.24 |
| (5,5) | $r = 2$ | 10 | 82 | 28.97 |
| (7,7) | $r = 2$ | 10 | 90 | 33.02 |

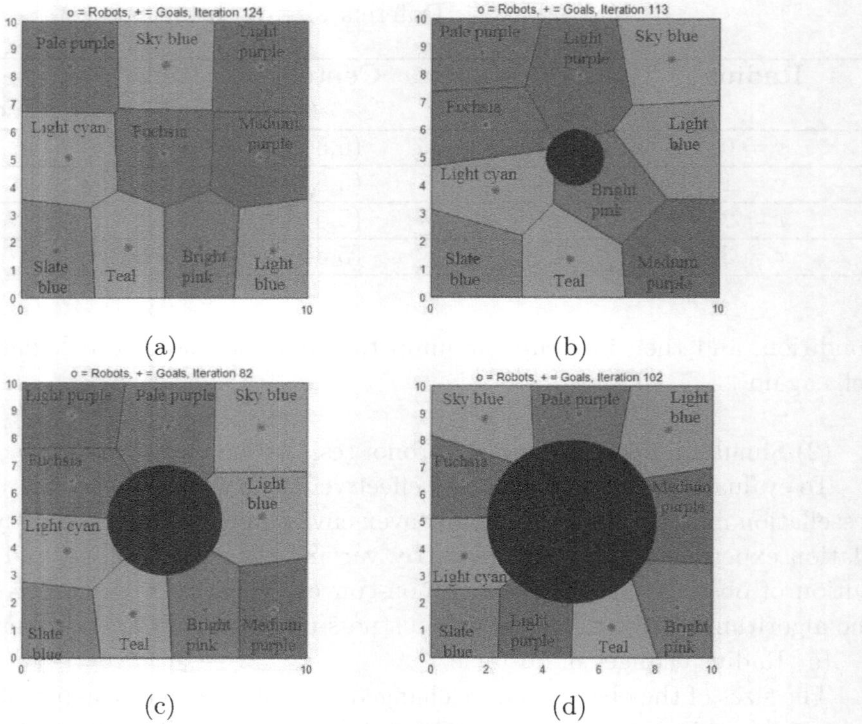


Figure 5.12: Different sizes of obstacles: (a) $r$= 0; (b) $r$= 1; (c) $r$= 2; (d) $r$= 3.

nearly 1/3 of the total area, and the amount of calculation increases dramatically, resulting in increased overall iteration time.

(b) Different locations of obstacle

In the virtual environment, in addition to obstacles being of different sizes, their positions are usually random. The experiment reported here verified the effectiveness of the proposed algorithm when obstacles occur at different positions. The $10m \times 10m$ environment was again used in this experiment. For $r = 2$, the positions of the center of the obstacle were selected as (3,3), (5,5), (7,7). The experimental results are shown in Fig 5.14 and the cost function is shown in Fig. 5.15. The minimum values were obtained when the number of iterations was 107, 82, and 90, respectively. The experimental data are presented in Table 5.2.

Fig. 5.14 and Table 5.2 show that the algorithm can ensure a reasonable partition of the area for different locations of an obstacle. Fig. 5.15 shows the graphs of the cost function for different locations of the obstacle. The number of iterations is the smallest when the obstacle was located at the center of the area. When the obstacle was at (3,3), which was close to the robots, the partition conditions of the Voronoi cell were not maintained; hence, the algorithm recalculated the centroids of the Voronoi cell, which increased the computation cost. The robots moved in the opposite direction when the obstacle was at (7,7). The obstacle was encountered toward the end of the area. Therefore, the fluctuation was not apparent for a few iterations at the beginning. In the range of 20 to 30 iterations, the cost function fluctuated wildly

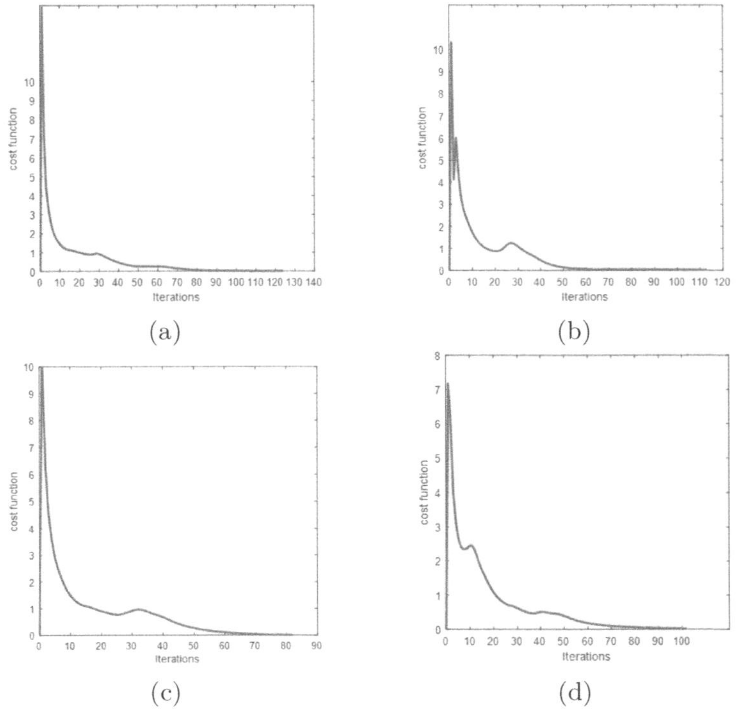


Figure 5.13: Cost function values for different sizes of obstacles: (a) $r = 0$; (b) $r = 1$; (c) $r = 2$; (d) $r = 3$.

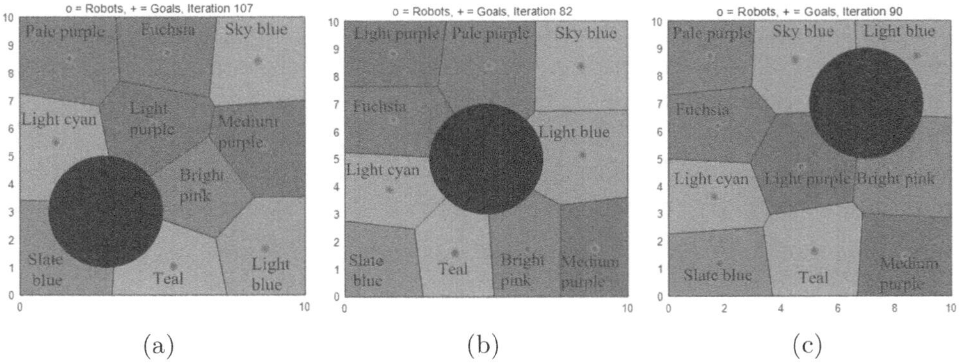


Figure 5.14: Different locations of obstacles: (a) Position (3,3); (b) position (5,5); (c) position (7,7).

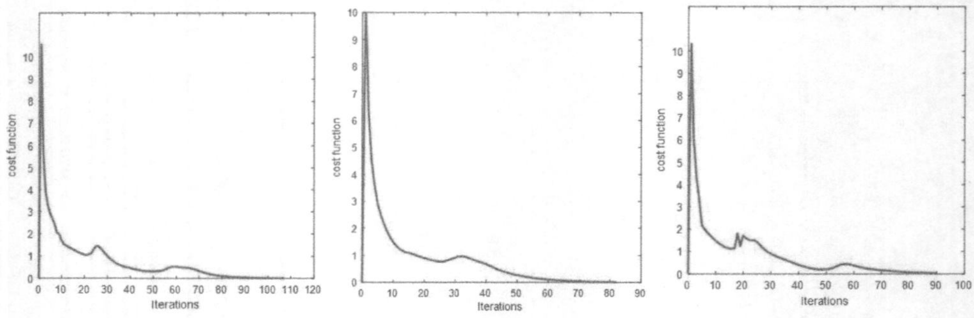


(a) (b) (c)

Figure 5.15: Cost function values for different obstacle locations: (a) Position (3,3); (b) Position (5,5); (c) Position (7,7).

but eventually stabilized. However, in these two cases, the robots overcame several situations where there were obstacles in the environment. When the obstacle was at (5,5), the iterations were smaller than in the other cases because the initial positions of the robots were significantly different from the desired locations. It can be seen that as the obstacle position moved closer to the center, the iteration time required was longer.

(c) Different quantities of obstacles

The following simulation focuses on different quantities of obstacles. It can be seen from Figs. 5.16 and 5.17, and Table 5.3 that as the number of obstacles increased, the number of iterations for achieving the steady state increased. Fig. 5.17c shows that for iterations from 1 to 90, owing to the obstacle position and the initial position of the swarm, the swarm robotics system repeatedly calculated the centroid positions in the lower right corner area, resulting in the robots staying in this area for a long time. Therefore, the cost function between 30 and 90 iterations was zero. When the 90th iteration recalculated the robot's Voronoi cell and the center of mass, the position of the center of mass was iterated again, resulting in a large error. The system stabilized at 188 iterations. The iteration time was the shortest in the presence of one obstacle, and increased as the number of obstacles increased.

In the above three simulations, although the size and spatial distribution of obstacles varied, the results demonstrate that the obstacle-constrained Voronoi tessellations maintained effective control over the swarm robots in non-convex environments.

Table 5.3: Different numbers of obstacles

| **Starting point** | **No. of robots** | **No. of obstacles** | **Radius** | **Iterations** | **Time consumed(s)** |
|---|---|---|---|---|---|
| (1.1) | 10 | 1 | $r = 2$ | 82 | 28.97 |
| (1,1) | 10 | 2 | $r = 2$ | 147 | 47.38 |
| (1,1) | 10 | 4 | $r = 1$ | 180 | 52.68 |

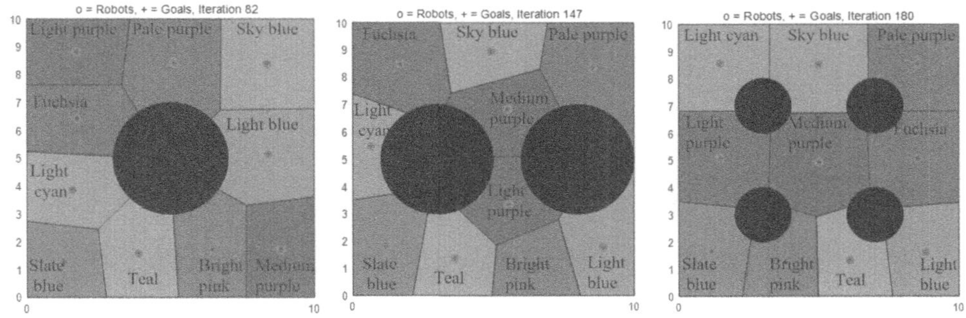


Figure 5.16: CVT for different numbers of obstacles: (a) $n = 10$; (b) $n = 50$; (c) $n = 100$.

Moreover, its control efficiency remained stable and was not significantly affected by changes in obstacle size or distribution.

### Voronoi Tessellations Method Based on Separating Hyperplanes

In addition to the traditional obstacle constrained Voronoi tessellations, it is also possible to generate Voronoi cells that do not overlap with obstacles by constructing a separation hyperplane, thereby controlling the robot to achieve obstacle avoidance within the task area. Given a two-dimensional task area $Q \subseteq \mathbb{R}^2$, the position points of two adjacent robots are $\boldsymbol{p}_i$ and $\boldsymbol{p}_j$, respectively. The hyperplane between the two robots can be represented as:

$$\boldsymbol{a}_{ij}^T \boldsymbol{q} = b_{ij}, \tag{5.33}$$

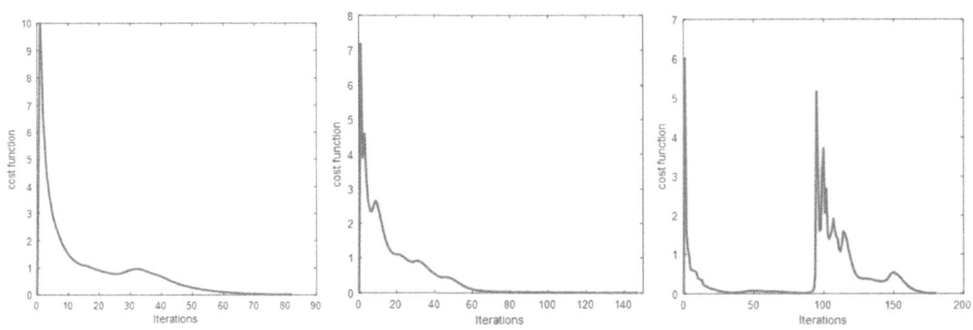


Figure 5.17: Cost function values for different numbers of obstacles: (a) $N = 1$; (b) $N = 2$; (c) $N = 4$.

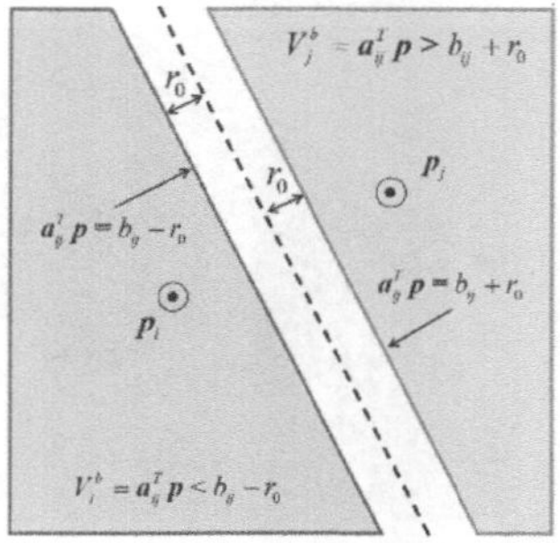


Figure 5.18: Adding buffer bands to Voronoi cells based on separating hyperplanes.

where $\boldsymbol{a}_{ij}^T$ is the normal vector that determines the direction of the hyperplane, and $b_{ij}$ is the distance from the origin to the hyperplane, which can be expressed as:

$$\begin{cases} \boldsymbol{a}_{ij}^T = \frac{\boldsymbol{p}_i - \boldsymbol{p}_j}{\|\boldsymbol{p}_i - \boldsymbol{p}_j\|} \\ b_{ij} = \frac{\boldsymbol{a}_{ij}^T(\boldsymbol{p}_i + \boldsymbol{p}_j)}{2}. \end{cases} \tag{5.34}$$

Due to the fact that the Voronoi unit separates the intersection of multiple point hyperplane sets, the Voronoi cells can be constructed through hyperplane tessellations. According to the definition of separating hyperplanes in (5.2), the Voronoi cell between two adjacent robots $\boldsymbol{p}_i$ and $\boldsymbol{p}_j$ can be expressed as:

$$\begin{cases} v_i^b = \left\{\boldsymbol{q} \in Q \mid \boldsymbol{a}_{ij}^T \boldsymbol{q} < b_{ij}\right\} \\ v_j^b = \left\{\boldsymbol{p} \in Q \mid \boldsymbol{a}_{ij}^T \boldsymbol{q} > b_{ij}\right\}. \end{cases} \tag{5.35}$$

On this basis, in order to improve the collision avoidance performance between robots, a buffer zone is introduced on the basis of the standard Voronoi cell. Based on (5.35), considering the actual radius $r_0$ of the robot, the distances from the origin to the hyperplane are modified to $b_{ij} - r_0$ and $b_{ij} + r_0$, respectively. The Voronoi units between robots $\boldsymbol{p}_i$ and $\boldsymbol{p}_j$ can be expressed as:

$$\begin{cases} v_i^b = \left\{\boldsymbol{q} \in Q \mid \boldsymbol{a}_{ij}^T \boldsymbol{q} < b_{ij} - r_0\right\} \\ v_j^b = \left\{\boldsymbol{q} \in Q \mid \boldsymbol{a}_{ij}^T \boldsymbol{q} > b_{ij} + r_0\right\}. \end{cases} \tag{5.36}$$

Formula (5.36) shrinks the edge of the Voronoi cell $V_i^b$ based on the radius $r_0$ of the robot in the actual scene to ensure that two adjacent robots do not collide when approaching the edge of the Voronoi cell. The specific construction method is shown in Fig. 5.18.

To further consider the obstacle avoidance performance between robots and obstacles, a portion of the Voronoi unit containing obstacles is removed using the hyperplane method. Define robot $\boldsymbol{p}_i$ and obstacle $\boldsymbol{O} = \{\boldsymbol{o}_1, \boldsymbol{o}_2, \ldots, \boldsymbol{o}_n\}$, where $\boldsymbol{o}_i$ represents the $i$ vertex of the obstacle. Defining the position $\boldsymbol{p}_i$ of the robot and the vertex $\boldsymbol{o}_i$ of the obstacle as two different sample types in the separation hyperplane dataset, the binary classification sample dataset can be obtained as $(F, y)$. Therefore, the determination equation for defining the separation hyperplane between the robot and the obstacle can be obtained as follows:

$$y = \boldsymbol{a}_{io}^T \boldsymbol{F} + b_{io}$$
$$\begin{cases} y > 0, & \boldsymbol{F} = \boldsymbol{o}_i \\ y < 0, & \boldsymbol{F} = \boldsymbol{p}_i, \end{cases} \tag{5.37}$$

where $\boldsymbol{a}_0$ is the normal vector of the hyperplane between the robot and the obstacle, and $b_{io}$ is the distance from the origin to the hyperplane between the robot and the obstacle. $\boldsymbol{F}$ is the coordinate of the test point, $y$ is the corresponding label. When $y > 0$, it indicates that the test point is on the side of the obstacle, and when $y < 0$, it indicates that the test point is on the side of the robot.

Considering the two points $\boldsymbol{x}'$ and $\boldsymbol{x}''$ in the hyperplane, one can obtain that:

$$\boldsymbol{a}_{io}^T \boldsymbol{x}' = -b_{io}, \quad \boldsymbol{a}_{io}^T \boldsymbol{x}'' = -b_{io}. \tag{5.38}$$

From this, the distance from any point $\boldsymbol{x}$ to the hyperplane can be calculated as:

$$\text{dist}\left(\boldsymbol{x}, \boldsymbol{a}_{io}, b_{io}\right) = \frac{\left|\boldsymbol{a}_{io}^T \boldsymbol{x} + b_{io}\right|}{\|\boldsymbol{a}_{io}\|}. \tag{5.39}$$

Given $y^2 > 0$, Can use $y$ to replace the absolute value symbol, Substituting $y$ into formula (5.39) yields:

$$\text{dist}\left(\boldsymbol{x}, \boldsymbol{a}_{io}, b_{io}\right) = \frac{y\left(\boldsymbol{a}_{io}^T \cdot \boldsymbol{F} + b_{io}\right)}{\|\boldsymbol{a}_{io}\|}. \tag{5.40}$$

According to (5.40), the separation hyperplane problem can be summarized as an optimization problem.

$$J = \arg \max_{\boldsymbol{a}_{io}, b_{io}} \left\{ \frac{1}{\|\boldsymbol{a}_{io}\|} \min_i \left[ y_i \left( \boldsymbol{a}_{io}^T \boldsymbol{F} + b_{io} \right) \right] \right\}. \tag{5.41}$$

When $|y| > 1$, it can be determined that $y^2 > 1$. Formula (5.41) can be rewritten as:

$$J = \arg \max_{\boldsymbol{a}_{io}, b_{io}} \left\{ \frac{1}{\boldsymbol{a}_{io}} \right\}$$
$$\text{s.t. } y_i \left( \boldsymbol{a}_{io}^T \cdot F + \boldsymbol{b}_{io} \right) \geq 1. \tag{5.42}$$

By using the Lagrange multiplier method, (5.42) can be further written as:

$$J = \arg \max_{\boldsymbol{a}_{io} \boldsymbol{b}_{io}} \left\{ \frac{1}{2} \boldsymbol{a}_{io}^2 \right\}$$
$$\text{s.t. } y_i \left( \boldsymbol{a}_{io}^T \cdot F + \boldsymbol{b}_{io} \right) \geq 1. \tag{5.43}$$

The separation hyperplane between the robot and the obstacle can be obtained by optimizing (5.43), and solved using support vector machine method:

$$a_{io} \boldsymbol{q} = b_{io} - r_0. \tag{5.44}$$

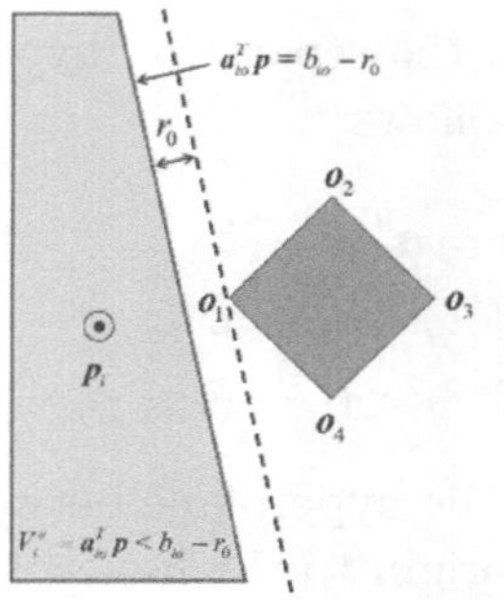


Figure 5.19: Constructing Voronoi cells based on separated hyperplanes in obstacle environments.

By combining (5.36) and (5.44), the complete definition of the formula for generating Voronoi cells from separated hyperplanes can be obtained as follows:

$$v_i^{b,s} = \Big\{ \boldsymbol{q} \in Q : \boldsymbol{a}_{ij}^T \boldsymbol{q} < b_{ij} - r_0 \\ \text{and } \boldsymbol{a}_{io}^T \boldsymbol{q} < b_{io} - r_0 \Big\} . \tag{5.45}$$

(1) Construct a Voronoi cell based on a separated hyperplane in obstacle environment, as shown in Fig. 5.19.

In the practical application of multi-robot formation, the separation hyperplane method is used to generate the respective Voronoi cells of the robots, and the centroid Voronoi tessellations is used to control the robots to move. Specifically, the centroid of the Voronoi area where the robot is located is used as the tracking target point, and the Euclidean distance between the robot's position point and the density centroid of the Voronoi cell is used as the tracking error. When the tracking error is less than the convergence threshold of the formation, it indicates that the formation has reached the convergence state. Otherwise, it continues to iterate until the error meets the requirements. The specific algorithm process is as follows:

(2) Simulation verification of Voronoi tessellations based on separating hyperplanes.

To verify the effectiveness of the Voronoi tessellations based on separated hyperplanes in non-convex environments, two sets of simulation experiments were designed in MATLAB environment to evaluate the performance and robustness of the algorithm under different conditions by changing the number and spatial distribution of obstacles. Set the robot task environment size to $100m \times 100m$, the number of robots to 6, the actual radius of the robots to $1m$, and the density function of the centroid Voronoi tessellations to:

$$50 \cdot \exp\left( - \left( \frac{(x-80)^2}{200} + \frac{(y-80)^2}{200} \right) \right) . \tag{5.46}$$

This density function defines the expected region of target distribution in the task space, and the image distribution is shown in Fig. 5.20, presenting high-density features concentrated in the upper right corner of the region.

**Algorithm 3** Centroid Voronoi tessellations based on separating hyperplanes.

**Require:** Robot initial position $\boldsymbol{p}_i$, task area $Q \subset \mathbb{R}^N$, environmental density function $\rho(\boldsymbol{q})$, cost function threshold $\varepsilon$, obstacle vertex coordinates $\boldsymbol{o}_i$ in the environment;
**Procedure:**
**while** The cost function is greater than $\varepsilon$ **do**
obtain the current position information $\{\boldsymbol{p}_i\}_{i=1}^{n}$ of the robot
constructing Voronoi cells by separating hyperplanes based on the current position of the robot
**for** $i = 1, \ldots, n$ **do**
calculate the centroid $\boldsymbol{c}_{v_i}$ of each robot Voronoi cell based on the density function $\rho(\boldsymbol{q})$
drive the robots to their respective centroid positions $\boldsymbol{p}_i = \boldsymbol{c}_{v_i}$
**end for**
**end while**

a. Simulation verification in two obstacle environments

First, simulate two obstacle environments, where the robot's trajectory is represented by curves of different colors, with black dots indicating the actual position of the robot. Set the center coordinates of the obstacle one (50,60) with a side length of $10m$; the center coordinates (60,20) of the obstacle two with a side length of $5m$. Define the cost function threshold as 0.1, the actual radius of the robot $r_0 = 1$ m, i.e. the safe distance between robots is $2m$, and the safe distance between the robot and the obstacle is $1m$. The specific simulation effect is shown in Fig. 5.21.

Figs. 5.21a, 5.21b, and 5.21c, show the simulation results of the initial distribution, 19 iterations, and 4 iterations of the robot, respectively. It can be seen that the six robots ultimately completed the centroid Voronoi tessellations. Fig. 5.21d shows the cost function of the multi-agent system. It can be seen from the figure that when the number of iterations of the multi robot system reaches 44, the cost function of the system decreases to 0.1. Fig. 5.21e shows the minimum distance between robots during the iteration process, where different colored curves represent the minimum distance between the corresponding intelligent agent and other robots. It can be seen

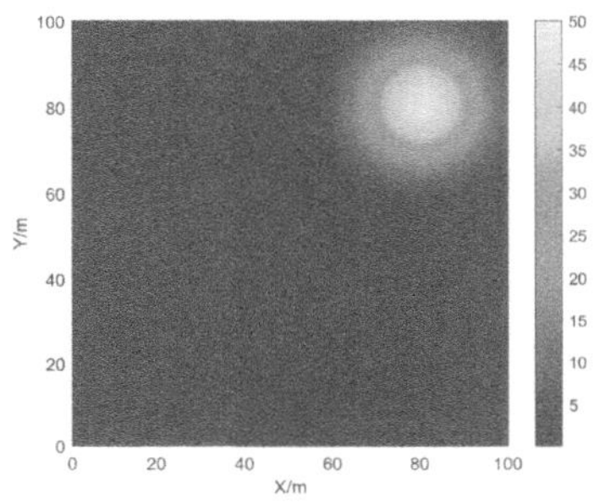


Figure 5.20: Density distribution of task environment.

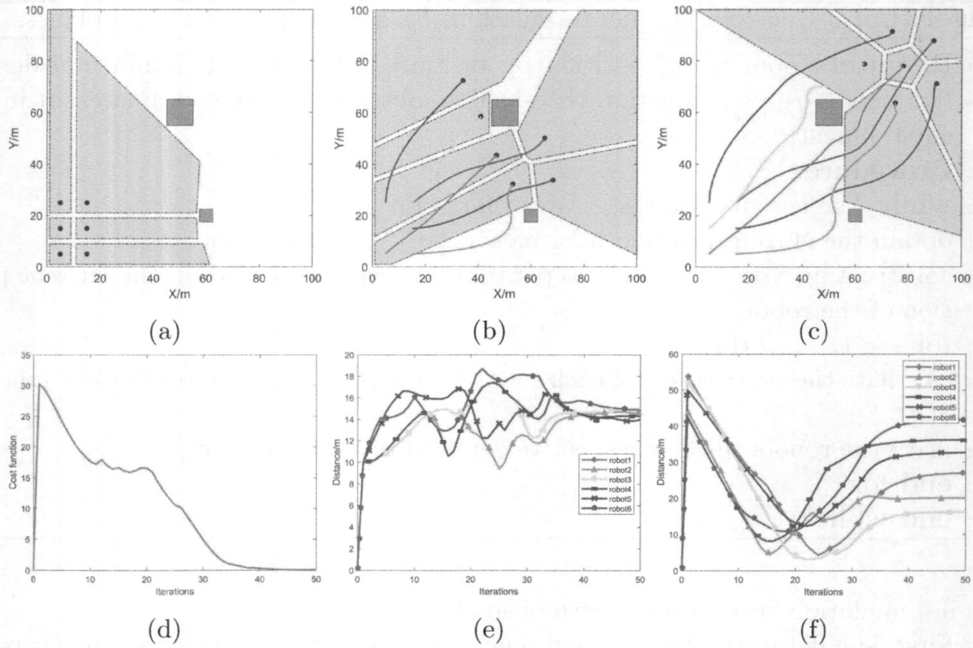

Figure 5.21: Simulation verification in two obstacle environments: (a) $k = 1$; (b) $k = 19$; (c) $k = 44$; (d) system cost function; (e) minimum distance between robots; (f) minimum distance between robot and obstacle.

from the figure that the distance between all robots is greater than the safe distance of 2m, ensuring collision avoidance performance between robots. Fig. 5.21f, shows the minimum distance indicator between the robot and the obstacle, indicating that the distance between the intelligent agent and the obstacle is always greater than the set safety value of 1m.

b. Simulation verification in five obstacle environments

Next, we will simulate five obstacle environments. Set the center coordinates of obstacle 1 to be (50,60), with a side length of $10m$; obstacle 2 to be (60,30), with a side length of $5m$; obstacle 3 to be (20,70), with a side length of $8m$; obstacle 4 to be (85,40), with a side length of $7m$; obstacle 5 to be (30,40), with a side length of $8m$. Define the threshold value of the robot cost function as 0.1, the actual radius of the robot $r_0 = 1$m, that is, the safe distance between the robots is $2m$, and the safe distance between the robot and the obstacle is $1m$. The specific simulation effect is shown in Fig. 5.22.

Figs. 5.22a, 5.22b, and 5.22c show the simulation results of the initial distribution, 18 iterations, and 48 iterations of the robot, respectively. It can be seen that the six robots ultimately completed the centroid Voronoi tessellations. Fig. 5.22d shows the cost function of the multi robot system. It can be seen from the figure that when the multi-robot system reaches 48 iterations, the cost function of the system drops to 0.1. Fig. 5.22e shows the minimum distance between robots during the iteration process, where different colored curves represent the minimum distance between the

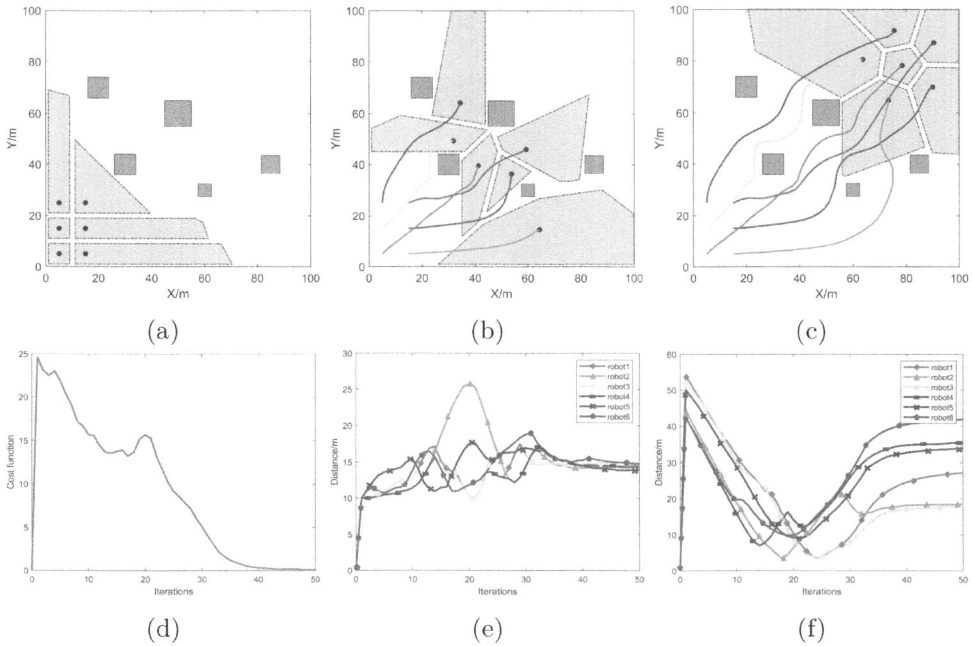


Figure 5.22: Simulation verification in five obstacle environments: (a) $k = 1$; (b) $k = 19$; (c) $k = 44$; (d) system cost function; (e) minimum distance between robots; (f) minimum distance between robot and obstacle.

corresponding robots and other robots. It can be seen from the figure that the distance between all robots is greater than the safe distance of 2m, ensuring the collision avoidance performance between robots. Fig. 5.22f shows the minimum distance indicator between the robot and the obstacle, indicating that the distance between the robot and the obstacle is always greater than the preset safety value.

From the two simulation results, it can be seen that in environments with different numbers and distributions of obstacles, the Voronoi tessellations based on separated hyperplanes can always effectively ensure the safe operation of multi robot systems, demonstrating good environmental adaptability and system robustness.

### 5.3.2 The Centroid Voronoi Tessellations Method with Obstacle Avoidance Strategies

An alternative paradigm to modifying cell boundaries is to maintain the standard Voronoi tessellation but enhance the robot's control law to handle obstacles. This "algorithm fusion" approach treats the Voronoi centroid as a high-level goal, while a low-level local planner is responsible for navigating around obstacles encountered along the path to that goal. This method can be advantageous as it preserves the simple convex geometry of the standard Voronoi cells, simplifying centroid calculations. This section explores the integration of two popular local planners with CVT:

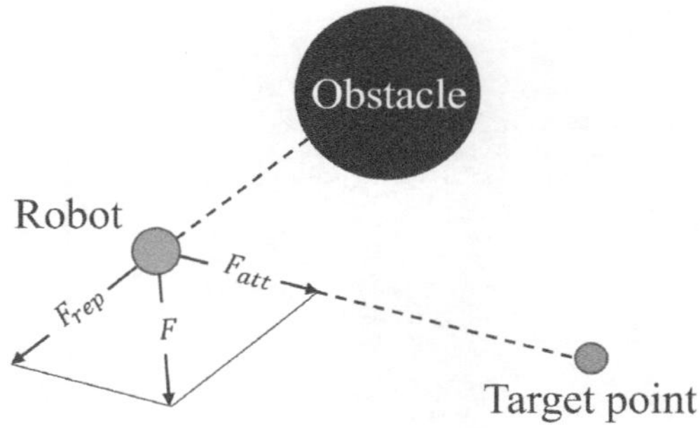


Figure 5.23: Schematic diagram of artificial potential field.

the Artificial Potential Field (APF) method and the Dynamic Window Approach (DWA).

**Centroid Voronoi Tessellations Based on Improved Artificial Potential Field**

(1) Principle of artificial potential field method

The artificial potential field method is a spatial virtual force proposed by Khatib inspired by the repulsive and attractive forces between homogeneous and heterogeneous charges in an electric field [22]. Obstacles generate repulsive fields, while target points generate attractive fields. Robots can avoid obstacles by bypassing them under the superposition of these two potential fields. The principle of artificial potential field is shown in Fig. 5.23, where the robot is represented by gray circles. The robot is simultaneously affected by repulsive and attractive forces, and ultimately moves in the direction of the resultant force, moving away from obstacles and reaching the target point.

Assuming the position of the robot is $\boldsymbol{x}$, and the obstacles and target points are $\boldsymbol{X}_O$ and $\boldsymbol{X}_g$, respectively, the repulsive potential field generated and the gravitational potential field generated by the target point are represented as $U_{rep}$ and $U_{att}$, respectively. The repulsive potential field generated by the obstacle, the closer the point to the obstacle the greater the force in the repulsive potential field, and vice versa, is mathematically defined as:

$$U_{rep}(x) = \begin{cases} \frac{\varepsilon}{2}\left(\frac{1}{d(x,X_o)} - \frac{1}{d_o}\right)^2, & d\left(x, X_o\right) < d_0 \\ 0 & , \quad d\left(x, X_O\right) > d_0 \end{cases}, \tag{5.47}$$

where $\varepsilon$ is the repulsion gain coefficient; $d\left(\boldsymbol{x}, \boldsymbol{X}_O\right)$ indicates the distance from that discrete point to the center of the obstacle; $d_o$ is the minimum radius of the repulsive action. The repulsive force is the negative gradient of the repulsive field, denoted as:

$$F_{rep}(q) = \begin{cases} \varepsilon\left(\frac{1}{d(x,X_o)} - \frac{1}{d_o}\right)\frac{1}{d(x,X_o)^2} & , d\left(x, X_O\right) < d_0 \\ 0 & , d\left(x, X_O\right) > d_0 \end{cases}. \tag{5.48}$$

The attractive potential field is generated by the target point.

$$U_{att} = \frac{1}{2}\eta d\left(x, X_g\right)^2, \tag{5.49}$$

where $\eta$ is the gravitational potential field gain coefficient, $d\left(\boldsymbol{x}, \boldsymbol{X}_g\right)$ is the distance between the robot and the target point, and the attractive force is also the negative gradient of the gravitational potential function, and the combined force on the robot is the combined force of (5.47) and (5.50), as shown in (5.51).

$$F_{att}(q) = -\nabla U_{att}(q) = \varepsilon d\left(x, X_g\right), \tag{5.50}$$

$$F = F_{att} + F_{rep}. \tag{5.51}$$

Although the traditional APF allows the robot to avoid obstacles in the process of advancement, there are two unavoidable drawbacks [23] : the local optimum and the target unreachability problem. The local minimum refers to the situation when the robot, the target point, and the obstacle are in a straight line, and the robot is subjected to the attractive force and the repulsive force of equal magnitude and opposite direction at a certain place so that the combined force is 0. The unreachable target refers to the situation when there is an obstacle near the target point, and the robot is subjected to the increasing repulsive force and decreasing attractive force in the process of approaching the target point, which makes the robot unable to reach the target point. In addition to these two cases, once the target point is generated inside the obstacle in the process of CVT, it will also make the robot unable to reach the centroid. Therefore, this paper proposes improvements to address the above problems by firstly eliminating the problems in APF, and secondly constraining the generated centroidal position to improve the obstacle avoidance performance of the coverage system.

(2) Improved artificial potential field method

(a) Improve on unreachable targets

Target unreachability refers to the situation when the target point of the robot is located near an obstacle in the traditional APF when the repulsive force at the point is infinite, and the repulsive force on the robot in the process of going to the target point is much larger than the attractive force on the robot, which leads to the robot not being able to reach the point, as shown in Fig. 5.24.

In Fig. 5.24a, the red star represents the starting point, the blue star represents the target point of the robot, the black circle represents the obstacle, the motion trajectory of the robot is represented by the green hollow circles, and the black curve in Fig. 5.24b the position error, which represents the distance between the robot and the target point. As seen from the two figures above, the robot is never able to reach the target point due to the infinitive repulsive force. To solve this problem, in this chapter, the repulsive potential field near the obstacle is replaced by:

$$U_{rep}(x) = \begin{cases} \frac{\varepsilon}{2}\left(\frac{1}{d_m} - \frac{1}{d_o}\right)^2 d^n, d\left(x, X_O\right) < d_m \\ \frac{\varepsilon}{2}\left(\frac{1}{d(x, X_o)} - \frac{1}{d_o}\right)^2 d^n, d_O < d\left(x, X_O\right) < d_m, \end{cases} \tag{5.52}$$

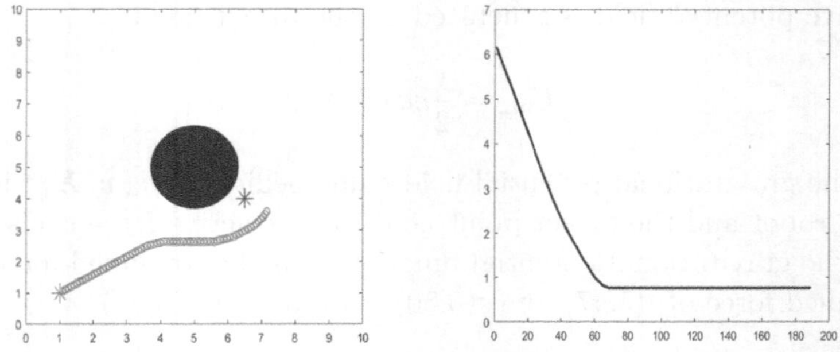

(a) (b)

Figure 5.24: Target unreachable: (a) Target near the obstacle; (b) position error.

where $d$ is the distance between the robot and the target point; $d_m$ is the limiting distance, after reaching this distance, the repulsive force will enter a constant value, preventing the case of infinite repulsive force near the obstacle. Then the repulsive force near the obstacle is given by:

$$F_{rep}(x) = -\nabla U_{rep}(x) = \begin{cases} f_1 & d(x, X_O) < d_m \\ f_2 + f_3 & d_m < d(x, X_O) < d_0, \\ 0 & d(x, X_O) < d_0, \end{cases} \tag{5.53}$$

$$\begin{cases} f_1 = \frac{n\varepsilon}{2}\left(\frac{1}{d_m} - \frac{1}{d_0}\right)^2 d^{n-1} \\ f_2 = \varepsilon\left(\frac{1}{d(x,X_O)} - \frac{1}{d_0}\right)\frac{1}{d^2(x,X_O)} d^n \\ f_3 = \frac{n\varepsilon}{2}\left(\frac{1}{d(x,X_O)} - \frac{1}{d_0}\right)^2 d^{n-1}. \end{cases} \tag{5.54}$$

Then the modified repulsive force is shown in below. We can see that for the comparison before the correction, the robot finally reaches the target point and the position error is reduced to 0.

(b) Improving local optimal problems

Although the improved repulsive function can solve part of the local optimum problem, when the robot is subjected to attractive force and repulsive force of exactly

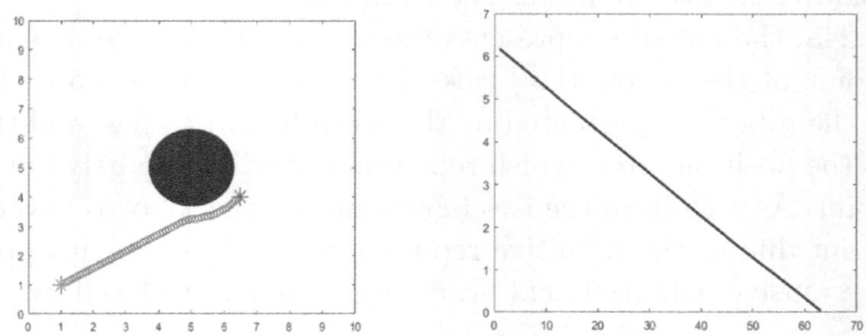

(a) (b)

Figure 5.25: The improvement of target unreachable: (a) Target near the obstacle; (b) position error.

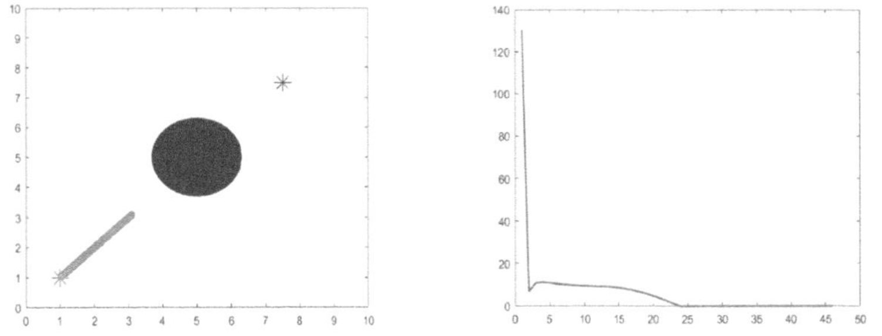

(a) (b)

Figure 5.26: Local optimum: (a) Target collinear with robot; (b) position error.

opposite direction and equal size, the robot is caught in the local optimum at this time, and the robot cannot escape from the region only by the improved repulsive potential field, as shown in Fig. 5.26.

Let the condition that the robot is caught in a local optimum be

$$\begin{cases} |F_{att} + F_{rep}| \leq 0 + \varepsilon \\ |\angle F_{rep} - \angle F_{att}| = \pi + \delta, \end{cases} \tag{5.55}$$

where $x$ and $X_g$ are the coordinates of the robot and the target point, respectively, $\angle F_{rep}$ and $\angle F_{att}$ are the angles between the robot and the attractive and repulsive forces, respectively, and $\varepsilon$, $\delta$ are the arbitrary minima of the combined forces and angles. When the condition is satisfied, the robot is determined to be in a local optimum. At this time, the robot starts from its own position and searches eight points within its detection radius, and selects the best point of the resultant force as the next target point, so as to jump out of the local optimum region. The coordinates of the robot in the detection process are shown in (5.56), where $t_s$ is the detection radius of the robot

$$\begin{cases} t_{i\_}x = x\_x + t_s \sin \frac{(i-1)\pi}{4} \\ t_{i\_}y = x\_y + t_s \cos \frac{(i-1)\pi}{4} \end{cases} (i = 1, 2, \ldots 8)\ . \tag{5.56}$$

The motion trajectory of the improved method is shown in Fig. 5.27a, and it can be seen that the robot eventually escapes from the local optimum under the action of (5.56), and the distance between the robot and the target point is finally reduced to 0 in Fig. 5.27b.

(c) Add centroidal repulsion field

In the obstacle environment, when multi-robot covered using the CVT combined with the improved APF described above, although collisions between robots and between robots and obstacles can be avoided during robot tracking of the centroid when the final position of the centroid is generated inside the obstacle, the entire Voronoi coverage appears to fail to converge because the robots cannot reach the centroidal position. As Fig. 5.28 shows 1 robot performing Voronoi coverage of the environment in an obstacle environment, black circles represent obstacles, the blue

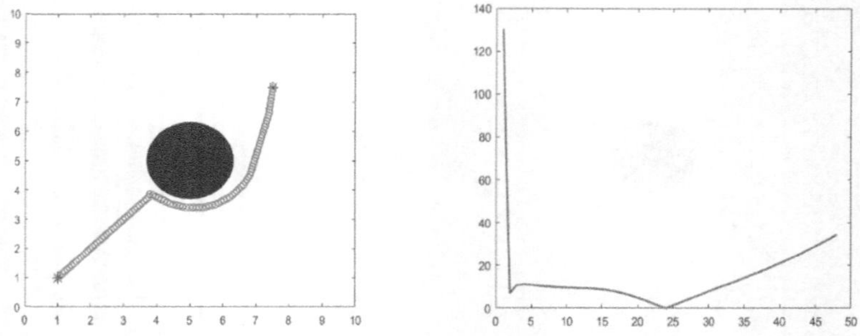

(a) (b)

Figure 5.27: The improvement of local optimum: (a) Target collinear with robot; (b) position error.

circle is the robot, and the white crossing is the centroid of the Voronoi cell. When the centroid is finally generated inside the obstacle, it makes the whole system appear to be unable to converge. For this reason, here we propose centroid repulsion field (CRF) located inside the obstacle, once the centroid is generated inside the obstacle, a new centroid will be generated outside the obstacle due to the presence of CRF, defined as follows:

**Definition of CRF:** When the centroid is generated at a distance from the center of the outer circle of the obstacle is less than the radius of the outer circle, it will generate a repulsive force in the direction of the line from the center of the outer circle to the position of the centroid, which will "push out" the centroid inside the obstacle, so that it is generated outside.

By definition, the closer the location of the outer circle, the greater the repulsive force. The force field in line with the center of the circle as the extreme value of the Gaussian distribution, the cross-section of the normal distribution function, the confidence interval is shown in Table 5.4, $\mu$ is the extreme value point, $\sigma$ is the standard deviation, according to the normal distribution of the $3\sigma$ criterion, the

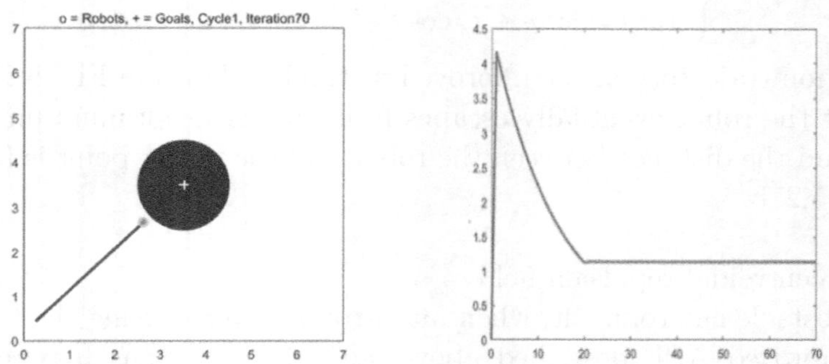


(a) (b)

Figure 5.28: The centroid is generated inside the obstacle: (a) Centroid inside obstacle; (b) position error.

Table 5.4: Confidence interval of normal distribution

| Value interval | $\mu \pm \sigma$ | $\mu \pm 2\sigma$ | $\mu \pm 3\sigma$ |
|---|---|---|---|
| Probability | 68.3% | 95.4% | 99.7% |

probability distribution of the centroid falling in different regions inside the obstacle can be obtained, as shown in Table 5.4

The interval $(\mu - 3\sigma, \mu + 3\sigma)$ is used to denote the obstacle diameter $R$. Therefore, it can be expressed numerically as:

$$\sigma = \frac{R_i}{3}. \tag{5.57}$$

Therefore, if the centroid is generated inside the obstacle, its new centroid generation equation under the constraint of CRF can be expressed as:

$$Z^* = Z + A \cdot e^{-\left(\frac{(Z-X)^2}{2\sigma^2}\right)} \cdot I = Z + A \cdot e^{-\left(\frac{9(Z-X)^2}{2R_i^2}\right)} \cdot I, \tag{5.58}$$

where $X$ is the coordinate of the center of the obstacle, $Z$ is the position of the centroid falling inside the obstacle, $Z^*$ is the position of the centroid after the constraint, $I$ is the unit vector of the centroid pointing to the center of the obstacle circle, expressed as:

$$I = \frac{Z - X}{|Z - X|}. \tag{5.59}$$

Fig. 5.29 shows the motion trajectory of the robot after adding CRF. By comparing with Fig. 5.28, it can be seen that CRF can effectively prevent the centroid from being generated inside the obstacle, thus minimizing the position error of the robot.

(d) Work flow of centroid Voronoi tessellations based on improved artificial potential field

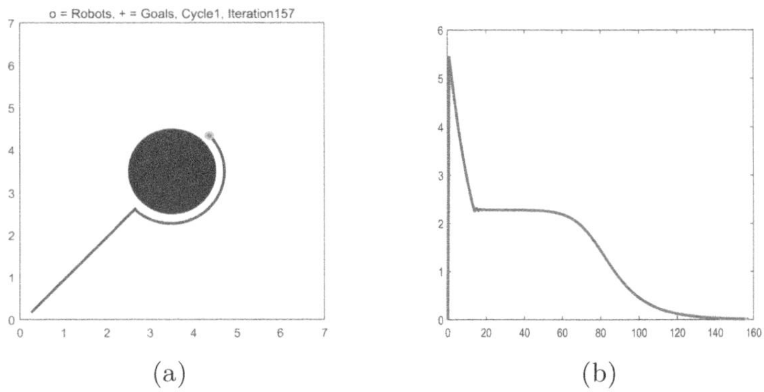


Figure 5.29: Motion of the robot after adding CRF: (a) Centroid outside obstacle; (b) position error.

This subsection combines the improved APF with the CVT and constrains the position of the centroid during its division to avoid the generation of the centroid inside the obstacle. The pseudo-code of the algorithm is shown in Algorithm 4.

**Algorithm 4** CVT coverage algorithm based on an improved APF.

**Initialization:** Initialize $k$ robots position $\{p_i\}_{i=1}^{k}$; mission area $Q \in \mathbb{R}^2$, density function $\rho(q), q \in Q$
**Procedure:**
**Calculate the Voronoi cell** $\{v_i\}_{i=1}^{k}$**, and the centroid of each Voronoi cell** $\{Z_v\}_{i=1}^{k}$ **while** $((\mathrm{err}\,(p_i) = p_i - Z_{v_i}) \gg \varepsilon)$ & & $\min\,(f(p))$ **do**
Each robot drive to centroid $p_i \rightarrow Z_{v_i}$
Generate a new Voronoi cell $p_i = Z_{v_i}$
Calculate the centroid of the new Voronoi element $\{Z_v\}_{i=1}^{k}$ $(\rho(q) = 1)$
**IF** the centroid of the cell inside the obstacle $(Z_{v_i} - X \leq R\_ob)$
Calculate the force according the CRF
Generate new position of the centroid
**else if** the distance between robot and obstacle $< d_0$
Calculate the $F_{att}$ and $F_{rep}$
**end if**
Calculate the $F_{att}$ and $F = F_{att} + F_{rep}$
**end while**

(3) Simulation verification of centroid Voronoi tessellations based on improved artificial potential field

In this section, we simulate the above-proposed algorithm in MATLAB. First, we compare the APF before and after the improvement, combine the improved APF with CVT, and verify the feasibility of the CRF on this basis. The number of robots in the simulation is 4, the working area is 10 m × 10 m, the obstacles present in the environment are black circles, and set the density function inside its area as a constant density function.

(a) Improvement of APF

In an obstacle environment, if there is an obstacle near the centroid of the Voronoi cell, the conventional APF cannot make the robots reach that target point. As shown in Fig. 5.30, the robots are represented by light blue circles, and four robots are started from the lower left corner to their respective target points, and their motion trajectories are represented by blue, yellow, red, and purple, respectively. In the environment shown in Fig. 5.30a, 4 black circular obstacles are arranged near the final convergence position of the centroid, and due to the infinite repulsive force near the obstacles, the robots are unable to reach their respective target points, it can be seen from Fig. 5.30b that the position error between the robots and the centroid cannot be reduced to 0, resulting in the whole coverage system failing to converge.

Fig. 5.31 shows the combination of the improved APF and the CVT. It can be seen that in the same task environment, the robot reaches the target point near the obstacle separately, and the position error with each Voronoi cell centroid is reduced to 0.

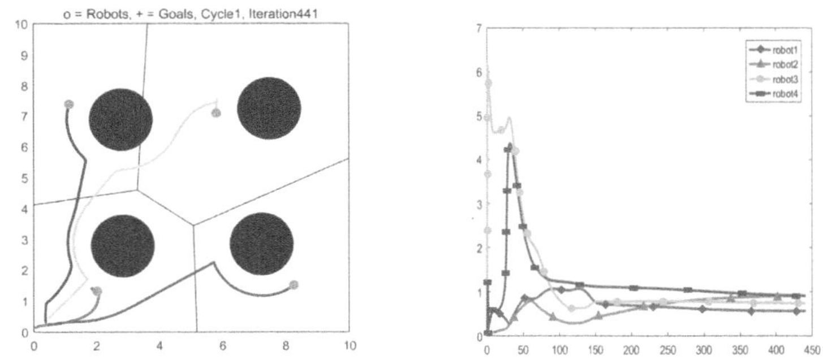


(a) (b)

Figure 5.30: CVT in combination with the conventional APF: (a) Robot final position; (b) position error.

(b) Add CRF to centroid Voronoi tessellations

After adding the improved APF, the robot can avoid the obstacles in the environment due to the virtual repulsive force to achieve the effect of obstacle avoidance, so that the position error is reduced to 0. However, in some special scene, if the centroid is generated inside the obstacle, the coverage system will be in another trouble, which is called centroid unreachable, as is shown in Fig. 5.32. In Fig. 5.32a, the white crossing sign represents the centroid, we can see the three of four centroids are generated inside the obstacles, at this time the robot cannot reach the target position, and Fig. 5.32b shows the error, the error cannot converge to 0.

To solve this centroid unreachable problem, we assume that the trajectory of the centroid in the CVT is a smooth straight line. After that, we add the Centroidal Repulsive Field (CRF) inside the obstacle to prevent the centroid generated with in it. Fig. 5.33 shows the performance of CRF in the same scene as in Fig. 5.32.

Fig. 5.33a shows the CVT coverage of the robot after adding the CRF, Fig. 5.33b shows the error of the robot position and the centroidal position, which shows that the

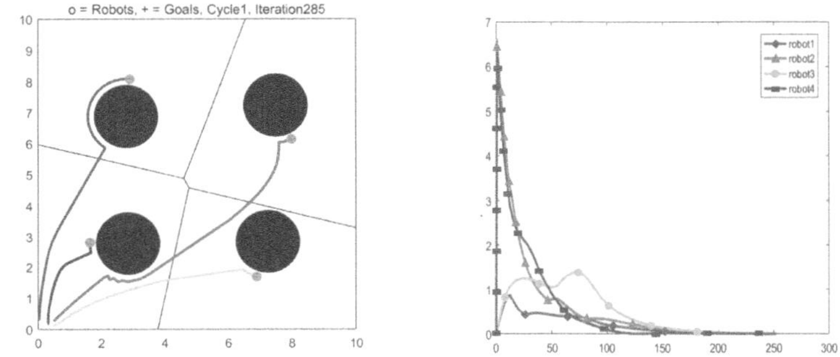


(a) (b)

Figure 5.31: CVT combined with the improved APF: (a) Robot final position; (b) position error.

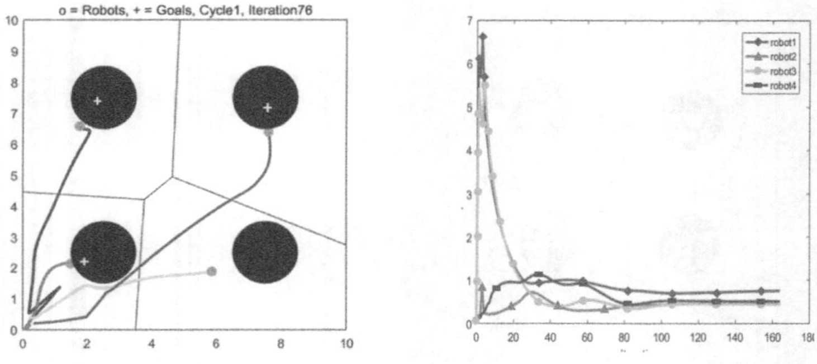


(a) (b)

Figure 5.32: Centroids generated inside the obstacle: (a) Centroid inside obstacle; (b) position error.

error finally converges to 0 asymptotically compared to the coverage scheme without adding the CRF.

(c) The performance of the whole algorithm

To further verify the feasibility of the algorithm proposed in this paper, we added two different experimental scene. In the first scene, we put 4 obstacles with a setting position, as shown in Fig. 5.34.

In Fig. 5.34a, as in the previous scenes, the four robots start from the lower left corner of the scene and go to their respective target points. Fig. 5.34b shows their position error, and the position error relative to the centroidal position increases slightly in the beginning due to the need to bypass the obstacle, but eventually all reach the target points and the errors tend to 0.

Fig. 5.35a shows another scene, in which the position of obstacles are randomly assigned. Although the process is different, the result in Fig. 5.35b is similar to Fig. 5.34 that all robots reached their target in the Voronoi cell, the coverage system converges eventually.

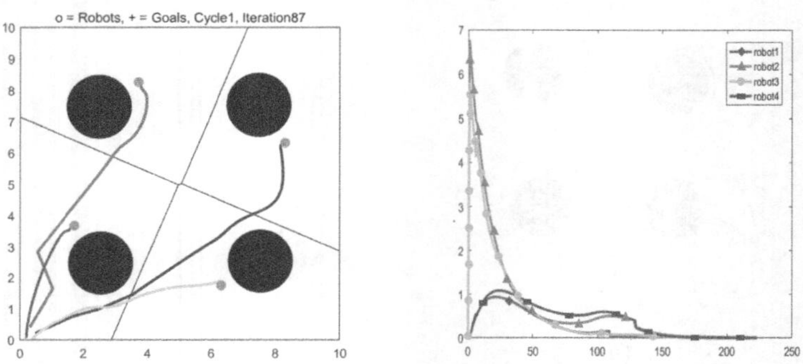


(a) (b)

Figure 5.33: CVT coverage after adding the CRF: (a) Robot final position; (b) position error.

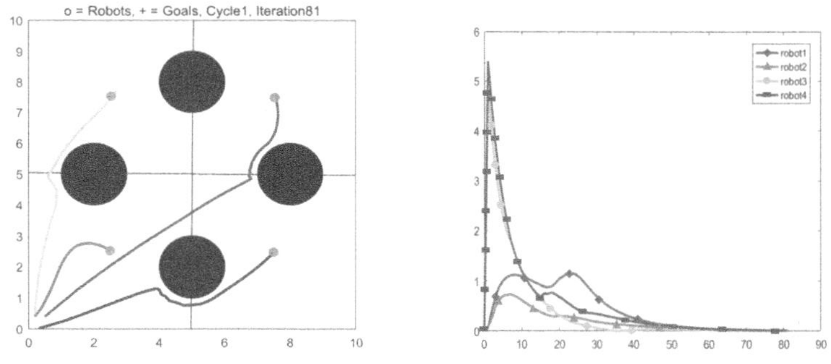


(a) (b)

Figure 5.34: CVT coverage in obstacle environment with setting positions: (a) Robot motion trajectory; (b) position error.

To further verify the feasibility of the algorithm proposed in this chapter, six randomly located obstacles are randomly distributed in Fig. 5.35a, and according to the distribution position of the obstacles and the motion trajectory of the robot, it can be seen that the improved APF method can effectively help the robot escape from the local optimum and help the robot reach the target point near the obstacles, and due to the existence of CRF, the centroid is not generated inside the obstacles, which improves the efficiency of the CVT coverage, and it can be seen through Fig. 5.35b that the distance between each robot and the corresponding centroid finally converges to 0, which verifies the effectiveness of the algorithm proposed in this chapter. Compared with the traditional Voronoi coverage, the advantage of our method are shown in Table 5.5.

**Centroid Voronoi Tessellations Based on Improved Dynamic Window**

(1) Principle of dynamic window approach

(a) Sampling space of dynamic window approach

The dynamic window approach [24] [25] selects the optimal motion trajectory from countless combinations of linear velocity and angular velocity, while considering

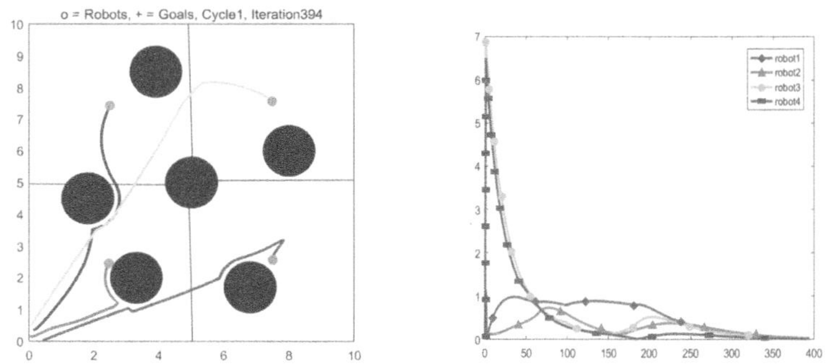


(a) (b)

Figure 5.35: CVT coverage in obstacle environment with random positions: (a) Robot motion trajectory; (b) position error.

Table 5.5: Comparison between Voronoi coverage methods

| Environments | Obs-free environment | Centroid constrain | Obstacle environment |
|---|---|---|---|
| Voronoi coverage | ✓ | X | X |
| Improved coverage method | ✓ | ✓ | ✓ |

the performance and environmental constraints of the robot itself, which requires limiting the sampling speed. Usually, the speed sampling of the dynamic window approach consists of three parts: the set of achievable robot speed obtained through the set maximum and minimum speed limits, the set of speed constraints obtained by the robot's own motor performance limitations, and the set of collision distance based speeds set to ensure safety between the robot and obstacles.

In terms of the maximum and minimum speed constraints for robots, after setting the maximum and minimum speeds of the robot, the window for sampling speed can be represented as:

$$V_m = \{(v, \omega) \mid v_{\min} \leq v \leq v_{\max}, \omega_{\min} \leq \omega \leq \omega_{\max}\}, \tag{5.60}$$

where $V_m$ represents the set of velocity constraints, $V_{min}$ and $V_{max}$ represent the minimum and maximum linear velocities of the robot, and $\omega_{min}$ and $\omega_{max}$ represent the minimum and maximum angular velocities of the robot, respectively.

In terms of speed constraints limited by motor performance, the maximum and minimum acceleration constraints are added to the speed space of the robot, so that the acceleration and deceleration windows of the robot within one sampling period can be expressed as:

$$V_d = \{(v, \omega) \mid v_c - \dot{v}_b \Delta t \leq v \leq v_c + \dot{v}_a \Delta t, \omega_c - \dot{\omega}_b \Delta t \leq \omega \leq \omega_c + \dot{\omega}_a \Delta t\}, \tag{5.61}$$

where $V_d$ represents the velocity window under the constraint of braking distance, and $dis(v, \omega)$ is the closest distance of the robot to the obstacle within the trajectory set composed of linear velocity and angular velocity.

The velocity space set of the robot should simultaneously satisfy the above three velocity constraints. Therefore, the intersection of the three different velocity constraints is taken as the final sampling space, defined as:

$$V_r = V_s \cap V_d \cap V_a. \tag{5.62}$$

(b) Trajectory evaluation function of the dynamic window approach

After determining the velocity sampling space, we shall simulate the possible trajectory of the robot by combining the line velocity and angular velocity in the velocity space. To determine the optimal trajectory, it is necessary to design a trajectory evaluation function to select the most reasonable trajectory. To establish the trajectory evaluation function, following criteria are considered the robot can reach the target point at a faster speed and can avoid possible obstacles by changing its heading. The trajectory evaluation function of the dynamic window approach is generally defined as:

$$G(v, \omega) = \sigma(\alpha \cdot \text{heading}(v, \omega) + \beta \cdot \text{dist}(v, \omega), + \gamma \cdot \text{vel}(v, \omega)), \tag{5.63}$$

where $\sigma$ represents the parameters of smooth normalization processing, heading$(\nu, \omega)$ represents the heading evaluation function that measures the deviation between the current heading of the robot and the target position, dist$(\nu, \omega)$ represents the distance evaluation function from the robot to the obstacle, vel $(\nu, \omega)$ represents the current speed of the robot, and $\alpha$, $\beta$, and $\gamma$ represent the weight parameters of the three evaluation functions, respectively.

Among the three evaluation functions, the heading evaluation function mainly measures the degree of deviation between the heading of the end point of the robot trajectory and the target point, as shown in (5.64). $\theta$ represents the deviation angle. The smaller the $\theta$ angle, the smaller the angle between the robot's heading and the target point, and the robot tends to move toward the target point.

$$\begin{aligned} \text{heading } (v, \omega) &= 180^\circ - \theta \\ \theta &= \arctan \frac{y_p - y_c}{x_p - x_c}, \end{aligned} \tag{5.64}$$

where $\left(\boldsymbol{x}_p, \boldsymbol{y}_p\right)$ represents the coordinate point at the end of the trajectory, and $(\boldsymbol{x}_c, \boldsymbol{y}_c)$ represents the coordinate of the target point.

The obstacle distance evaluation function represents the distance between the end of the trajectory generated by the robot and the obstacle. The main purpose of this function is to detect the distance between obstacles and robots to assign safety measures to each trajectory point. The farther the robot is from the obstacle, the higher the trajectory safety level. Otherwise, the trajectory safety level is low and should not be selected as the next target point.

$$\begin{aligned} \text{dist}(v, \omega) &= \begin{cases} d, d < d_{\max} \\ d_{\max}, d \geq d_{\max} \end{cases} \\ d &= \sqrt{(x_c - x_o)^2 + (y_c - y_o)^2}. \end{aligned} \tag{5.65}$$

where $(\boldsymbol{x}_c, \boldsymbol{y}_c)$ represents the position point of the obstacle, $d$ represents the shortest distance from the simulate trajectory of the robot to the obstacle, and $d_{max}$ represents the preset distance threshold.

Finally, the main function of the speed evaluation function is to assess the current speed of the robot, usually directly using the absolute value of the robot's speed as the value of the speed evaluation function. The higher the speed evaluation value, the faster the robot can approach the target trajectory. Reasonable selection of points on the trajectory can ensure that the robot reaches the target point quickly.

$$\mathrm{vel}(v,\omega) = |v_g|, \tag{5.66}$$

where $v_g$ represents the current speed value of the robot.

The trajectory evaluation function is the weighted sum of three evaluation functions, and the values of the three evaluation functions belong to different dimensions. In order to avoid excessive weight of a certain item in the evaluation function, which may result in unreasonable trajectories, it is necessary to normalize it before weighing it. The main idea of normalization is to normalize each value to an interval by dividing each individual indicator by the sum of the indicators. The normalization calculation formulas for the three evaluation functions are:

$$normal\_heading(i) = \frac{heading(i)}{\sum_{i=1}^{n} heading(i)}, \tag{5.67}$$

$$normal\_dist(i) = \frac{dist(i)}{\sum_{i=1}^{n} dist(i)}, \tag{5.68}$$

$$normal\_vel(i) = \frac{vel(i)}{\sum_{i=1}^{n} vel(i)}, \tag{5.69}$$

where $n$ represents all possible trajectories within the simulation period, and $i$ represents the currently evaluated trajectory.

(2) An improved dynamic window approach

(a) Add virtual obstacles

The use of dynamic window approach also needs to consider the problem of unreachable centroid in non-convex areas. When the robot moves near an obstacle, two situations may occur: one is that the centroid of the robot is generated inside the obstacle, and the CVT focuses on reducing the distance error between the centroid point and the robot position as the main evaluation index, guiding the intelligent agent to always move toward the centroid located inside the obstacle, ultimately leading to collision phenomenon. Another situation is that the center of mass of the robot is generated too close to the obstacle. Although the center of mass is not generated inside the obstacle, during the process of approaching the center of mass, the dynamic window approach controls the intelligent agent to make significant turns in order to avoid the obstacle, resulting in many unnecessary trajectories and additional energy consumption.

Therefore, to solve the above problems, first, a virtual centroid is introduced when executing the CVT, and the density of the obstacle location is set to 0. Then, the complement of the obstacle area in the Voronoi unit is taken to obtain the Voronoi cell without obstacles:

$$\begin{cases} \mathcal{V}_i = \{q \in \Omega \mid \|p_i - q\| \leq \|p_j - q\|\} \\ \mathcal{V}^{o,i} = \complement_{\mathcal{V}_i} O_i. \end{cases} \tag{5.70}$$

The centroid position constrained by (4.12) will not be generated inside the obstacle, resulting in an improved robot trajectory compared to the original algorithm. However, there are still cases where the centroid point is too close to the obstacle. To solve this problem, virtual obstacles are introduced to create a larger zero density area, so that the centroid of the Voronoi cell is generated outside the virtual obstacles. The non-convex environment constrained CVT division under virtual obstacles is as follows:

$$\begin{cases} \mathcal{V}_i = \{q \in \Omega \mid \|p_i - q\| \leq \|p_j - q\|\} \\ O_i = \{\|q - p_o\| \leq \lambda\} \\ \mathcal{V}^{o,i} = \complement_{V_i} O_i, \end{cases} \tag{5.71}$$

where $O_i$ represents the obstacle area, $\mathcal{V}_i$ represents the standard Voronoi area, $p_o$ represents the position of the obstacle, $\lambda$ represents the inflation factor, $\lambda \in [1, 1.5]$, and $\mathcal{V}^{o,i}$ represents the part of Voronoi cells $\mathcal{V}_i$ that does not contain obstacles.

(b) Adaptive mechanism for dynamic window parameter weights

The dynamic window approach considers factors such as the dynamic constraints of the robot and the detection environment information. Multiple sets of different possible sampling trajectories are constructed by combining linear velocity and angular velocity. Then, based on the set evaluation function, the trajectory at the optimal linear velocity and angular velocity is selected as the next motion trajectory of the agent. The principle of dynamic window approach can be summarized as follows:

**Step 1:** Determine the current state of the intelligent agent, including information such as position, velocity, and orientation.

**Step 2:** Calculate the set of windows that generate feasible actions based on the dynamic and environmental constraints of the robot.

**Step 3:** For each window, calculate its potential reward based on a predefined evaluation function and evaluate the actions within the window candidate set.

**Step 4:** Select the action with the highest potential reward as the next action for the robot.

**Step 5:** Repeat the above steps until the goal is achieved or the best path is found.

The dynamic window approach has real-time performance and can quickly perform path planning, avoiding dependence on the global map. It can also dynamically adapt to environmental changes and find the best path in complex environments. Compared with local path planning algorithms such as artificial potential field approach, dynamic window approach considers the dynamic constraints and mechanical characteristics constraints of the robot itself [26], making the path planned by this approach more in line with real scenarios and able to meet actual control needs.

Although the dynamic window approach has certain advantages in local path planning, it cannot accurately predict the movement and changes of obstacles in highly dynamic and uncertain environments [27], resulting in possible collisions between robots and obstacles. Second, the dynamic window approach relies on the evaluation function to select the best action, but the weight of the evaluation function in the traditional dynamic window approach is fixed, and in some complex scenarios, it cannot select the optimal path and falls into a local extremum. Therefore, it is necessary to add an adaptive adjustment mechanism to the weight parameters of the dynamic window approach to improve the overall safety and efficiency of the multi-robot system. The adaptive adjustment mechanism is introduced below. The formula for dynamically adjusting speed weights is given by

$$\gamma = \begin{cases} \gamma_m + \mu \cdot \frac{1}{1+e^{-\frac{d_s}{d_m}}}, & d_m < d_s \\ \gamma_m, d_m > d_s \end{cases}, \tag{5.72}$$

where $d_m$ represents the Euclidean distance from the robot to the obstacle, $d_s$ represents the distance threshold for executing the dynamic window method, and $\gamma_m$ represents the maximum speed weight threshold. $\mu$ is the parameter for adjusting the second term of the formula, and $\mu < 0$.

When $d_s < d_m$, the centroid Voronoi partitioning algorithm is used to control the motion of the robot. When $d_s > d_m$, the robot enters the obstacle avoidance phase, and the speed weight is controlled by (5.71). As the distance between the robot and the obstacle decreases, the speed weight $\gamma$ continuously decreases.

The relationship between the weight coefficients of the yaw angle evaluation function and the input is as follows:

$$\alpha = \begin{cases} \alpha_m - \beta \cdot \frac{1}{1+e^{-\frac{d_s}{d_m}}}, & d_m < d_s \\ \alpha_m, d_m > d_s \end{cases}, \tag{5.73}$$

where $\alpha$ represents the yaw angle weight, $\alpha_m$ is the maximum yaw angle weight, and $\beta$ is the adjustment parameter. When $d_s > d_m$, the weight of the yaw angle decreases continuously as the distance from the robot to the obstacle decreases. The overall work flow of the algorithm is as follows.

(3) Simulation verification of centroid Voronoi tessellations based on improved dynamic window

This section uses MATLAB software to simulate and analyze the proposed method, verifying the performance of the algorithm after adding virtual obstacles and adaptive weight adjustment mechanism. The two-dimensional simulation experimental environment is 10 m × 10 m, set to 7 robots, and the density function is a uniform density function.

(a) Control of centroid Voronoi formation with virtual obstacles added

In this simulation, the center positions of the obstacles are set at points $(2m, 8m)$, $(5m, 4m)$, $(8m, 2m)$ and the radius of the obstacles. From Fig. 5.36, it can be seen that after adding virtual obstacles, the robot's motion trajectory is relatively reasonable, and there is no oscillation or collision near the obstacles. However, as shown in Fig. 5.36d, the cost function of the system failed to converge to a lower value because unreasonable velocity and heading weights prevented the robot from converging to the target centroid in a timely manner. The red dashed line in Fig. 5.36f represents the safety distance. Setting the safety distance to $0.25m$, the minimum distance between the 7 robots and the obstacle is always greater than the safety distance, indicating

**Algorithm 5** Obstacle avoidance algorithm for centroid Vino formation based on adaptive dynamic window method.

**Require:** $n$ intelligent agents' position information $\{p_i\}_{i=1}^{n}$, task area $D \in R^2$, formation distance convergence threshold $p$, density information $\rho$
**Procedure:**
**while** The minimum distance $d <$ the intelligent agent and the centroid in the formation is less than the minimum convergence threshold $p$ of the multi-agent formation
**if** The distance $d_m <$ the intelligent agent and the obstacle is less than the distance threshold $d_s$ for executing the dynamic window method
Generate velocity sampling space $V_r$ based on the position information of the intelligent agent
Simulate trajectories for each set of speeds
Using the evaluation function $G(v, \omega)$ under dynamic weights to select the optimal trajectory
**if** Successfully avoided obstacles
Using centroid Voronoi formation control to drive intelligent agents
**else**
Reselect the optimal trajectory in the velocity space $V_r$
**end**
**else** Select the next trajectory point as the target point $p_{i+1}$
According to the density information $\rho$, divide the Voronoi cells $\mathcal{V}_i$ of each agent and calculate the corresponding centroid $c_{\mathcal{V}_i}$
Drive the intelligent agent to the centroid position $p_i = c_{\mathcal{V}_i}$
**end**
**end while**

that the strategy of adding virtual obstacles can ensure the safety of the robots, that is, the robots will not collide with obstacles.

(b) Improved centroid Voronoi formation control under dynamic window

Set the number of obstacles in the simulation experiment environment to 4, with their centers located at $(3m, 2.5m)$, $(7.5m, 2.5m)$, $(3m, 7.5m)$, $(7.5m, 7.5m)$, and the radius of the obstacles $R = 1m$. The simulation experiment of formation obstacle avoidance using the improved adaptive dynamic window method is shown in Figs. 5.37, 5.37f shows that the algorithm terminates at the 160th iteration, while Fig. 5.37a to 5.37c demonstrate that all robots in the formation task are able to avoid obstacles and there are no issues with large turns or excessive speed that prevent convergence to the target point.

The changes in speed and heading weights of Robot 7 during formation obstacle avoidance are shown in Figs. 5.37d and 5.37e, indicating that the weights of the dynamic window method vary with the distance between the robot and the obstacle. That is, when the robot approaches the obstacle, it has lower speed and heading weights, and as the distance between the obstacle and the robot increases, the two weights also increase accordingly. Fig. 5.37g shows that the area of the Voronoi unit

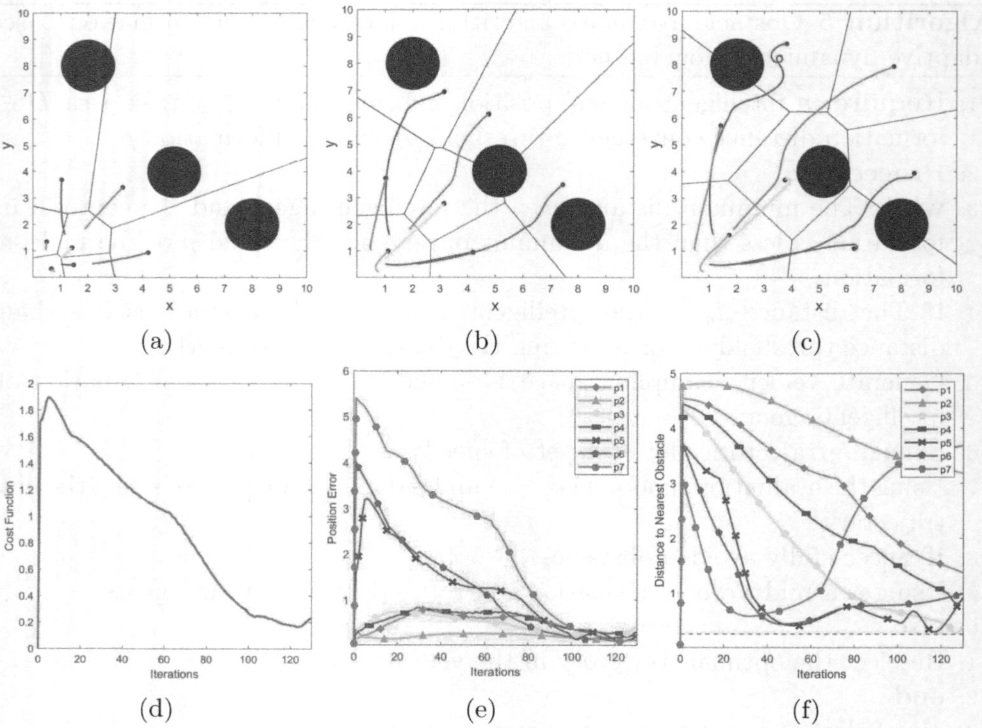


Figure 5.36: Add simulation of virtual obstacles: (a) $k = 15$; (b) $k = 63$; (c) $k = 102$; (d) system cost function; (e) distance error; (f) minimum distance between robot and obstacle.

tends to be equal, indicating that the area occupied by each robot is approximately equal, which conforms to the CVT coverage characteristics of equal distribution under a constant density function. The position error iteration chart in Fig. 5.37h shows that the final distance between the centroid and the robot is relatively small. Fig. 5.37i illustrates that during the formation obstacle avoidance process, the distance between each robot and the obstacle is always greater than $0.25m$.

To analyze the performance of the improved algorithm, 20 repeated experiments were conducted. Table 5.6 shows the simulation results analysis of the adaptive dynamic window method. From the table, it can be seen that the average number of iterations for the improved algorithm is 158, which is an increase of about 6.96% compared to the high heading and speed weight case. However, the safety distance of the improved algorithm is relatively large, which can ensure the safety of the robot. Compared with the low heading and speed weight situation, the algorithm's iteration times are reduced by 10.23%, the minimum distance between the robot and obstacles is 0.27m, and the total path length is reduced by about 5.56%.

The results in Table 5.6 highlight a classic trade-off in motion planning: aggressive parameters (high weights) lead to faster completion times and shorter paths but compromise safety (smaller $D_{min}$), while conservative parameters are safer but less efficient. The proposed adaptive weight mechanism successfully navigates this

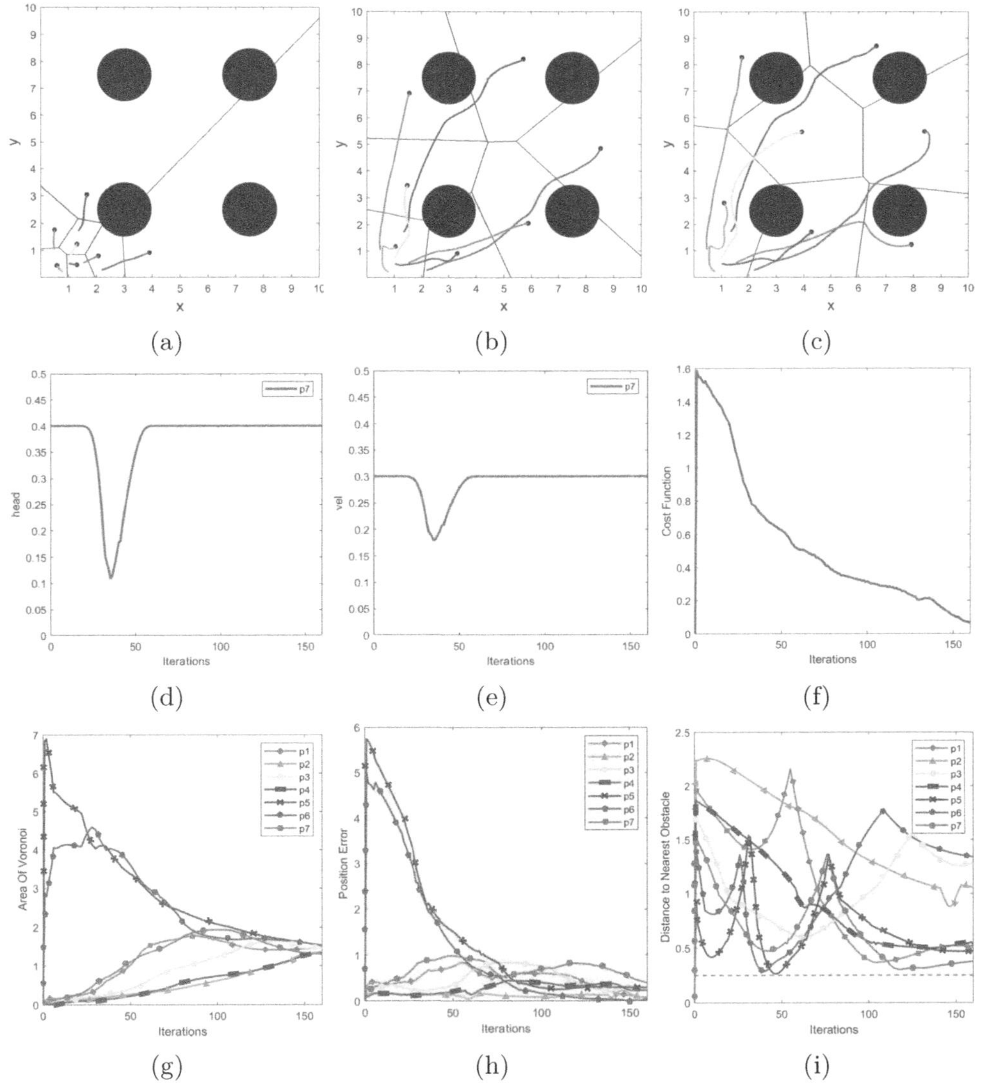


Figure 5.37: Add simulation of virtual obstacles: (a) $k = 15$; (b) $k = 91$; (c) $k = 160$; (d) heading weight; (e) speed weight; (f) System cost function; (g) Voronoi cells area; (h) distance error; (i) minimum distance between robot and obstacle.

Table 5.6: Comparison of various indicators before and after improvement

| **Weight parameters** $(\alpha, \gamma)$ | $D_{\min}/m$ | $D_t/m$ | *step* | **Time** / $s$ |
|---|---|---|---|---|
| (0.1,0.1) | $0.83 \pm 0.2$ | $48.42 \pm 0.15$ | $176 \pm 2$ | $151.84 \pm 3.2$ |
| $(\alpha_d, \gamma_d)$ | $0.27 \pm 0.06$ | $45.73 \pm 0.31$ | $158 \pm 3$ | $136.43 \pm 4.5$ |
| (0.4,0.4) | $0.13 \pm 0.03$ | $40.51 \pm 0.12$ | $147 \pm 3$ | $115.69 \pm 3.8$ |

trade-off, achieving near-optimal efficiency while robustly maintaining a larger safety margin, demonstrating its superior balance of performance and safety.

## 5.4 Chapter Summary

In summary, this chapter focuses on the safety collaborative control problem of multi robot systems, and constructs a collision avoidance mechanism framework based on Voronoi tessellations. Starting from the physical constraints and positioning errors between robots, we introduce buffer uncertainty Voronoi units and dynamic buffer area tessellations algorithms, effectively solving the problem of traditional CVT algorithms ignoring robot physical size and perception errors. By modeling the state distribution of each robot through particle filtering and confidence ellipse, the uncertainty can be further quantified, and an adaptive Voronoi boundary was constructed to improve the collision avoidance ability and convergence of the multi-robot system in dynamic environments.

In addition, in response to the obstacle avoidance requirements between robots and the environment, this chapter proposes two types of extension methods: one is to actively bypass obstacle areas by adjusting the Voronoi boundary, including introducing obstacle space constraints and separating hyperplane constructions, to achieve geometric processing of complex task environments. The second is to integrate the artificial potential field method (APF) and dynamic window approach (DWA) into the CVT framework, effectively solving classic problems such as local optima and unreachable targets issues while maintaining the continuity of robot motion. Through simulation experiments on two-dimensional and three-dimensional scenes, these improved strategies demonstrate good robustness and performance in environments with dense obstacles and irregular terrain, providing theoretical support and algorithm foundation for multi-robot execution of specific tasks in our subsequent chapters.

## Bibliography

[1] Guanghong Yang, Liwei An, and Can Zhao. Collision/obstacle avoidance coordination of multi-robot systems: A survey. In *Actuators*, volume 14, page 85. MDPI, 2025.

[2] Jihao Huang, Jun Zeng, Xuemin Chi, Koushil Sreenath, Zhitao Liu, and Hongye Su. Velocity obstacle for polytopic collision avoidance for distributed multi-robot systems. *IEEE Robotics and Automation Letters*, 8(6):3502–3509, 2023.

[3] Kang Xu, Jiejie Xing, Wenbin Sun, Peng Xu, and Ranbing Yang. Multi-robot collision avoidance method in sweet potato fields. *Frontiers in Plant Science*, 15:1393541, 2024.

[4] Meng Zhou, Zihao Wang, Jing Wang, and Zhengcai Cao. Multi-robot collaborative hunting in cluttered environments with obstacle-avoiding Voronoi cells. *IEEE/CAA Journal of Automatica Sinica*, 11(7):1643–1655, 2024.

[5] Zhenhua Pan, Chengxi Zhang, Yuanqing Xia, Hao Xiong, and Xiaodong Shao. An improved artificial potential field method for path planning and formation control of the multi-UAV systems. *IEEE Transactions on Circuits and Systems II: Express Briefs*, 69(3):1129–1133, 2021.

[6] Songtao Xie, Junyan Hu, Parijat Bhowmick, Zhengtao Ding, and Farshad Arvin. Distributed motion planning for safe autonomous vehicle overtaking via artificial potential field. *IEEE Transactions on Intelligent Transportation Systems*, 23(11):21531–21547, 2022.

[7] Ryandika Afdila, Fahmi Fahmi, and Arman Sani. Distributed formation control for groups of mobile robots using consensus algorithm. *Bulletin of Electrical Engineering and Informatics*, 12(4):2095–2104, 2023.

[8] Yang FengYing, Ahmad Din, Liu HuiChao, Muhammad Babar, and Shafiq Ahmad. Decentralized consensus in robotic swarm for collective collision and avoidance. *IEEE Access*, 12:72143–72154, 2024.

[9] Shi-Lu Dai, Ke Lu, and Jun Fu. Adaptive finite-time tracking control of nonholonomic multi-robot formation systems with limited field-of-view sensors. *IEEE Transactions on Cybernetics*, 52(10):10695–10708, 2021.

[10] Yunping Liu, Cheng Chen, Yan Wang, Tingting Zhang, and Yiguang Gong. A fast formation obstacle avoidance algorithm for clustered UAVs based on artificial potential field. *Aerospace Science and Technology*, 147:108974, 2024.

[11] Zhengtian Wu, Jinyu Dai, Baoping Jiang, and Hamid Reza Karimi. Robot path planning based on artificial potential field with deterministic annealing. *ISA Transactions*, 138:74–87, 2023.

[12] Shiliang Shao, Jin Zhang, Ting Wang, Achyut Shankar, and Carsten Maple. Dynamic obstacle-avoidance algorithm for multi-robot flocking based on improved artificial potential field. *IEEE Transactions on Consumer Electronics*, 70(1):4388–4399, 2023.

[13] Jie Luo, Zhong-Xun Wang, and Kang-Lu Pan. Reliable path planning algorithm based on improved artificial potential field method. *IEEE Access*, 10:108276–108284, 2022.

[14] Haiyan Tu, Yizhao Deng, Qiyang Li, Mingjun Song, and Xiujuan Zheng. Improved RRT global path planning algorithm based on bridge test. *Robotics and Autonomous Systems*, 171:104570, 2024.

[15] Huixia Zhang, Yadong Tao, and Wenliang Zhu. Global path planning of unmanned surface vehicle based on improved A-Star algorithm. *Sensors*, 23(14):6647, 2023.

[16] Hongqiang Sang, Yusong You, Xiujun Sun, Ying Zhou, and Fen Liu. The hybrid path planning algorithm based on improved A* and artificial potential field for unmanned surface vehicle formations. *Ocean Engineering*, 223:108709, 2021.

[17] Yuming Ning, Tuanjie Li, Cong Yao, Wenqian Du, and Yan Zhang. HMS-RRT: A novel hybrid multi-strategy rapidly-exploring random tree algorithm for multi-robot collaborative exploration in unknown environments. *Expert Systems with Applications*, 247:123238, 2024.

[18] Shangding Gu, Jakub Grudzien Kuba, Yuanpei Chen, Yali Du, Long Yang, Alois Knoll, and Yaodong Yang. Safe multi-agent reinforcement learning for multi-robot control. *Artificial Intelligence*, 319:103905, 2023.

[19] Kunal Garg, Songyuan Zhang, Oswin So, Charles Dawson, and Chuchu Fan. Learning safe control for multi-robot systems: Methods, verification, and open challenges. *Annual Reviews in Control*, 57:100948, 2024.

[20] Yiran Ding. Distributed multi-robot collision avoidance using the Voronoi-based method. In *Journal of Physics: Conference Series*, volume 1948, page 012015. IOP Publishing, 2021.

[21] Daniel Hennes, Daniel Claes, Wim Meeussen, and Karl Tuyls. Multi-robot collision avoidance with localization uncertainty. In *Proceedings of the 11th International Conference on Autonomous Agents and Multiagent Systems-Volume 1*, pages 147–154, 2012.

[22] Oussama Khatib. Real-time obstacle avoidance for manipulators and mobile robots. In *Proceedings. 1985 IEEE International Conference on Robotics and Automation*, volume 2, pages 500–505. IEEE, 1985.

[23] Kai Cao, Yangquan Chen, Song Gao, and Jiajia Gao. Path planning of RRT* mobile robot guided by eddy current artificial potential field. *Journal of Frontiers of Computer Science & Technology*, 15(4), 2021.

[24] Xu Zhenyang and Yuan Wei. Mobile robot path planning based on fusion of improved A* algorithm and adaptive DWA algorithm. In *Journal of Physics: Conference Series*, volume 2330, page 012003. IOP Publishing, 2022.

[25] Shiqi Wang, Yiyi Hu, Zhenni Liu, and Lijun Ma. Research on adaptive obstacle avoidance algorithm of robot based on DDPG-DWA. *Computers and Electrical Engineering*, 109:108753, 2023.

[26] Zenan Lin, Ming Yue, Guangyi Chen, and Jianzhong Sun. Path planning of mobile robot with PSO-based APF and fuzzy-based DWA subject to moving obstacles. *Transactions of the Institute of Measurement and Control*, 44(1):121–132, 2022.

[27] Zhaozhen Jiang, Wenlong Wang, Wenqi Sun, and Lianglong Da. Path planning method for mobile robot based on a hybrid algorithm. *Journal of Intelligent & Robotic Systems*, 109(3):47, 2023.

CHAPTER 6

# Multi-Robot Source Search and Coverage Control Based on Voronoi Tessellations

## 6.1 Introduction

In recent years, with the rapid development of computer technology, sensor technology, and artificial intelligence, multi-robot cooperative systems have been widely applied in various fields, including search and rescue [1–3], inspection [4,5], and pollution source localization [6,7]. Multi-robot systems can efficiently accomplish complex tasks through self-organized collaborative control, which is often unattainable by a single robot. Compared to a single robot, a multi-robot swarm offers distinct advantages: it enables cooperative operation, information sharing, and complementary capabilities to efficiently accomplish complex tasks [8–10]. This synergy is particularly crucial for achieving efficient and accurate area coverage and source exploration in unknown and complex environments. Consequently, it is essential to develop rational and robust exploration strategies that allow robots to autonomously and safely determine optimal trajectories.

Extensive research has been conducted in the fields of coverage and source localization in unknown environments [11–17]. A review of the literature reveals that multi-robot source localization studies commonly employ techniques such as density gradient methods [18–22], information entropy [23–25], or Bayesian approaches [26–30]. The density gradient method utilizes local gradient information for target localization, adapting to real-time environmental changes without relying on prior map data. Jabeen et al. [31] proposed an adaptive search algorithm for pollution sources, which estimates the gradient of odor plumes to locate target pollution sources. Francis et al. [32] introduced a hybrid approach combining gradient and probabilistic search for source localization, where gradient-based navigation guides robot movement while probabilistic search estimates pollutant locations. This integration reduces localization time and improves both efficiency and accuracy. However, gradient-based methods are prone to local optima, particularly in complex environments, which may

DOI: 10.1201/9781003714569-6

compromise localization performance. To address these limitations, some researchers have adopted deep reinforcement learning (DRL) to guide robot motion. Instead of depending on prior environmental knowledge, robots interact with the environment, learning optimal source localization strategies through exploration and feedback. Li et al. [33] proposed a DRL-based indoor odor source localization method, framing the problem as a Markov Decision Process (MDP) with carefully designed state space, action space, and dense reward functions to mitigate sparse reward challenges. The Proximal Policy Optimization (PPO) algorithm was then applied to generate optimal robot decisions. Similarly, Zhao et al. [34] developed a DRL-based source search method using Deep Q-Networks (DQN) to derive an optimal search policy. While effective, this approach suffers from high computational costs during model training. Hu et al. [35] investigates the multi-robot collaborative exploration problem in unknown environments by integrating Voronoi diagrams with deep reinforcement learning strategies. This approach dynamically partitions Voronoi regions based on robotic movements and achieves obstacle avoidance through deep reinforcement learning. While demonstrating effective search performance, it faces challenges including high sample complexity and prolonged training duration. Beyond learning-based approaches, heuristic optimization methods have also been explored to further improve global search efficiency and reduce computational complexity. Intelligent optimization algorithms emulate collective behaviors of bird flocks or fish schools, enabling global optimization through inter-agent cooperation. Feng et al. [36] proposes a flow-independent multi-robot olfaction method employing particle swarm optimization for odor source tracking, which expands search scope via extremum perturbation beacons to prevent local optima entrapment. However, such heuristic algorithms exhibit limitations in convergence speed, environmental field modeling accuracy, and interpretability. Zhu et al. [37] utilizes particle filtering and information entropy for plume tracking, designing an objective function based on target candidates' entropy gain and re-exploration scores, while eliminating pseudo-sources through artificial potential fields. Zhai et al. [38] addresses robot health degradation in pollution source coverage control by developing an energy-optimized management strategy combined with Centroidal Voronoi Tessellation (CVT) for formation cost function design, ultimately enhancing coverage robustness. Abdulghafoor et al. [39] presents a Gaussian Mixture Model (GMM)-based distributed adaptive coverage algorithm that online-learns environmental models through prior information, guiding robots to explore high-uncertainty regions for complete source distribution reconstruction. Luo et al. [40] employs GMM for environmental modeling, directing robotic exploration and reconstruction via Voronoi diagrams. Although capable of reconstructing unknown environments, this method suffers from high computational complexity when processing large-scale datasets, demanding substantial storage and computational resources that prolong training periods.

This chapter analyzes multi-robot source localization methods in unknown environments, summarizing their respective limitations. Building upon previous analyses of Voronoi characteristics and their multi-robot formation applications, we conclude that integrating Voronoi diagrams with multi-robot source localization strategies can overcome the deficiencies of existing methods, enabling multi-robot systems to accomplish source localization tasks across diverse environments.

## 6.2 Source Localization Algorithm Based on Variational Sparse Gaussian Process

### 6.2.1 Sparse Gaussian Process Regression

Gaussian Process Regression (GPR) represents a non-parametric regression methodology grounded in Gaussian processes, capable of establishing mapping relationships between input and output variables. This approach enables the inference of output variable values based on a given set of input data.

The fundamental assumption of GPR posits that the regression function follows a Gaussian distribution, formally expressed as:

$$f(x) \sim \mathcal{GP}\left(m(z),\, k(z, z')\right). \tag{6.1}$$

In this framework, $m(z)$ represents the mean function, while $k(z, z')$ denotes the covariance kernel function of the GPR model, also referred to as the kernel function. The radial basis function is adopted as the kernel function for GPR:

$$k(z, z') = M \exp\left(-\frac{\|z_i - z_i'\|^2}{2l^2}\right). \tag{6.2}$$

Where $M$ and $l$ are hyperparameters, where $M$ represents the amplitude scale and $l$ represents the length scale.

This covariance function quantifies the correlation between two input vectors $z_i$ and $z_i'$ , where a stronger correlation corresponds to a smaller distance between the vectors. Selecting optimal hyperparameters is crucial for ensuring the model's fitting performance, making hyperparameter optimization a key challenge. Typically, the Maximum Likelihood Estimation (MLE) method is employed to construct the likelihood function, and the hyperparameters. To optimize the hyperparameters $\xi = M, l, \sigma_n^2$, where $\xi$ comprise the amplitude scale $M$, length scale $l$, and noise variance $\sigma_n^2$, the maximum likelihood estimation (MLE) approach is adopted. The marginal likelihood is expressed as:

$$p(y|Z, \xi) = \frac{1}{(2\pi)^{N/2}|K_y|^{1/2}} \exp\left(-\frac{1}{2} y^\top K_y^{-1} y\right), \quad K_y = K(Z, Z) + \sigma_n^2 I. \tag{6.3}$$

Taking the logarithm gives the log-marginal likelihood:

$$\log p(y|Z, \xi) = -\frac{1}{2} y^\top K_y^{-1} y - \frac{1}{2} \log |K_y| - \frac{N}{2} \log(2\pi), \tag{6.4}$$

which is maximized with respect to the hyperparameters to ensure optimal model fitting.

Given the training data $Z$ and test data input $Z_*$, the predictive distribution of GPR follows:

$$p(f_*|X, y, X_*) = \mathcal{N}(m(X_*), \Sigma(X_*)), \tag{6.5}$$

where,

$$m(Z_*) = K(Z_*, Z)\big[K(Z, Z) + \sigma_n^2 I\big]^{-1} y \tag{6.6}$$

$$\sigma_*^2(Z_*) = K(Z_*, Z_*) - K(Z_*, Z)\big[K(Z, Z) + \sigma_n^2 I_n\big]^{-1} K(Z, Z_*), \tag{6.7}$$

where $y$ represents the training targets, $K$ is the covariance matrix, and $\sigma_n^2$ is the noise variance.

In conventional Gaussian process regression, predictions are made by computing the inverse of the covariance matrix in the input space. However, the computational and storage requirements become prohibitively expensive for large datasets. This study employs the Fully Independent Training Conditional (FITC) approximation to implement sparse Gaussian process regression, introducing inducing points to reduce computational costs, thereby effectively lowering computational complexity while maintaining sufficient predictive accuracy.

Considering the reference dataset inputs $Z_* = \{z'_m\}_{m=1}^M$ and corresponding targets $\overline{f} = \{f_m\}_{m=1}^M$ , the conditional distribution can be derived as follows:

$$\mathbf{p}(y \mid z, Z_*, \overline{f}) = \mathcal{N}\left(y \,\middle|\, k_z^\top K_M^{-1} \overline{f},\ K_z - k_z^\top K_M^{-1} k_z + \sigma^2\right), \tag{6.8}$$

where $[K_M]_{mm'} = K(z'_m, z'_{m'})$, $[k_z]_m = K(z'_m, z)$. $\sigma^2$ denotes the variance.

Setting Gaussian priors:

$$\mathbf{p}(\bar{f} \mid Z_*) = \mathcal{N}(\bar{f} \mid \mathbf{0}, K_M), \tag{6.9}$$

and using Bayesian inference method, the posterior probability distribution of the sparse model is given by:

$$\mathbf{p}(\overline{f}|\mathcal{D}, Z_*) = \mathcal{N}\left(\overline{f} \,\middle|\, K_M Q_M^{-1} K_{MN} (\Phi + \sigma^2 I)^{-1} y,\ K_M Q_M^{-1} K_M\right), \tag{6.10}$$

where the reference dataset is denoted as $\mathcal{D}$.

$$Q_M = K_M + K_{MN}\left(\Lambda + \sigma^2 I\right)^{-1} K_{NM}. \tag{6.11}$$

The model parameters and hyperparameters are iteratively updated using new training inputs $z_*$, with the predictive distribution obtained by marginalizing over both the marginal likelihood function and the model's posterior probability distribution.

$$p(y_* \mid z_*, \mathcal{D}, Z_*) = \int p(y_* \mid z_*, Z_*, \bar{f})\, p(\bar{f} \mid \mathcal{D}, Z_*)\, \mathrm{d}\bar{f}. \tag{6.12}$$

The posterior mean $m_*$ and variance $\sigma_*^2$ are employed to estimate both the expected output values and the associated uncertainty for new inputs.

$$m_* = k_*^T Q_M^{-1} K_{MN} \left(\Phi + \sigma^2 I\right)^{-1} y, \tag{6.13}$$

$$\sigma_*^2 = K_{**} - k_*^T \left(K_M^{-1} - Q_M^{-1}\right) k_* + \sigma^2, \tag{6.14}$$

where $k_*$ is the kernel vector between the test point and the reference points, and $K_{**}$ is the covariance of the test points themselves.

In contrast to conventional Gaussian Process Regression (GPR) methods, the Fully Independent Training Conditional (FITC) approximation decomposes the full model into multiple independent sub-models. This decomposition reduces the model's dependency on specific input dimensions and enhances robustness to input variations. The computational complexity of training is reduced to $O(MN^2)$, yielding significant improvements in both time efficiency and memory consumption.

### 6.2.2 Upper Confidence Bound-Centroidal Voronoi Tessellations

Taking an environment with pollutant source leakage as an example, the conventional density-based Greedy-Centroidal Voronoi Tessellation (G-CVT) algorithm employs a greedy strategy to continuously direct robots toward high-concentration locations. However, this approach frequently causes robots to get trapped in local high-concentration regions, ultimately failing to achieve complete area coverage for inspection tasks.

To enhance multi-robot environmental exploration and coverage performance, we propose here a novel multi-robot coverage framework termed Upper Confidence Bound-Centroidal Voronoi Tessellations (UCB-CVT). Building upon sparse Gaussian process regression theory, this method extends the conventional CVT algorithm by incorporating a regularization term based on predictive uncertainty. As shown in (6.15), it evaluates target locations through a weighted sum of uncertainty and density metrics.

$$f(\boldsymbol{q}) = \varepsilon(\boldsymbol{q}) + \sigma^2 \ln \sigma^2, \tag{6.15}$$

where $\varepsilon(\boldsymbol{q})$ and $\sigma^2$ denote the predicted concentration value and posterior predictive variance at points within the Voronoi region, respectively.

The proposed formulation improves the robotic motion strategy by enabling a trade-off between exploration and exploitation, preventing convergence to local concentration extremes while continuously reducing predictive model uncertainty. Fig. 6.1 illustrates the changes in CVT centroids before and after considering uncertainty, where black dots represent the original centroid positions of the robots (typically more focused on high-concentration areas), while green dots indicate new positions adjusted through the UCB algorithm's balancing.

The extended CVT cost function is shown in (6.16).

$$\mathcal{H}_e = \sum_{i=1}^{n} \int_{\mathcal{V}_{b,i}} \|\boldsymbol{q} - \boldsymbol{p}_i\|^2 f(\boldsymbol{q})\, \mathrm{d}q. \tag{6.16}$$

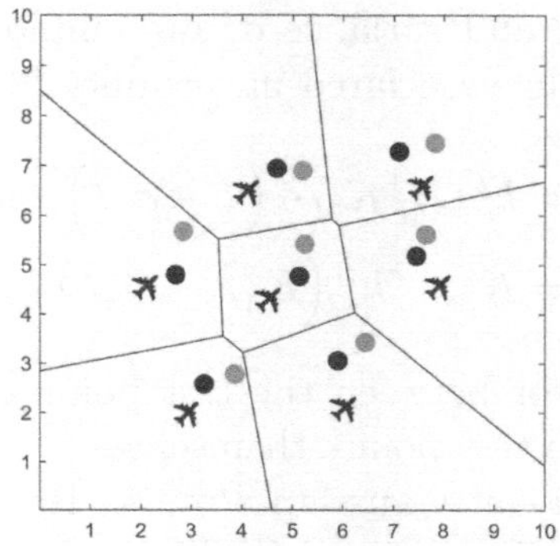


Figure 6.1: Environment coverage considering uncertainty.

The density function in the control strategy is derived from an information-theoretic estimation based on predictive information, with the centroid coordinates of each Voronoi cell updated as follows:

$$\boldsymbol{q}_* = \frac{\int_{\mathcal{V}_{b,i}} \boldsymbol{q} f(\boldsymbol{q})\,\mathrm{d}\boldsymbol{q}}{\int_{\mathcal{V}_{b,i}} f(\boldsymbol{q})\,\mathrm{d}\boldsymbol{q}} = \widehat{\boldsymbol{C}}_{\mathcal{V}_{b,i}}. \tag{6.17}$$

Unlike density-based CVT algorithms, the updated Voronoi centroids consist of two distinct components: the predicted density centroid and the uncertainty centroid.

$$\widehat{\boldsymbol{C}}_{\mathcal{V}_{b,i}} = \boldsymbol{C}_\epsilon + \boldsymbol{C}_\sigma. \tag{6.18}$$

where $\boldsymbol{C}_\epsilon = \dfrac{\int_{\mathcal{V}_{b,i}} \varepsilon(\boldsymbol{q})\boldsymbol{q}\,\mathrm{d}\boldsymbol{q}}{\int_{\mathcal{V}_{b,i}} \varepsilon(\boldsymbol{q})\,\mathrm{d}\boldsymbol{q}}$, $\boldsymbol{C}_\sigma = \dfrac{\int_{\mathcal{V}_{b,i}} (\sigma^2 \ln\sigma)\boldsymbol{q}\,\mathrm{d}\boldsymbol{q}}{\int_{\mathcal{V}_{b,i}} \sigma^2 \ln\sigma\,\mathrm{d}\boldsymbol{q}}$.

After balancing density and uncertainty, the robots will move toward the new centroids, with the control law updated as:

$$\dot{\boldsymbol{p}}_{b,i} = k_p(\widehat{\boldsymbol{C}}_{\ell_{b,i}} - \boldsymbol{p}_{b,i}), \tag{6.19}$$

where $k_p$ denotes a constant adjusting the step size weight, and $\boldsymbol{p}_{b,i}$ represents the current position of the robot $i$.

### 6.2.3 Distributed Exploration Strategy

During environmental exploration missions, area information detection is achieved by equipping robots with specific sensors. When the onboard sensors fail to detect any usable prior information, traditional formation control based on CVT methods becomes ineffective, making it difficult to maintain formation patterns. Therefore, a Voronoi diagram-based cooperative search strategy is proposed to guide robots in continuously exploring unknown regions, thereby reducing the effect of environmental uncertainty. This strategy enables each robot to make autonomous decisions based on local sensing data and limited neighbor communication, eliminating the need for a centralized coordinator.

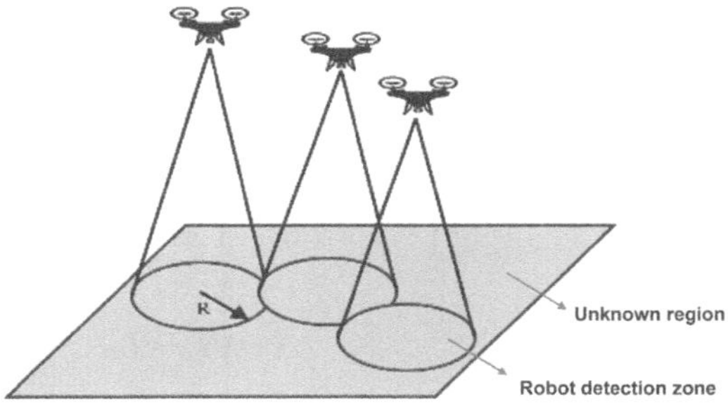


Figure 6.2: The robot search area.

Due to environmental noise and the inherent sensing range limitations of the sensors, the robotic exploration capability gradually degrades as the Euclidean distance between the task point $q$ in the search area and the robot increases. As shown in Fig. 6.2, the blue area represents the sensor detection range, while the green irregular region denotes unknown areas. To characterize the sensor's exploration capability, we define $R$ as the maximum sensing range of the robotic sensors and establish the search capability function as:

$$f(r) = k \exp\left(-\frac{\|\boldsymbol{p}_i - \boldsymbol{q}\|^2}{2\sigma^2}\right), \tag{6.20}$$

where $k$, $\sigma$ represents two constants, and $r$ denotes the robot search radius. Specifically, when the distance exceeds the maximum search radius $r \leq R$, the robot search capability becomes zero.

Upon completing each iteration, robots executing the search task will update the global uncertainty Shannon entropy distribution as follows:

$$\begin{cases} F_{m+1} = \sum\limits_{i=1}^{n} F_m \min\left\{1 - f(\|\boldsymbol{p}_i - \boldsymbol{q}\|)\right\} \\ F_m = -\sum\limits_{i=1}^{n} \phi_m(\boldsymbol{q}) \log \phi_m(\boldsymbol{q}), \end{cases} \tag{6.21}$$

where $m$ represents the number of searches, $F_m$ denotes the uncertainty entropy distribution, $\phi_m$ indicates the uncertainty density at a given point $\boldsymbol{q}$, and $\phi_m = 1$ signifies the maximum uncertainty at that point $\boldsymbol{q}$.

In regional search tasks, since pollution source information cannot be detected, the objective of the distributed exploration strategy is to minimize environmental uncertainty within the task area. The variation in the uncertainty density function is defined as follows:

$$\begin{aligned} \Delta F_m(\boldsymbol{q}) &= F_m(\boldsymbol{q}) - F_{m+1}(\boldsymbol{q}) \\ &= \left(1 - \min_{i \in I_n}\left\{1 - f(\|\boldsymbol{p}_i - \boldsymbol{q}\|)\right\}\right) F_m(\boldsymbol{q}). \end{aligned} \tag{6.22}$$

The cost function for the CVT algorithm is defined as follows:

$$\begin{aligned}\mathcal{H}_v &= \int_{\mathcal{Q}} \Delta F_m(\boldsymbol{q})\,\mathrm{d}\boldsymbol{q} \\ &= \int_{\mathcal{V}_i} \left(1 - \min_{i \in I_n}\left\{1 - f(\|\boldsymbol{p}_i - \boldsymbol{q}\|)\right\}\right) F_m(q)\,\mathrm{d}\boldsymbol{q} \\ &= \sum_{i=1}^{n} \int_{\mathcal{V}_{o,i}} \left(k e^{-\frac{\|\boldsymbol{p}_i - \boldsymbol{q}\|^2}{2\sigma^2}}\right) F_m(\boldsymbol{q})\,\mathrm{d}\boldsymbol{q},\end{aligned} \tag{6.23}$$

where $\mathcal{V}_{o,i}$ represents a circular search region centered at robot $\boldsymbol{p}_i$ with radius $r$.

The gradient of the cost function at any point $\boldsymbol{q}$ is given by:

$$\begin{aligned}\frac{\partial \mathcal{H}_v}{\partial p_i} &= \sum_{i=1}^{n} \int_{\mathcal{V}_{o,i}} k e^{-\frac{\|\boldsymbol{p}_i - \boldsymbol{q}\|^2}{2\sigma^2}} \left[-\frac{1}{\sigma^2}(\boldsymbol{p}_i - \boldsymbol{q}) F_m(q)\right] \mathrm{d}q \\ &= -\int_{\mathcal{V}_{o,i}} \widetilde{F}_m(q)(\boldsymbol{p}_i - \boldsymbol{q})\,\mathrm{d}q \\ &= -\frac{1}{\sigma^2} \widetilde{M}_{\mathcal{V}_{o,i}} \left(\boldsymbol{p}_i - \widetilde{\boldsymbol{C}}_{\mathcal{V}_{o,i}}\right),\end{aligned} \tag{6.24}$$

where $\widetilde{M}_{\mathcal{V}_{o,i}}$ and $\widetilde{\boldsymbol{C}}_{\mathcal{V}_{o,i}}$ represent the mass and centroid, respectively, of the uncertainty entropy distribution over the robot's search region $\mathcal{V}_{o,i}$. When the cost function $\mathcal{H}_v$ is minimized, the robot reaches the centroid position $\boldsymbol{C}_{o,i}$ of the search region.

After obtaining the uncertainty density function distribution, the CVT algorithm directs the robot to move toward the new centroid position, with the control law expressed as:

$$\dot{\boldsymbol{p}}_{o,i} = k_m \left(\widetilde{\boldsymbol{C}}_{\mathcal{V}_{o,i}} - \boldsymbol{p}_{o,i}\right), \tag{6.25}$$

where $p_{o,i}$ denotes the robot's current position.

Furthermore, define the target point set $\Omega = \{\boldsymbol{P} \mid \boldsymbol{p}_i = \boldsymbol{C}_i,\ \boldsymbol{C}_i \in \widetilde{\boldsymbol{C}}_{\mathcal{V}_{o,i}}, \widehat{\boldsymbol{C}}_{\mathcal{V}_{b,i}}\}$. When the condition $\|C_i - p_i\| \to 0$ is satisfied, this state is termed as a the convergent configuration. Set the single-movement distance threshold for the robot as $D$, with each robot's target point updated as follows:

$$\begin{cases} \boldsymbol{p}_{i+1} = \boldsymbol{p}_i + d\dfrac{\overline{\boldsymbol{p}_i \boldsymbol{C}_i}}{\|\boldsymbol{p}_i - \boldsymbol{C}_i\|}, & \|\boldsymbol{p}_i - \boldsymbol{C}_i\| \geq d \\ \boldsymbol{p}_{i+1} = \boldsymbol{p}_i + \overline{\boldsymbol{p}_i \boldsymbol{C}_i}, & \|\boldsymbol{p}_i - \boldsymbol{C}_i\| < d \end{cases}, \tag{6.26}$$

where $\dfrac{\overline{\boldsymbol{p}_i \boldsymbol{C}_i}}{\|\boldsymbol{p}_i - \boldsymbol{C}_i\|}$ represents the unit vector between the robot and its corresponding centroid.

The pseudocode for the online exploration and coverage algorithm is presented in Algorithm 6:

**Algorithm 6** Multi-robot online coverage algorithm based on Centroid Voronoi Control

**Require:**
Initial positions of $n$ robots $\{\boldsymbol{p}_i\}_{i=1,\ldots,n}$ and uncertainty radius $\{r_i\}_{i=1,\ldots,n}$, Task coverage area $D \in \mathbb{R}^2$, Formation minimum convergence threshold $\boldsymbol{p}$.
**Procedure:**
**while** The minimum distance $d$ between any robot and its centroid is less than $\boldsymbol{p}$ **do**
    Obtain robot position $\boldsymbol{p}_i$ and uncertainty radius $r_i$
    Calculate distance $\|\boldsymbol{q} - \boldsymbol{p}_i\|$ to points in task area
    Construct buffered Voronoi cell $\mathcal{V}_{b,i}$ for each robot
    **if** A robot $p_i$ detects a concentration below threshold $T$ **then**
        Compute uncertainty volume $\mathbf{F}_m(\boldsymbol{q})$
        Calculate Voronoi cell mass $\widetilde{M}_{\mathcal{V}_i}$ and uncertain centroid $\widetilde{\boldsymbol{C}}_{\mathcal{V}_{b,i}}$
        Drive robot to centroid position $\boldsymbol{p}_i = \widetilde{\boldsymbol{C}}_{\mathcal{V}_{o,i}}$
    **else**
        Perform global centroidal Voronoi partition for $n$ robots at $\boldsymbol{p}_i$
        Obtain environmental prior information
        Compute FITC sparse model and its predictive variance
        Update evaluation formula $f(\boldsymbol{q})$ combining density and uncertainty
        Calculate coverage cost function $\mathcal{H}_e$ and weighted centroid $\hat{\boldsymbol{C}}_{\mathcal{V}_d}$
        Adjust step weight $k_p$ and drive robot for coverage
        Combine historical information to get explored point set $X = \{x_n\}_{n=1}^N$
    **end if**
**end while**

## 6.3 Simulation Analysis of Source Search and Coverage

The effectiveness of the proposed algorithm was verified through MATLAB 2018a, with two comparative simulation experiments conducted. The simulation environment was configured with a size of $10m \times 10m \times 10m$. To prevent excessive robot velocity, the maximum number of iterations per cycle was set to 120 and the maximum movement step length was limited to 1 m. Considering the physical dimensions of the robots, their uncertainty radius was configured as 0.2 m, resulting in a minimum collision distance of 0.4 m. To balance task efficiency and sampling accuracy, the maximum robot velocity per iteration was constrained to 1.5 $m/s$ with a sampling time interval of 1 seconds.

In order to evaluate the accuracy of reconstructing the leaking source field, three evaluation metrics are proposed:

(1) The distance error calculates the difference between the predicted concentration maximum and the true concentration field maximum:

$$DE = \sqrt{(x_p - x_t)^2 + (y_p - y_t)^2}, \tag{6.27}$$

where $(x_p - y_p)$ represents the coordinates of the concentration extremum point in the true field, and $(x_t - y_t)$ represents the coordinates of the concentration extremum point in the predicted field.

(2) The extremum error calculates the concentration difference between the predicted field and the true field at their respective concentration maximum points:

$$EE = p_*^{\max} - t_*^{\max}, \tag{6.28}$$

where $p_*^{\max}$ represents the concentration extremum of the predicted field, and $t_*^{\max}$ represents the concentration extremum of the true field.

(3) The average error is used to measure the accuracy of the predictive model:

$$RMSE = \sqrt{\frac{1}{n}\sum_{i=1}^{n}(p_* - t_*)^2}, \tag{6.29}$$

where $p_*$ represents the predicted point concentration value, and $t_*$ represents the true value.

### 6.3.1 Online Coverage in Multi-Source Pollution Scenarios

The sampling environment is assumed to contain two pollution sources located at coordinates (4, 4) and (8, 7), respectively, with the true density function following a composite Gaussian distribution.

$$\rho_{xy} = 3e^{-0.7\left[(x-4)^2+(y-4)^2\right]} + 4e^{-0.4\left[(x-8)^2+(y-7)^2\right]}. \tag{6.30}$$

The multi-source simulation environment is illustrated in Fig. 6.3a, where (8, 7) represents the global maximum point and (4, 4) denotes a local extremum. The initial Voronoi partition for the seven robots is shown in Fig.6.3b.

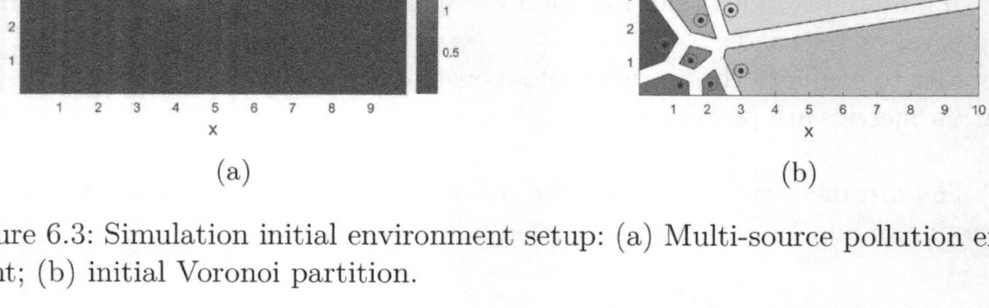


(a) (b)

Figure 6.3: Simulation initial environment setup: (a) Multi-source pollution environment; (b) initial Voronoi partition.

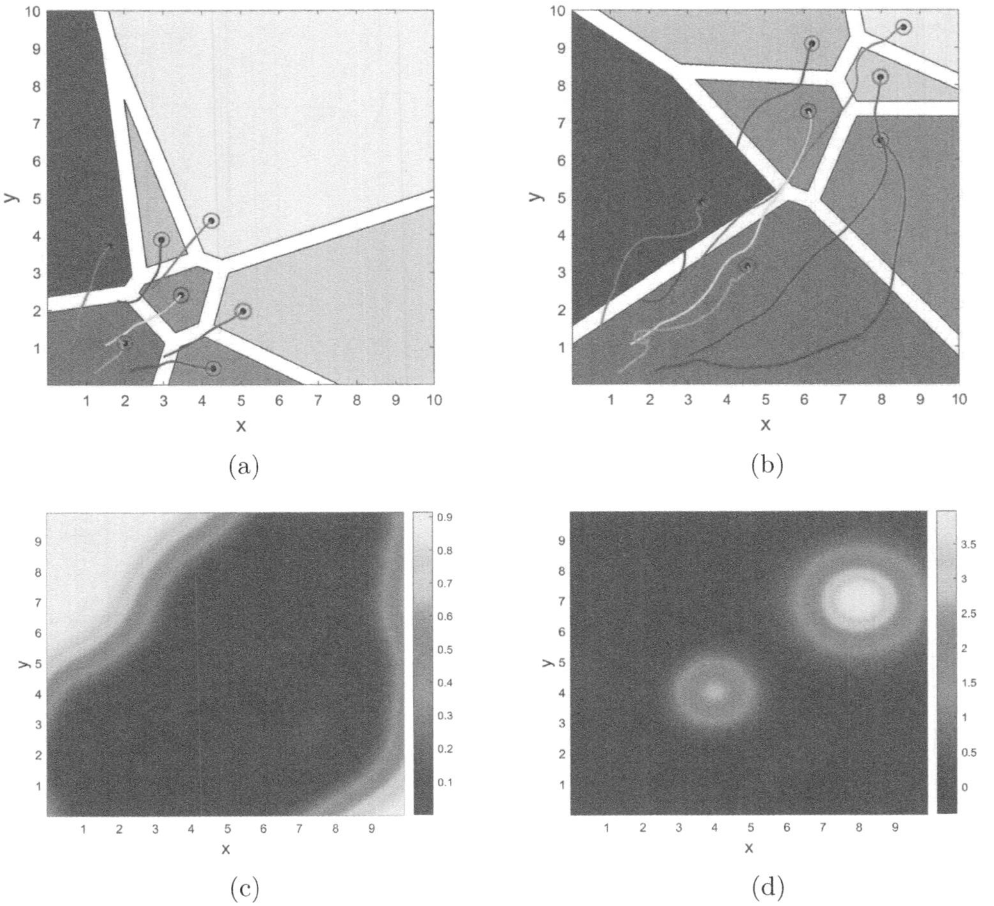


Figure 6.4: Simulation of UCB-CVT algorithm under multiple source density function: (a) $k = 17$; (b) $k = 95$; (c) uncertainty distribution; (d) prediction of field sources.

In Figs. 6.4a and 6.4b, trajectories of different robots are depicted by distinct colored curves, while irregular polygons represent their respective Voronoi cells.

As demonstrated in Fig. 6.4, robots under the coverage algorithm thoroughly explored the pollution sources, achieving an 88.1 percent coverage rate of the task area. Consequently, Fig. 6.5c shows minimal residual uncertainty distribution, and Fig. 6.5d confirms the algorithm's accurate localization of both pollution sources.

Fig. 6.5 reveals that the baseline G-CVT algorithm exhibits significantly higher mean and peak value errors in its predictive model. In contrast, the proposed algorithm directs robots to prioritize uncertain regions, reducing the prediction error by 85.25 percent compared to the true field data. Positional and peak value errors are reduced by 97.92 percent and 98.60 percent, respectively. Fig. 6.5d verifies that inter-robot distances consistently exceed the collision threshold throughout iterations.

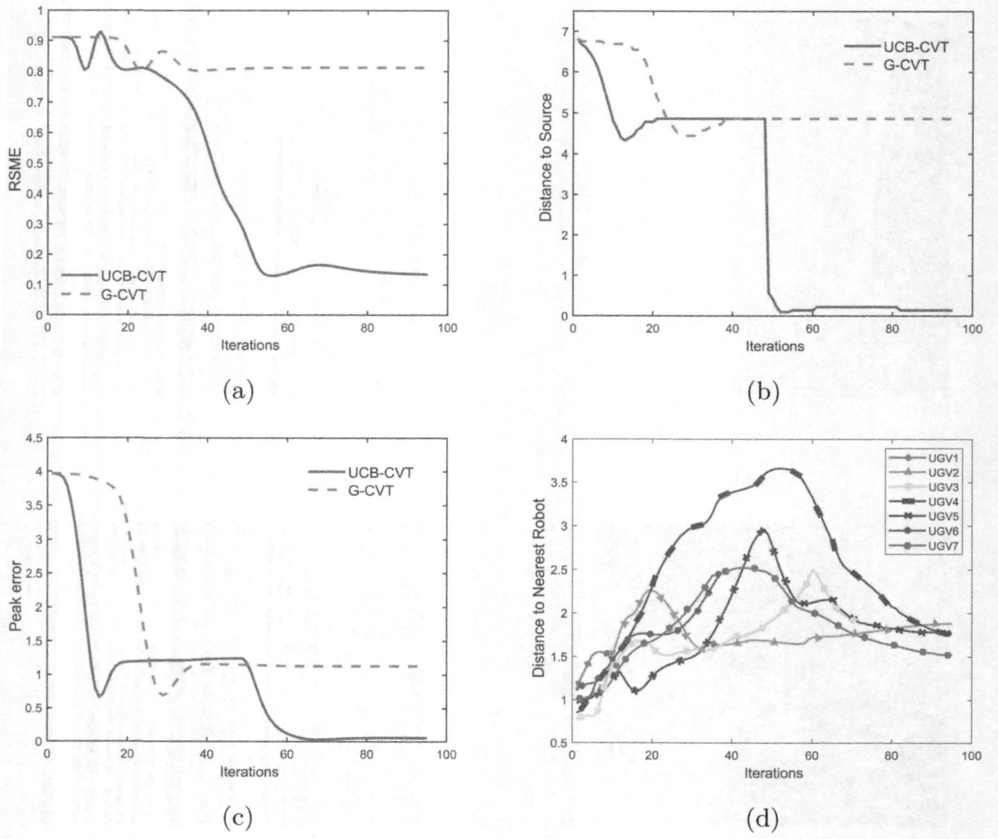


Figure 6.5: Performance of two algorithms in different evaluation metrics: (a) Root mean squared error; (b) extremum point localization distance error; (c) extremum point localization concentration error; (d) minimum inter-robot distance.

A runtime comparison between G-CVT and UCB-CVT algorithms is presented in Table 6.1. The UCB-CVT algorithm completes computations in 239.53 seconds, with an average iteration time of 2.52 seconds, whereas G-CVT requires 6.15 seconds per iteration, demonstrating significant efficiency gains. This simulation validates that the UCB-CVT method dynamically adjusts the uncertainty regularization term based on sparse GPR-predicted variance, enabling robots to escape local optima and expand coverage. Meanwhile, it substantially reduces computational time.

Table 6.1: Runtime of two online coverage algorithms

| Coverage algorithm | Runtime (s) |
|---|---|
| G-CVT | 583.79 |
| UCB-CVT | 239.53 |

(a)

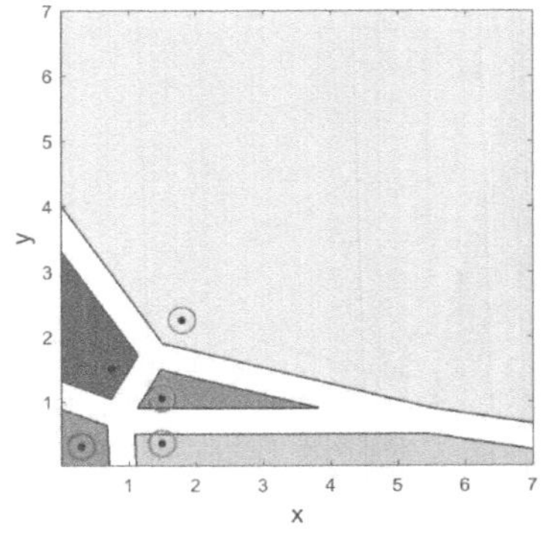


(b)

Figure 6.6: Simulation initial environment setup: (a) Single-source pollution environment; (b) initial Voronoi partition.

### 6.3.2 Online Exploration and Coverage Without Prior Environmental Information

In this experimental scenario, the pollution source is located far from the robots, meaning that the robots lack usable prior environmental information during initial iterations. The multi-robot system must explore the unknown environment to reduce environmental uncertainty.

The experiment employs five robots, with the true environmental field's density function given by:

$$\rho_{xy} = 6e^{-0.6\left[(x-6)^2+(y-5)^2\right]}. \tag{6.31}$$

Fig. 6.6a displays the source distribution of the experimental field, while Fig. 6.6b shows the initial Voronoi partition formed by the five robots.

From Fig. 6.7, it can be observed that in the initial stage, since all environmental information is unknown, the robots tend to perform exploration tasks. As the robots move within the mission area and unexplored regions continuously decrease, the robots become more inclined to perform coverage tasks. That is, the environmental density and uncertainty information are estimated using the sparse Gaussian process regression method, and the CVT algorithm is used to complete the partitioning of the mission area and the assignment of task points. As shown in Fig.6.7c, the robots achieve a coverage rate of 96.7 percent for the explored area.

Without usable prior information, robots without the exploration algorithm cannot utilize gradient information to complete the partitioning of the mission area or establish a field source model. Therefore, traditional algorithms fail in this scenario. The comparison results of the three evaluation metrics are shown in Fig.6.8, where the G-CVT based online coverage algorithm performs poorly, with the robots unable to complete the exploration task. In contrast, the proposed algorithm can ensure the

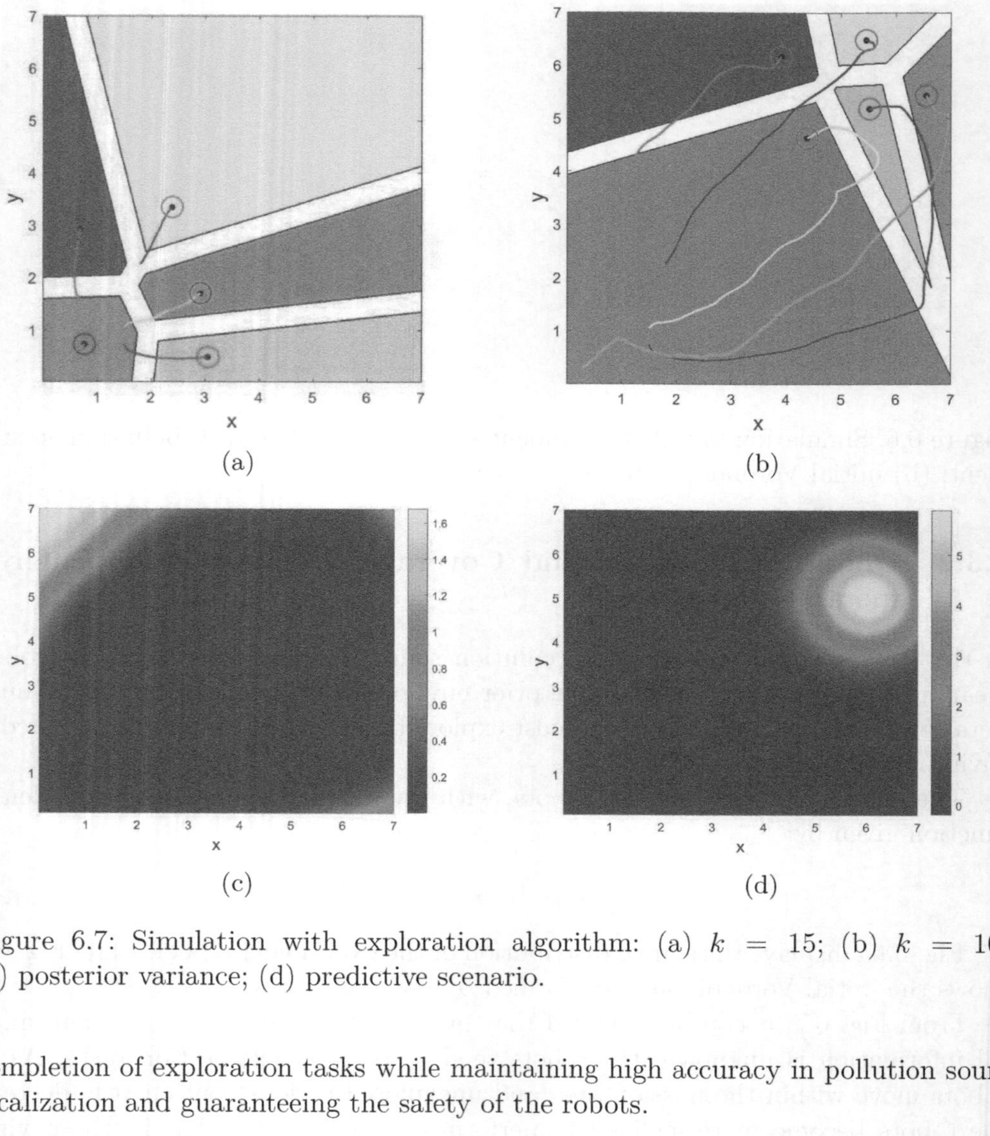

Figure 6.7: Simulation with exploration algorithm: (a) $k = 15$; (b) $k = 109$; (c) posterior variance; (d) predictive scenario.

completion of exploration tasks while maintaining high accuracy in pollution source localization and guaranteeing the safety of the robots.

### 6.3.3 Online Coverage in Three-Dimensional Multi-Source Scenarios

Since most practical application scenarios are in three-dimensional space, to further illustrate the generality of the algorithm, the experimental scenario is extended to three-dimensional space in this subsection. It is assumed that there are two leakage sources in the coverage area, located at coordinates (2.5, 2.5, 2.5) and (5.5, 5.0, 4.0) respectively. As shown in (6.32), the actual density function is a composite Gaussian distribution.

$$\begin{aligned}\rho_{xyz} &= 5e^{-0.5\left[(x-2.5)^2+(y-2.5)^2+(z-2.5)^2\right]} \\ &\quad + 10e^{-0.3\left[(x-5.5)^2+(y-5)^2+(z-4)^2\right]}.\end{aligned} \tag{6.32}$$

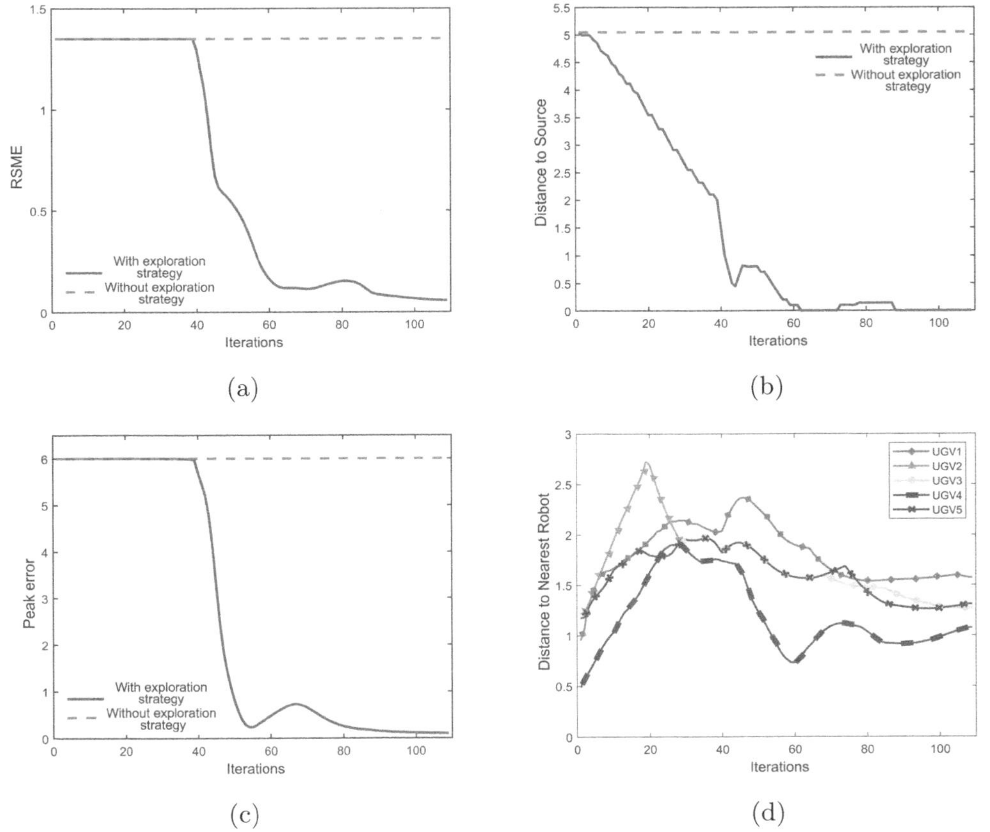


Figure 6.8: Performance of two algorithms in different evaluation metrics: (a) Root mean square error; (b) extremum point localization distance error; (c) extremum point localization concentration error; (d) minimum inter-robot distance.

The following adjustments were made to the simulated environment: the scene size was set to 7 m × 7 m × 7 m, the point (5.5, 5.0, 4.0) is the position with maximum concentration. Consequent to the change in scene size, and to optimize for task efficiency and sampling precision, the safety clearance for robots from obstacles was defined as $0.4m$, while the robots' maximum iterative speed was set to $0.25m/s$. Additionally, three obstacles were added at two-dimensional coordinates (5.0,1.5), (1.2,3.5), and (4.0,5.0), and the number of robots is five.

Fig. 6.9a to 6.9f show the simulation results of multi-robot online coverage tasks completed using the G-CVT algorithm, while Fig. 6.10a to 6.10f show the results using the UCB-CVT algorithm. As can be seen from Fig. 6.9b, the traditional G-CVT algorithm may collide with obstacles during movement, resulting in safety problems for the robot. In the subsequent simulation in Fig. 6.9c, the traditional G-CVT algorithm returned after covering a certain distance, and could not accurately cover the location of the pollution source. From Fig. 6.10a, it can be observed that the UCB-CVT algorithm introduces a buffer zone between each Voronoi cell during the

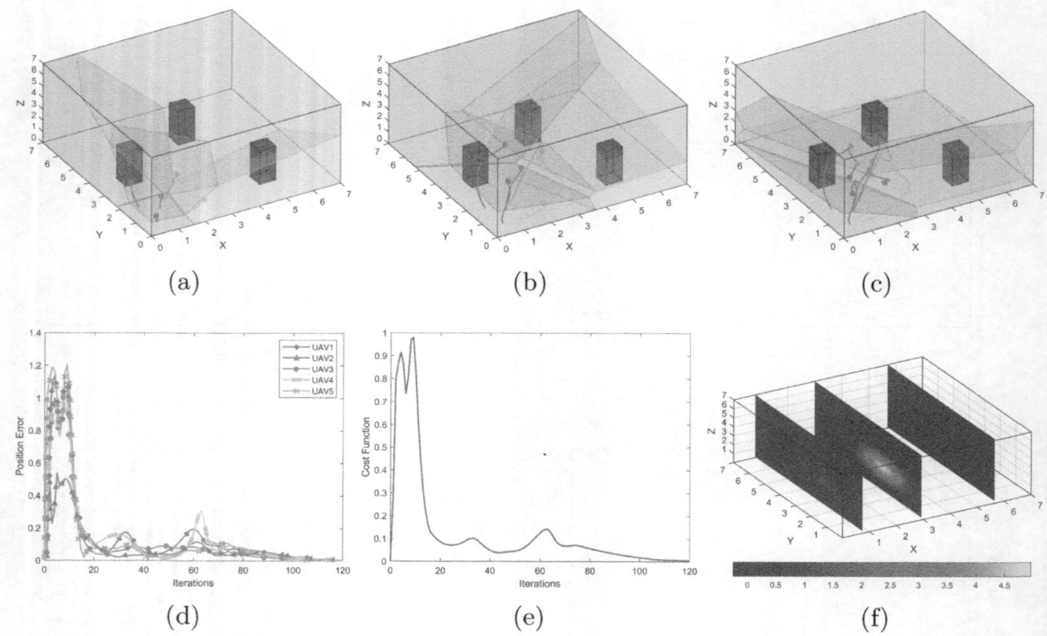


Figure 6.9: Multi-UAV coverage and source localization process under the G-CVT: (a) $k = 5$; (b) $k = 30$; (c) $k = 120$; (d) position error with respect to the centroid; (e) cost function; (f) predicted source field.

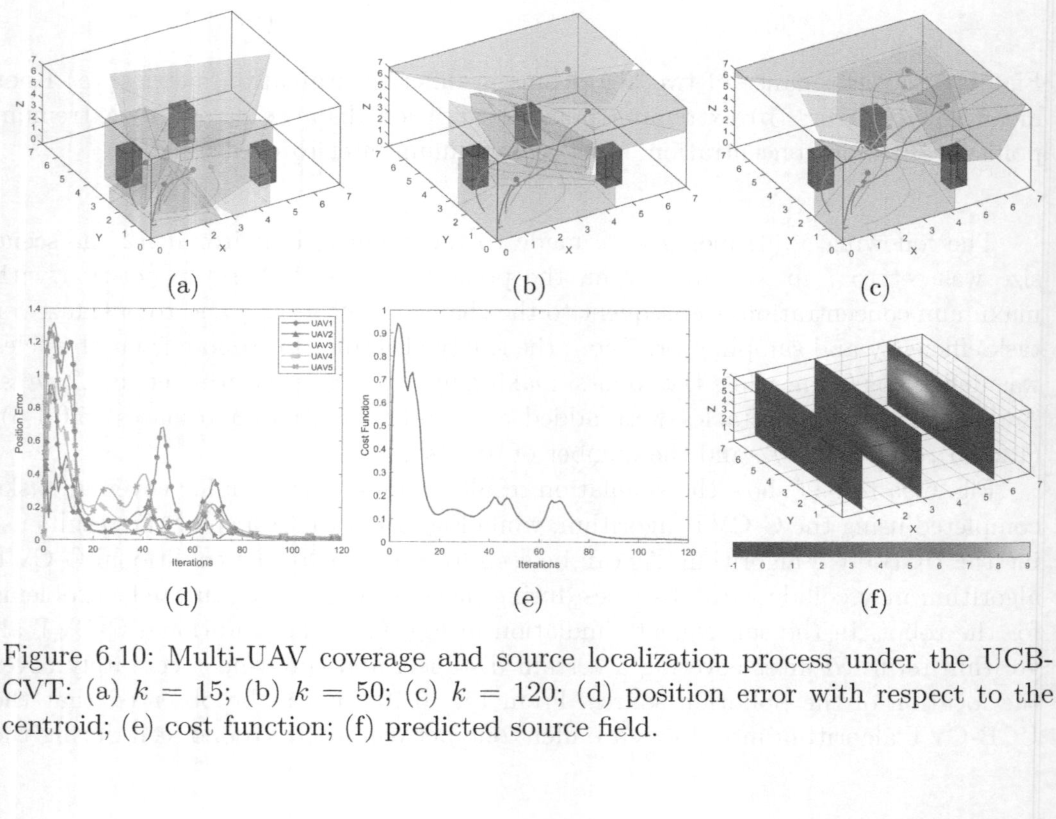


Figure 6.10: Multi-UAV coverage and source localization process under the UCB-CVT: (a) $k = 15$; (b) $k = 50$; (c) $k = 120$; (d) position error with respect to the centroid; (e) cost function; (f) predicted source field.

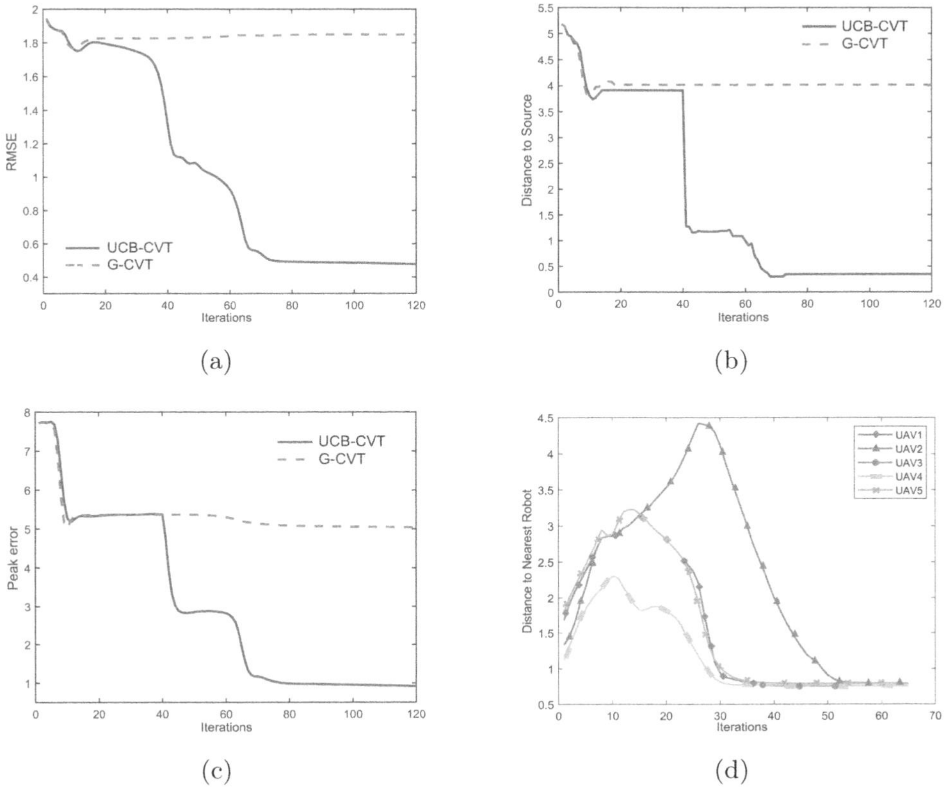


Figure 6.11: Performance of two algorithms in different evaluation metrics: (a) Model accuracy metrics; (b) distance to source; (c) peak error; (d) minimum distance between robots.

initial Voronoi cell generation stage to avoid collisions between robots. In addition, it can be seen from the subsequent simulation Figs. 6.10b and 6.10c that the robot has been moving toward the location of the pollution source, so that the robot can locate the location of the pollution source more accurately in the future. Figs. 6.9f and 6.10f respectively depict the concentration prediction maps of leakage sources under the G-CVT and UCB-CVT algorithms. The traditional G-CVT algorithm fails to accurately locate the two pollution sources, while the UCB-CVT algorithm can cover more areas and ultimately successfully identify the two different pollution sources.

Fig. 6.11a shows the root mean squares error of the models established by the two algorithms under the source described in (6.29). Compared with the real field data, at the end of the iteration, the error of the model established by the UCB-CVT algorithm decreased by 75.44 percent, while the error of the model established by the G-CVT algorithm only decreased by 4.96 percent due to the robots being trapped in a local extremum. Therefore, in terms of the accuracy metrics of the established models, the UCB-CVT algorithm has significant advantages. Figs. 6.11b and 6.11c shows that the traditional G-CVT algorithm performs poorly and cannot accurately locate the two pollution sources. In contrast, the algorithm proposed in this paper

Table 6.2: Runtime and overall path distance of two online coverage control algorithms

| Coverage control algorithm | G-CVT | UCB-CVT |
|---|---|---|
| Runtime (s) | 673.22 | 272.63 |
| Path distance (m) | 43.5 | 48.3 |

guides the robots to explore uncertain areas as much as possible. Compared with the real field data, the positioning distance error and concentration error of the extremum points are reduced by 93.31 percent and 88.14 percent, respectively.

Table 6.2 shows the simulation running time and total path distance of the two algorithms. It can be seen that the total iteration time of the UCB-CVT algorithm is 272.63 seconds, with an average single iteration time of 2.27 seconds, while that of the G-CVT algorithm is 5.61 seconds. When using these two algorithms, there is almost no difference in the total path distance, which further indicates that the proposed UCB-CVT algorithm can move approximately the same path distance in a short time and ensure precise positioning of the pollution source. However, the traditional G-CVT algorithm not only takes longer time and moves approximately the same distance but also cannot guarantee the positioning of the pollution source, which further proves the positioning effectiveness of our proposed algorithm.

## 6.4 Toward Distributed Gaussian Process Regression

This chapter employs SGPR to ensure predictive accuracy while significantly reducing computational overhead. However, the current approach still relies on centralized data processing and model training. In real-world multi-robot systems, such centralized architectures face critical limitations in scalability and real-time responsiveness due to constraints such as limited communication bandwidth, asynchronous data acquisition, and robot heterogeneity.

To address these issues, Distributed Gaussian Process Regression (DGPR) presents a highly promising direction for future research. DGPR allows each robot to maintain and update its own local Gaussian process model and exchange statistical summaries—such as inducing points, covariance matrices, or predictive outcomes—with neighboring agents, thereby enabling efficient collaborative modeling. This decentralized scheme effectively reduces communication load and centralized computation pressure, enhances system robustness under communication delays or dropouts, and supports asynchronous updates and parallel modeling. It is especially suitable for large-scale multi-robot systems requiring collaborative learning and distributed task execution in dynamic environments.

Recent research on DGPR has proposed several representative approaches:

(1) Consensus-Based Distributed Fusion.

In this approach, each robot trains a local GPR model independently and iteratively exchanges model information with neighbors via consensus algorithms, gradually converging toward globally consistent predictions. This method is well-suited for structured communication networks but requires stability and convergence guarantees in the network topology.

(2) Local Experts and Mixture of Experts.

The workspace is partitioned into subregions, with each robot acting as a "local expert" responsible for modeling its designated area. Other robots can query these expert models or integrate predictions from multiple experts via weighted mixtures. This approach provides strong regional adaptability and scalability, especially in spatially distributed modeling tasks.

(3) Sparse Information Compression and Prior Propagation.

Instead of sharing raw data, robots transmit compressed model representations such as inducing inputs, kernel hyperparameters, or low-rank approximations of the covariance matrix. Methods like FITC and PITC exemplify this strategy, enabling efficient approximation of global inference with reduced communication, making it particularly suited for bandwidth-limited or resource-constrained settings.

(4) Federated Gaussian Process Learning

Inspired by federated learning, this approach allows each robot to keep its local data private and only share model parameters or updates with a central aggregator across multiple communication rounds. Federated GPR is particularly applicable to scenarios with strict privacy or security requirements, such as water quality monitoring or nuclear pollution localization.

These DGPR approaches naturally complement Voronoi-based distributed coverage control. Voronoi tessellations possess inherent spatial locality and distributed computability—each robot only needs local neighbor information to perform regional decomposition and real-time updates. This makes it particularly well-suited to integration with distributed modeling frameworks. By combining the environmental modeling capabilities of DGPR with the task allocation properties of Voronoi-based control, a truly fully distributed system can be realized. In such a system, robots autonomously model their local environments, adaptively adjust their coverage behavior based on dynamically updated Voronoi cells, and collaboratively achieve source localization, environmental reconstruction, and path planning. This greatly enhances the system's scalability, adaptability, and responsiveness in complex tasks.

Future research should also focus on key challenges such as ensuring model consistency, designing efficient information-sharing protocols, and analyzing the convergence behavior of collaborative learning processes. Incorporating distributed modeling into the system is expected to facilitate the development of an efficient, flexible, robust, and real-time multi-robot source-seeking and coverage control framework, laying a solid foundation for intelligent autonomous operations in complex environments.

## 6.5 Chapter Summary

This chapter presents a Voronoi-based multi-robot online coverage and source localization framework in unknown environments. By integrating CVT with SGPR, the proposed method enables efficient environment modeling while reducing computational complexity. Furthermore, the introduction of uncertainty regularization through the UCB mechanism enhances exploration efficiency and prevents convergence to local extrema. To address scenarios lacking prior environmental knowledge, a distributed entropy-based exploration strategy is incorporated, allowing robots to adaptively explore and model the pollution field while maintaining inter-robot safety via buffered Voronoi cells. Simulation results in both two-dimensional and three-dimensional environments demonstrate the proposed algorithm's superior performance in terms of source localization accuracy, reconstruction error, and computational efficiency compared to baseline CVT methods.

In addition, this chapter also outlines a perspective on DGPR, highlighting its advantages in asynchronous data modeling, communication efficiency, and system robustness. Owing to the naturally local and distributed properties of Voronoi partitioning, DGPR can be tightly coupled with distributed coverage control strategies. Future research will focus on the integration of distributed modeling and control, enabling scalable, real-time, and collaborative systems for complex multi-robot source-seeking and environmental reconstruction.

# Bibliography


[1] Gongcheng Wang, Weidong Wang, Pengchao Ding, Yueming Liu, Han Wang, Zhenquan Fan, Hua Bai, Zhu Hongbiao, and Zhijiang Du. Development of a search and rescue robot system for the underground building environment. *Journal of Field Robotics*, 40(3):655–683, 2023.

[2] Marcelo Elias Simon, Fabio Luiz Baldissera, Max Hering de Queiroz, and Felipe Gomes Cabral. Multi-robots coordination system for urban search and rescue assistance based on supervisory control theory. *Journal of Control, Automation and Electrical Systems*, 40(3):484–495, 2023.

[3] Yuwen Du. Multi-UAV search and rescue with enhanced A* algorithm path planning in 3D environment. *International Journal of Aerospace Engineering*, 2023:8614117, 2023.

[4] Xinyi Song, Ying Cheng, Qinglin Qi, and Xiaofu Zou. A path planning method for robot-aided aero-engine fleet inspection considering resource reuse strategy. In *2022 IEEE International Conference on Industrial Technology (ICIT)*, pages 1–6. IEEE, 2022.

[5] Graeme Best and Geoffrey A Hollinger. Decentralised self-organising maps for multi-robot information gathering. In *2020 IEEE/RSJ International Conference on Intelligent Robots and Systems (IROS)*, pages 4790–4797. IEEE, 2020.

[6] Tianpeng Zhang, Victor Qin, Yujie Tang, and Na Li. Source seeking by dynamic source location estimation. In *2021 IEEE/RSJ International Conference on Intelligent Robots and Systems (IROS)*, pages 2598–2605. IEEE, 2021.

[7] Meh Jabeen, Qing-Hao Meng, Tao Jing, and Hui-Rang Hou. Robot odor source localization in indoor environments based on gradient adaptive extremum seeking search. *Building and Environment*, 229:109983, 2023.

[8] David Gundana and Hadas Kress-Gazit. Event-based signal temporal logic synthesis for single and multi-robot tasks. *IEEE Robotics and Automation Letters*, 6(2):3687–3694, 2021.

[9] Boyu Zhou, Hao Xu, and Shaojie Shen. RACER: Rapid collaborative exploration with a decentralized multi-UAV system. *IEEE Transactions on Robotics*, 39(3):1816–1835, 2023.

[10] Jorge Pena Queralta, Qingqing Li, Fabrizio Schiano, and Tomi Westerlund. VIO-UWB-based collaborative localization and dense scene reconstruction within heterogeneous multi-robot systems. In *2022 International Conference on Advanced Robotics and Mechatronics (ICARM)*, pages 87–94, Guilin, China, 2022. IEEE.

[11] Vu Phi Tran, Matthew A Garratt, Kathryn Kasmarik, and Sreenatha G Anavatti. Dynamic frontier-led swarming: Multi-robot repeated coverage in dynamic environments. *IEEE/CAA Journal of Automatica Sinica*, 10(3):646–661, 2023.

[12] Ali El Romeh and Seyedali Mirjalili. Multi-robot exploration of unknown space using combined meta-heuristic Salp swarm algorithm and deterministic coordinated multi-robot exploration. *Sensors*, 23(4):2156, 2023.

[13] Shushman Choudhury, Jayesh K Gupta, Mykel J Kochenderfer, Dorsa Sadigh, and Jeannette Bohg. Dynamic multi-robot task allocation under uncertainty and temporal constraints. *Autonomous Robots*, 46(1):231–247, 2022.

[14] Faiza Gul, Adnan Mir, Imran Mir, Suleman Mir, Tauqeer Ul Islaam, Laith Abualigah, and Agostino Forestiero. A centralized strategy for multi-agent exploration. *IEEE Access*, 10:126871–126884, 2022.

[15] Alessandro Renzaglia and Lara Briñón-Arranz. Search and localization of a weak source with a multi-robot formation. *Journal of Intelligent & Robotic Systems*, 97(3):623–634, 2020.

[16] Frank Mascarich, Mihir Kulkarni, Paolo De Petris, Taylor Wilson, and Kostas Alexis. Autonomous mapping and spectroscopic analysis of distributed radiation fields using aerial robots. *Autonomous Robots*, 47(2):139–160, 2023.

[17] Xinxing Chen, Bo Yang, Jian Huang, Yuquan Leng, and Chenglong Fu. A reinforcement learning fuzzy system for continuous control in robotic odor plume tracking. *Robotica*, 41(3):1039–1054, 2023.

[18] Meng Zhao, Hui Lu, Shi Cheng, Siyi Yang, and Yuhui Shi. A multi-robot cooperative exploration algorithm considering working efficiency and working load. *Applied Soft Computing*, 128:109482, 2022.

[19] Le Hong, Weicheng Cui, and Hao Chen. A novel multi-robot task allocation model in marine plastics cleaning based on replicator dynamics. *Journal of Marine Science and Engineering*, 9(8):879, 2021.

[20] Tugay Alperen Karagüzel, Ali Emre Turgut, A. E. Eiben, and Eliseo Ferrante. Collective gradient perception with a flying robot swarm. *Swarm Intelligence*, 17(1):117–146, 2023.

[21] Mingrui Jiang, Yu Liao, Xun Guo, Hao Cai, Wenqing Jiang, Zhou Yang, Fei Li, and Fei Liu. A comparative experimental study of two multi-robot olfaction methods: towards locating time-varying indoor pollutant sources. *Building and Environment*, 207:108560, 2022.

[22] Lu Chang, Liang Shan, Weilong Zhang, and Yuewei Dai. Hierarchical multi-robot navigation and formation in unknown environments via deep reinforcement learning and distributed optimization. *Robotics and Computer-Integrated Manufacturing*, 83:102570, 2023.

[23] Hongro Jang, Minkyu Park, and Hyondong Oh. Improved socialtaxis for information-theoretic source search using cooperative multiple agents in turbulent environments. *Expert Systems with Applications*, 225:120033, 2023.

[24] Liwei Chen, Cong Zhou, Yu Wang, Yiran Zong, Tingting Lu, and Chunhua Chen. Autonomous search investigation for radioactive leaked source based on an updated infotaxis method during nuclear emergency rescue. *Nuclear Engineering and Design*, 416:112769, 2024.

[25] Liwei Chen, Cong Zhou, Yu Wang, Yiran Zong, Tingting Lu, and Chunhua Chen. Receding-horizon RRT-Infotaxis for autonomous source search in urban environments. *Aerospace Science and Technology*, 120:107276, 2022.

[26] Yang Liu, Christopher M Harvey, Frederick E Hamlyn, and Cunjia Liu. Bayesian estimation and reconstruction of marine surface contaminant dispersion. *Science of the Total Environment*, 907:167973, 2024.

[27] Hongyuan Jia and Hideki Kikumoto. Line source estimation of environmental pollutants using super-Gaussian geometry model and Bayesian inference. *Environmental Research*, 194:110706, 2021.

[28] Haitong Ma, Tianpeng Zhang, Yixuan Wu, Flavio P Calmon, and Na Li. Gaussian max-value entropy search for multi-agent Bayesian optimization. In *2023 IEEE/RSJ International Conference on Intelligent Robots and Systems (IROS)*, pages 10028–10035. IEEE, 2023.

[29] Christopher E Denniston, Oriana Peltzer, Joshua Ott, Sangwoo Moon, Sung-Kyun Kim, Gaurav S Sukhatme, Mykel J Kochenderfer, Mac Schwager, and Ali-akbar Agha-mohammadi. Fast and scalable signal inference for active robotic source seeking. In *2023 IEEE International Conference on Robotics and Automation (ICRA)*, pages 7909–7915. IEEE, 2023.

[30] Abdullah Al Redwan Newaz, Murtadha Alsayegh, Tauhidul Alam, and Leonardo Bobadilla. Decentralized multi-robot information gathering from unknown spatial fields. *IEEE Robotics and Automation Letters*, 8(5):3070–3077, 2023.

[31] Meh Jabeen, Qing-Hao Meng, Tao Jing, and Hui-Rang Hou. Robot odor source localization in indoor environments based on gradient adaptive extremum seeking search. *Building and Environment*, 229:109983, 2023.

[32] Adam Francis, Shuai Li, Christian Griffiths, and Johann Sienz. Gas source localization and mapping with mobile robots: A review. *Journal of Field Robotics*, 39(8):1341–1373, 2022.

[33] Hui Li, Jie Yuan, and Hao Yuan. An active olfaction approach using deep reinforcement learning for indoor attenuation odor source localization. *IEEE Sensors Journal*, 2024.

[34] Yong Zhao, Bin Chen, XiangHan Wang, Zhengqiu Zhu, Yiduo Wang, Guangquan Cheng, Rui Wang, Rongxiao Wang, Ming He, and Yu Liu. A deep reinforcement learning based searching method for source localization. *Information Sciences*, 588:67–81, 2022.

[35] Junyan Hu, Hanlin Niu, Joaquin Carrasco, Barry Lennox, and Farshad Arvin. Voronoi-based multi-robot autonomous exploration in unknown environments via deep reinforcement learning. *IEEE Transactions on Vehicular Technology*, 69(12):14413–14423, 2020.

[36] Qilin Feng, Hao Cai, Yibin Yang, Jiheng Xu, Mingrui Jiang, Fei Li, Xianting Li, and Chenchu Yan. An experimental and numerical study on a multi-robot source localization method independent of airflow information in dynamic indoor environments. *Sustainable Cities and Society*, 69:101897, 2020.

[37] Hongbiao Zhu, Yibo Wang, Chengjin Du, Quan Zhang, and Weidong Wang. A novel odor source localization system based on particle filtering and information entropy. *Robotics and Autonomous Systems*, 132:103619, 2020.

[38] Chao Zhai, Zhaoxu Wang, and Jie Dou. Multi-agent coverage control under health management. *Chinese Journal of Sensors and Actuators*, 35(7):902–912, 2022.

[39] Alaa Z Abdulghafoor and Efstathios Bakolas. Multi-agent distributed optimal control for tracking large-scale multi-target systems in dynamic environments. *IEEE Transactions on Cybernetics*, 54(5):2866–2879, 2024.

[40] Wenhao Luo, Changjoo Nam, George Kantor, and Katia Sycara. Distributed environmental modeling and adaptive sampling for multi-robot sensor coverage. In *Proceedings of the 18th International Conference on Autonomous Agents and Multi-Agent Systems*, pages 1488–1496. ACM, 2019.

CHAPTER 7

# Multi-Robot System Optimization and Health Management Based on Voronoi Tessellations

## 7.1 Introduction

THE proliferation of multi-robot systems in applications such as persistent surveillance, environmental monitoring, and logistics has underscored a critical challenge: ensuring long-term operational autonomy. While much research has focused on optimizing task performance under ideal conditions, the reality of sustained deployment involves inevitable wear and tear, energy depletion, and unforeseen failures, which significantly affect task completion and robot health status [1]. Therefore, as an intelligent control system, it is essential to consider these potential anomalies to maximize the overall robot performance and task completion.

In terms of optimization and management issues in formation missions [2], this chapter proposes a task allocation system based on health management which can predict the impact of various abnormalities on the task execution status through dynamic programming and selective measures to mitigate these effects. The authors in [3–5] investigated the impact of different robot driving capabilities in the team on the coverage task, using online and distributed methods to learn the changes in the relative driving performances of the robots, which can be a trade-off between time and the energy consumption of the robot, to ensure that the robots converged to the best configuration.

The study in this chapter further combines the CVT algorithm with multi-robot formation control [6]. First, we used the CVT algorithm to complete the Voronoi division of the task map and designed the controller to control the robot to track the specified Voronoi tessellation centroid. When the CVT algorithm converged, it guarantees that the robots form the specified desired formation in the specified area according to the optimal distribution. Second, considering that there are obstacles

DOI: 10.1201/9781003714569-7

in the real environments where robots perform tasks, this study also verified the formation maintenance experiment in an obstacle environment [7]. Under the premise of ensuring the uniform distribution of the global density function, the density value corresponding to the obstacle position was reduced to ensure that the centroid was not divided into the obstacle area. That is, the characteristics of the CVT algorithm were used to realize the repulsive force field that drives the robot to avoid obstacles. It was necessary to set up additional potential field functions to complete the collision avoidance and obstacle avoidance control of formation robots, which significantly improved the flexibility and scalability of the algorithm [8,9]. Finally, the health optimization management algorithm was used to maximize the endurance of unhealthy robots under the premise of minor changes to the original framework [10], so that the system could deal better with unforeseen damage such as mechanical failures and emergencies, making the formation more robust when performing tasks in different scenarios.

Specific contents of the formation algorithm were covered in past chapters, they will not be detailed in this chapter. Instead, we focus on introducing the formation simulation under the health optimization management algorithm.

In coverage tasks, Chen et al. [11] proposed a distributed control law for mobile robot networks, driving each robot toward the centroid of its Voronoi tessellation to cover the current task area. Chen et al. [12] used CVT to detect and control dispersed pollution sources in a region and subsequently proposed multi-robot coverage strategies for complex environments to address pollution source detection and treatment. Hatleskog et al. [13] studied a decentralized control scheme for Voronoi-based discrete-time multi-agent dynamic systems in multi-dimensional static convex polyhedral environments. They used local control laws to direct each robot toward the Chebyshev center of its time-varying Voronoi neighborhood, revealing the mechanism by which robot subsets reach consensus on neighbor internal distances. To avoid task duality in coverage path planning for multi-cooperative robots, Vishnu [14] employed Voronoi tessellation to achieve natural passive cooperation among robots. This approach combined Voronoi tessellation based on Euclidean distance and Voronoi tessellation based on Manhattan distance to address partition continuity in the presence of obstacles.

Most of the aforementioned work on multi-robot collaborative coverage control assumes ideal environmental and robot states. However, in the real world, robots may experience various health degradation issues during task execution, such as reduced sensing efficiency, communication loss or delay, insufficient energy, impaired mobility, and robot loss within the team. These factors make the ideal assumptions difficult to implement in practical applications [15–17].

To address these issues, this study investigates the problem of using a group of robots to cover a region of interest, where robots may have variable health conditions. A health optimization management algorithm is employed to enhance the system's ability to handle mechanical failures and unforeseen events with minimal changes to the original robot control framework. Additionally, robot health degradation is incorporated into the position optimization problem. W-CVT is used to partition

Voronoi regions, where each robot's health value serves as a weight. This elegantly translates a robot's physical condition into its share of task responsibility, allowing healthier robots to naturally assume larger roles, providing Voronoi tessellation tailored to the health characteristics of different robots. Distributed coverage control laws are then developed to ensure robots converge to optimal configurations. Finally, numerical simulations and experiments are conducted to validate the effectiveness of the proposed algorithm, followed by a summary of the findings.

## 7.2 Optimization Strategies Based on Health Management

### 7.2.1 Methods of Health Optimization Management Based on Formation

To improve the health of each robot and enhance the robustness and task completion of the formation, health degradation of each robot should be considered. Ensuring the survivability of a single robot will often have a significant impact on the task performance and future reliability of the overall formation, as factors such as low battery, high power consumption rate or unforeseen damage to physical structures can cause the robot to slow down in speed or task processing. Although these health problems caused by degraded operating conditions or malfunctions do not need to be dealt with immediately, continued operation will have a significant impact on the mission performance and future reliability of the robot itself and even on the formation.

In response to the above problems, the formation health optimization management algorithm was introduced to restrict the speed of the unhealthy robot's Voronoi tessellation, in order to make it move more conservatively, maximizing the endurance of the unhealthy robot. This makes the formation system more capable of dealing with unforeseen damage such as mechanical failures and emergencies, making it more robust when performing tasks in different scenarios.

In this algorithm, $u_i$ is the control input of the robot $i$. To better describe the health status of the robot, in response to the above problems, $h_i(t)$ is specially introduced, where $h_i(t)$ at time $t$ can be expressed as:

$$\begin{cases} h_i(t) = h_i(0) - d_i(t) \\ d_i(t) = \int_0^t k_{d,i} u_i^2 \, dt + f_{d,i}(t) \\ \dot{h}_i(t) \leq 0, \dot{f}_{d,i}(t) \geq 0, h_i(t), f_{d,i}(t) \in [0,1], \end{cases} \tag{7.1}$$

where $h_i(0)$ is the initial health of the robot, $d_i(t)$ is the health loss of the robot at each moment, $k_{d,i}$ is the usage gain coefficient, $f_{d,i}(t)$ is the negative impact of the external environment on the robot, such as damage caused by external collisions. The integral term $\int_0^t k_{d,i} u_i^2 \, dt$ encapsulates the cumulative effect of mechanical wear and tear due to motion. The squared control input $u_i^2$ serves as a proxy for the aggressiveness of maneuvers, which can expedite the consumption of energy resources. Assuming that the health of each robot decreases with the progress of its motion control and possible external damage, (7.1) is a function of its control accumulation and any external damage that may be encountered. Controlling the accumulation means that fast-moving and turning robots degrade their health faster than slow-moving or stationary robots, which can be simplified to a combination of remaining power and structural health in a physical sense.

The current health status of a multi-robot formation can be calculated using the average health level of all robots:

$$\begin{cases} h^*(t) = \frac{1}{n}\sum_{i=1}^{n} h_i(t) \\ \dot{h}^*(t) < 0, \quad h^*(t) \in [0,1]. \end{cases} \tag{7.2}$$

Assume that each robot can obtain the overall average health level $h^*(t)$ of all robots in the entire task area by communicating with neighboring robots within a limited range. The movement of each robot is controlled by a proportional controller, which is controlled by the proportional derivative controller of the formation health optimization management algorithm:

$$\begin{cases} u_i(t) = k_p\left[1 + k_{h,i}(t)\right](p_i - p_i^*) \\ e_i(t) = h_i(t) - h^*(t) \\ k_{h,i}(t) = k_{h_p} e_i(t) + k_{h_d}\dot{e}_i(t), \end{cases} \tag{7.3}$$

where $k_{h,i}(t)$ represents the gain obtained by the robot changing the mobile formation health optimization management system, $e_i(t)$ represents the difference between a robot and the overall average health level, $k_{h_p}$, and $k_{h_d}$ represent the gain coefficients of the formation health optimization management algorithm. Robot healthier than the average instantly becomes more aggressive, while a less healthy one becomes more conservative. The $k_{h_d}\dot{e}_i(t)$ acts as a damping factor, anticipating future health trends. For instance, if a robot's health is declining rapidly large negative $e_i(t)$, this term will further reduce its control gain, preemptively conserving its resources. This proactive control enhances the stability of the health-balancing mechanism.

Finally, combining the optimization of robot health management and the flexibility of the CVT algorithm, the remaining robots re-divide the original area after the robot, whose health value is reduced to zero, stops working (the robot cannot continue to work normally due to energy exhaustion, failure, or the deterioration of the external environment), to minimize the impact of this type of robot on the formation task.

Based on the traditional CVT construction algorithm, the insertion construction method was introduced, so that in the original CVT configuration, when the robot number is reduced, all the triangles whose circumscribed circle contains new points are recorded, the common edges that affect the construction of the new triangle are deleted, and then new edges are re-planned, until all points are inserted. The specific process is shown in Figs. 7.1a-d.

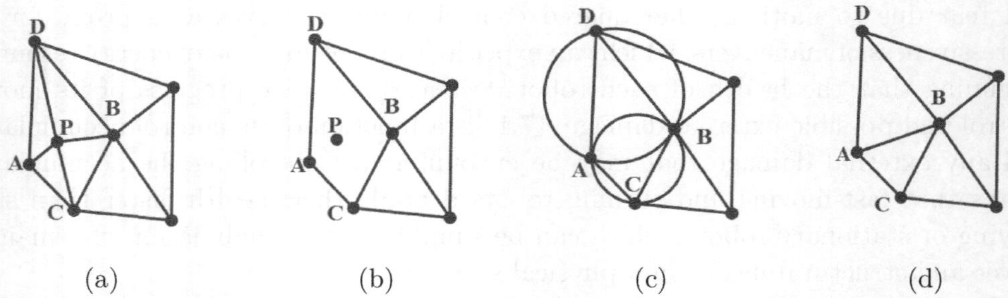


Figure 7.1: Insert construction method: (a) Robot P has a health value of zero; (b) delete the public edge related to P; (c) re-plan neighboring robots; (d) form a new distribution.

The specific control steps in this study were show in Algorithm 7.
When the formation is moving, the entire formation is recalculated to obtain the distribution in the environment. While controlling the robot to have the smallest possible impact on other robots, making the robots in the formation respond better to changes in the environment and improving the robustness and task execution ability of the formation in various environments.

---

**Algorithm 7** CVT-based multi-robot formation control and health optimization management

---

***Require***
The initial position of a group of robots $P : \{p_i\}_{i=1}^n, h_i(t), h_i(0)$;
Boundary information $\partial Q$ of task area $Q$, area density function $\rho(q)$;
***Procedure***
Calculate the centroid of the Voronoi tessellation $\{p_i^*\}_{i=1}^n$ and $\{V_i\}_{i=1}^n$ of the robot position $P$;
**while** $((p_i - p_i^*) \geq \varepsilon)$ && $\partial(E_V(p)) \neq 0$ **do**
  Calculate $d_i(t)$ and $h_i(t)$ for each robot;
  **if** $h_i(t) \leq 0$ **then**
    Use the insert construction method to re-form and divide the area;
  **end if**
  Calculate $h^*(t)$ and $u_i(t)$ for the entire formation;
  Drive the robot to the center-of-mass position $p_i = p_i^*$, $(i = 1, \ldots, n)$;
  Generate a new set of robot positions $P : \{p_i = p_i^*\}_{i=1}^n$;
  Calculate the new centroid $\{p_i^*\}_{i=1}^n$ and $\{V_i\}_{i=1}^n$ of the robot position set $P$;
**end while**

---

### 7.2.2 Weighted Centroidal Voronoi Tessellations Based on Coverage

Assume that a set of robots in a bounded convex region $Q \subset \mathbb{R}^2$ can be represented as $P = (p_1, \ldots, p_n)$, where the weight of robot $p_i$ is denoted by $w_i$, with $w_i \geq 0$, and

$w_i = h_i$. If $wv_i \cap wv_j = \partial wv_i$ for $i \neq j$, where $\partial wv_i$ is the boundary of the weighted Voronoi tessellation, meaning that two adjacent weighted Voronoi tessellation share only a common edge as their overlapping region, and if $\{wv_i\}_{i=1}^n = Q$, then a Voronoi tessellation of the region $Q$ can be generated as $WV = \{wv_i\}_{i=1}^n$, where the generator point $p_i$ corresponds to the weighted Voronoi tessellation $wv_i$, resulting in the weighted Voronoi tessellation:

$$wv_i = \left\{ x \in Q \mid \frac{\|q - p_i\|}{w_i} < \frac{\|q - p_j\|}{w_j},\ i, j = 1, 2, \ldots, n,\ j \neq i \right\}. \tag{7.4}$$

The objective of the multi-robot coverage task can be mathematically formulated as the minimization of a global cost function. This function, denoted as $F(P, V)$, quantifies the overall coverage quality for a given set of robot positions $P = \{p_1, \ldots, p_n\}$ and the corresponding partition $V = \{v_1, \ldots, v_n\}$ of the domain $Q$. A lower value of $F$ indicates a more rational or optimal deployment of the robots. The general form of this cost function is given by:

$$F(P, V) = \sum_{i=1}^{n} \int_{q \in v_i} \rho(q) f_{i,t}[q, h_i(t)] dq, \tag{7.5}$$

where $\rho(q)$ is the density function representing the importance of point $q$, and $f_{i,t}(q, h_i(t))$ is a local cost function associated with robot $i$ covering point $q$, which explicitly depends on the robot's health status $h_i(t)$.

To integrate the health status of each robot into this optimization problem, we define a specific form for the local cost function $f_{i,t}(q, h_i(t))$ as follows:

$$f_i(q, h_i(t)) = \frac{\|p_i - q\|^2}{h_i(t)}. \tag{7.6}$$

In the formula, $h_i(t)$ represents the health value of robot $i$ at time $t$. Assuming other variables remain constant, a smaller $h_i(t)$ leads to a higher coverage cost for the $i$-th robot (assuming $h_i(t) = 1$ corresponds to a fully healthy robot, $0 < h_i(t) < 1$ indicates a robot with health impairment, and $h_i(t) \leq 0$ corresponds to a malfunctioning robot). Therefore, the degradation of a robot's health affects the overall coverage performance. Additionally, the magnitude of a robot's health value can be determined through corresponding robot detection and fault diagnosis methods.

The coverage performance function with the above cost function is denoted by $F(P, V)$. When $h_i(t)$ is not identical, the optimal position of the robot is at the centroid of the coverage area, but at this time, the spatial division is no longer the traditional CVT, and should be planned according to the health value $h_i(t)$ of each robot. That is, when planning the Voronoi region, health value should be added as a weight, as a basis for judgment, transferring the traditional coverage optimization problem to a situation where health is changing over time for study.

Fig. 7.2 illustrates an example of CVT and W-CVT with a set of 10 weighted generator points. In Fig. 7.2a, the sub-cell areas covered individually by the ten robots are approximately equal in size, and the boundaries of each Voronoi tessellation lie

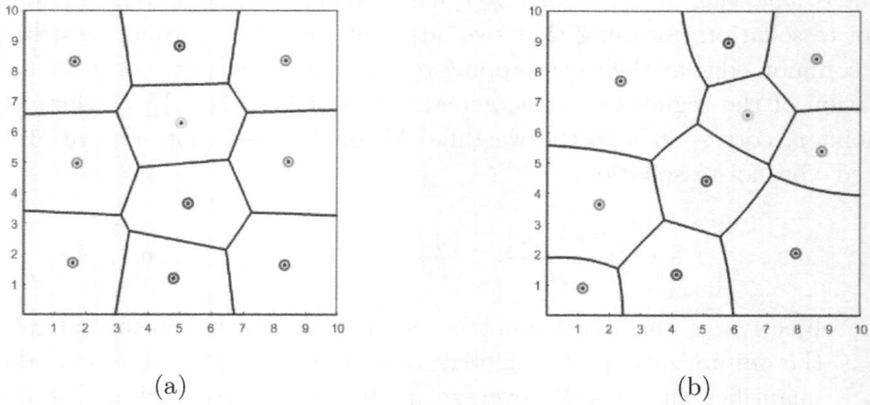


Figure 7.2: Weighted Voronoi diagram: (a) CVT; (b) W-CVT.

on the perpendicular bisector between adjacent robots, consistent with the CVT characteristics. In Fig. 7.2b, the robots in the top-left and bottom-right corners have larger $w_i$, resulting in larger cells in the final configuration compared to other robots. The robot in the bottom-left corner, with smaller $w_i$, occupies a relatively smaller cell, and the boundaries of each Voronoi tessellation are biased toward the robot with smaller $w_i$, consistent with the W-CVT characteristics. This section subsequently develops a distributed coverage algorithm for multi-robot systems using W-CVT.

This section uses a differential drive robot as an example. Assuming the wheeled robot is a rigid body that can rotate freely, its simplified motion model is shown in Fig. 7.3.

The center of mass of the wheeled robot is denoted as $C(x, y)$, and $\theta$ is the forward direction angle. The state information of the robot is represented by $P = (x, y, \theta)^\top$. $[v, w]$ are the linear and angular velocities of the robot, respectively. $2L$ is the distance

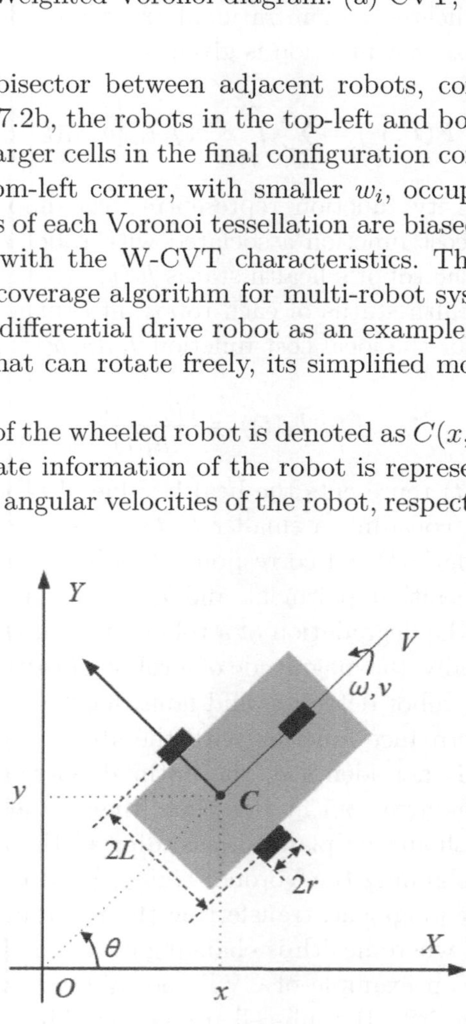


Figure 7.3: Robot kinematics model.

between the two wheels, and $2r$ is the diameter of each wheel. The kinematic state-space equation of the robot can be expressed as shown in (8.5).

$$\begin{cases} \dot{x} = v\cos\theta \\ \dot{y} = v\sin\theta \\ \dot{\theta} = \frac{v\tan\varphi}{2b}. \end{cases} \tag{7.7}$$

According to the robot kinematic model and control laws, to maintain zero dynamics, the control input should satisfy $\ddot{p}_i = 0$, i.e., $u_i = 0$. From this, the position control law for the robot can be derived as:

$$\begin{cases} \ddot{p}_i = u_i \\ u_i = -K_p(p_i - p_i^*) - K_d\dot{p}_i, \end{cases} \tag{7.8}$$

where $u_i$ represents the control input for the $i$-th robot. The robot's motion can be modeled using a double integrator dynamic model. This section focuses on the coordinated control of multi-robots rather than the posture control of individual robots. In this context, a more detailed dynamic model of the robot is required. Additionally, the dynamics of many robots can be feedback linearized into a double integrator form. This section derives a distributed coverage controller under the following assumptions: mobile robots can compute their weighted Voronoi tessellation in a distributed manner, and each robot can communicate with other robots in neighboring weighted Voronoi tessellation.

A position control law of the following form is proposed for the $i$-th robot:

$$\begin{cases} h^*(t) = \frac{1}{n}\sum_{i=1}^{n} h_i(t) \\ e_i(t) = h_i(t) - h^*(t) \\ k_{h,i}(t) = k_{hp}e_i(t) + k_{hd}\dot{e}_i(t) \\ u_i(t) = k_p\left[1 + k_{h,i}(t)\right](p_i^* - p_i) - k_d\dot{p}_i \end{cases}, \tag{7.9}$$

where $h^*(t) \in [0,1]$ is time-decreasing. $k_{h,i}(t)$ represents the gain obtained by the robot when changing its movement to cover the health optimization management system. $e_i(t)$ represents the difference between the health level of robot $i$ and the overall average health level $h^*(t)$. $k_{h_p}$ and $k_{h_d}$ are the gain coefficients of the coverage health optimization management algorithm.

The following theorem provides the convergence results. From (10), the Lyapunov function can be defined as:

$$\mathcal{V}(p_i) = \frac{1}{2}k_p\left[1 + k_{h,i}(t)\right] \parallel p_i - p_i^* \parallel + \frac{1}{2}k_d\dot{p}_i^2. \tag{7.10}$$

The health gain term, $k_{h,i}(t) = k_{hp}e_i(t) + k_{hd}\dot{e}_i(t)$, is a time-varying quantity. To ensure stability, it is necessary to prove that $1 + k_{h,i}(t) > 0$ always holds. Considering that the health value $h_i(t) \in [0,1]$ and the average health value $h^*(t) \in [0,1]$, the range of the health error $e_i(t) = h_i(t) - h^*(t)$ is within $[-1, 1]$. By appropriately selecting the gain coefficients $k_{hp}$ and $k_{hd}$ and constraining their values to a small positive range, we can ensure that $[1 + k_{h,i}(t)]$ is always positive.

Taking the time derivative of $V(p_i)$ and substituting the control law $u_i = \ddot{p}_i = -k_p[1 + k_{h,i}(t)](p_i - p_i^*) - k_d\dot{p}_i$, we obtain:

$$\dot{\mathcal{V}}(p_i) = k_p \left[1 + k_{k,i}(t)\right] (p_i - p_i^*)\dot{p}_i + k_d\dot{p}_i\ddot{p}_i. \tag{7.11}$$

By simplification, we can obtain:

$$\dot{\mathcal{V}}(p_i) = -k_d\dot{p}_i^2 \leq 0. \tag{7.12}$$

Since $\dot{V}(p_i)$ is negative semi-definite, according to LaSalle's Invariance Principle, the system trajectories will converge to the largest invariant set where $\dot{V}(p_i) = 0$. $\dot{V}(p_i) = 0$ implies that $\dot{p}_i = 0$, and consequently $\ddot{p}_i = 0$. Substituting this into the control law yields $k_p[1 + k_{h,i}(t)](p_i - p_i^*) = 0$. Because $k_p > 0$ and $1 + k_{h,i}(t) > 0$, it must be that $p_i = p_i^*$. This proves that each robot will converge to the centroid of its weighted Voronoi region, thus causing the entire system to reach the W-CVT configuration.

At each time step, robot $i$ detects its neighboring robots and exchanges position information with them. Then, robot $i$ constructs its W-CVT unit and calculates its centroid to construct the control input (7.6). The robot moves toward $p_i^*$ under this control input. All robots in the group simultaneously follow the above steps at each time step until the coverage cost function $F$ is minimized between two consecutive time steps $t$ and $t-1$, and the error between the robots is less than a specified threshold $\varepsilon$. For the specific algorithm process, see Algorithm 8.

---

**Algorithm 8** Distributed Coverage Algorithm Procedure

---

***Require***
A set of initial robot positions $P : \{p_i\}_{i=1}^n$, $h_i(t)$, $h_i(0)$;
Boundary information $\partial Q$ and density function $\rho(q)$ of the task area $Q$;
***Procedure***
Compute the centroids $\{p_i^*\}_{i=1}^n$ of weighted Voronoi tessellation and the weighted Voronoi regions $\{wv_i\}_{i=1}^n$ for the robot position set $P$;
**while** $\{\text{err}(p_i) = p_i - p_i^* \geq \varepsilon\}$ && $\min F$ **do**
    Compute $d_i(t)$ and $h_i(t)$ for each robot;
    Compute the system health $h^*(t)$ and control inputs $u_i(t)$ for the multi-robot coverage system;
    Drive robots to reach the centroid positions $p_i = p_i^*$, $(i = 1, \ldots, n)$;
    Update the robot position set $P : \{p_i = p_i^*\}_{i=1}^n$;
    Recompute the new centroids $\{p_i^*\}_{i=1}^n$ and weighted Voronoi regions $\{wv_i\}_{i=1}^n$ for the updated $P$;
**end while**

---

# 7.3 Simulation Results and Analysis

## 7.3.1 Simulation Analysis under Uniform Density Function

In this section, the performance of the health optimization management algorithm will be assessed via the health loss value when the robot group reaches the optimal configuration. In the implementation of health optimization management algorithms, the goal is to maintain the health level of all robots as high as possible and reduce disproportionate health degradation. The following is a simulation of four different scenarios.

Scenario 1: None of the robots in the group have any abnormal health behaviors. Scenario 1 represents the best-case scenario, in which no robot suffers from abnormal health behaviors, so the health optimization management algorithm is used to optimize the performance of each robot.

Scenario 2: The health degradation rate of a single robot increases. Scenario 2 may occur during a period of engine failure, reducing power efficiency.

Scenario 3: The health value of a single robot at the beginning of the task is lower than other members of the group. Scenario 3 may indicate that the battery is not fully charged at the beginning of the task.

Scenario 4: The health of a single robot decreases suddenly. Scenario 4 may be a small component failure, such as a structural fracture. Under these circumstances, the deterioration of the operational status or failure does not require immediate recovery, but continued operation may affect the overall future operational status and performance of the formation.

In these four scenarios, the four robots form a square formation as an example. During the formation, the health degradations described in the above four scenarios may be encountered. For each scenario, two simulations were performed: one for normal operation and the other for health optimization management.

Scenario 1: All Robots Are Healthy

Figs. 7.4a and 7.4b describe the results of the first case, where each robot is initialized with the same health level, and the health of each robot decreases normally with its movement. Fig. 7.4a depicts the health of each robot during formation control without health optimization management. When all the robots reach the final configuration, the health of robot 1 drops more than the rest, reaching the final configuration when $h_i(t) = 0.26$. In Fig. 7.4b, with optimized health management, the same robot extends its lifespan and does not reach the final configuration until $h_i(t) = 0.48$, increasing the lifespan by 22%. Notably, in Fig. 7.4b, the health degradation curves of all robots under HOM are more tightly clustered than in Fig. 7.4a. This visually demonstrates the effectiveness of the health-balancing mechanism, which prevents any single robot from bearing a disproportionate amount of work, thus promoting team-wide longevity.

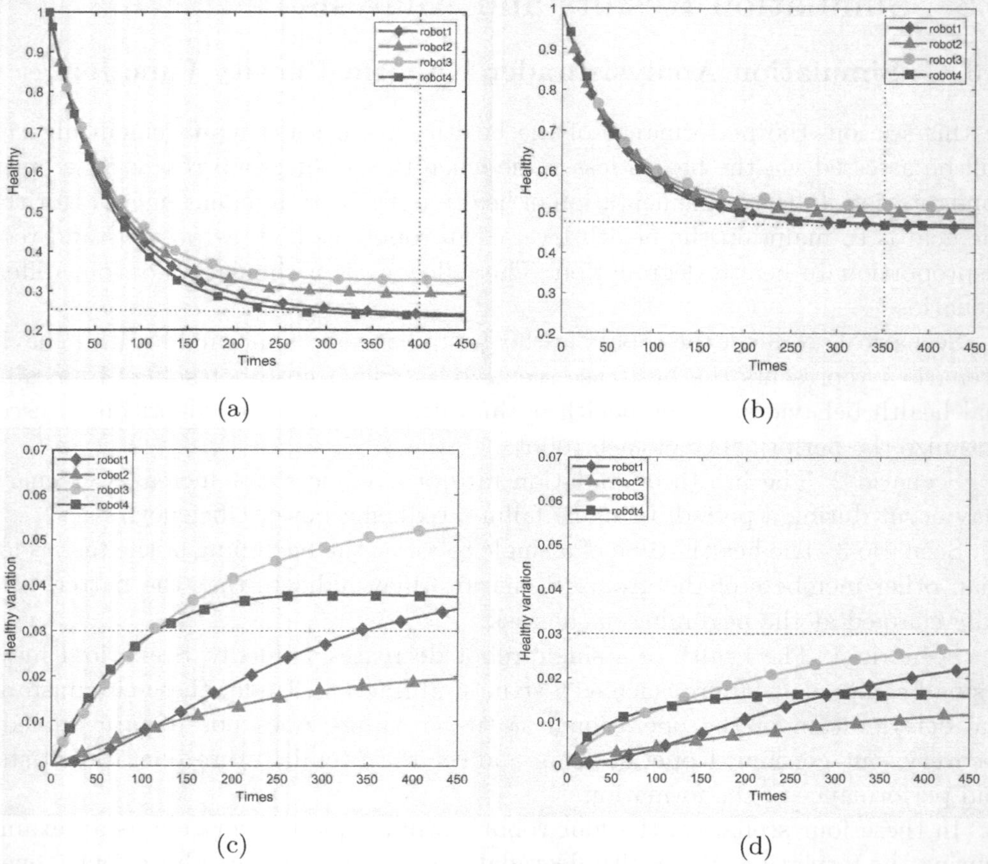


Figure 7.4: Comparison before and after adding HOM in the first scenario: (a) $h_i(t)$ of the robot under normal conditions; (b) $h_i(t)$ of the robot under HOM; (c) $e_i(t)$ under normal conditions; (d) $e_i(t)$ under HOM.

It can be seen from Fig. 7.4c that if there is no health optimization management, the differences in robot health will diverge over time, while in Fig. 7.4d, the differences are bounded.

Scenario 2: The initial power of a single robot is lower than that of other robots, resulting in a decrease in battery life

Figs. 7.5a and 7.5b describe the results in the scenario 2, where each robot is initialized with the same health level. Fig. 7.5a depicts the health of each robot during formation control without health optimization management. When all the robots reach the final configuration, the health of robot 4 drops more than the rest, reaching the final configuration when $h_i(t) = 0.04$. In Fig. 7.5b, with optimized health management, the same robot extends its lifespan and does not reach the final configuration until $h_i(t) = 0.38$, increasing the lifespan by 34%.

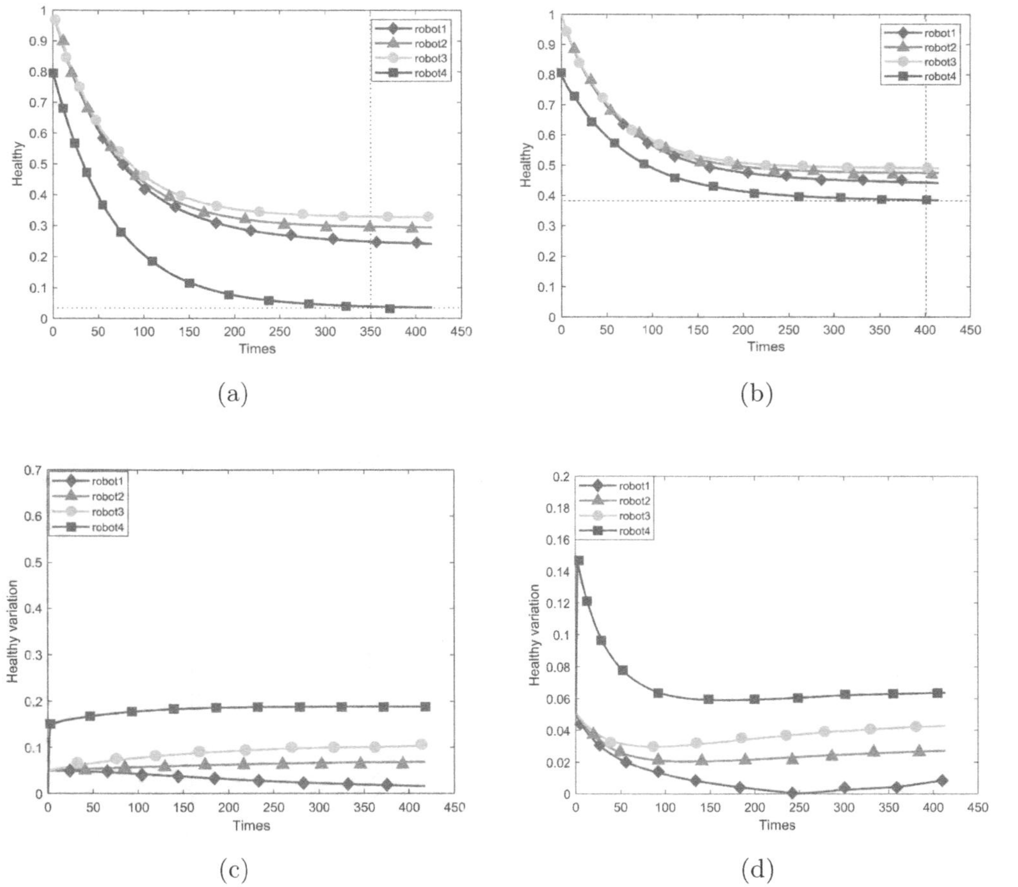


Figure 7.5: Comparison before and after adding HOM in the second scenario: (a) $h_i(t)$ of the robot under normal conditions; (b) $h_i(t)$ of the robot under HOM; (c) $e_i(t)$ under normal conditions; (d) $e_i(t)$ under HOM.

From Fig. 7.5c, if there is no health optimization management, the differences in robot health diverge greatly with the passage of time, while in Fig. 7.5d, the difference is small.

Scenario 3: The failure of a single robot component caused by the external environment, such as a damaged wheel that causes a sharp deterioration in the health of the robot

Figs. 7.6a and 7.6b describe the results of scenario 3, where each robot is initialized with the same health level. Fig. 7.6a depicts the health of each robot during formation control without health optimization management. When all the robots reach the final configuration, the health of robot 4 drops more than the rest, reaching the final configuration when $h_i(t) = 0.15$. In Fig. 7.6b, with optimized health management, the same robot extends its lifespan and does not reach the final configuration until $h_i(t) = 0.4$, increasing the lifespan by 25%.

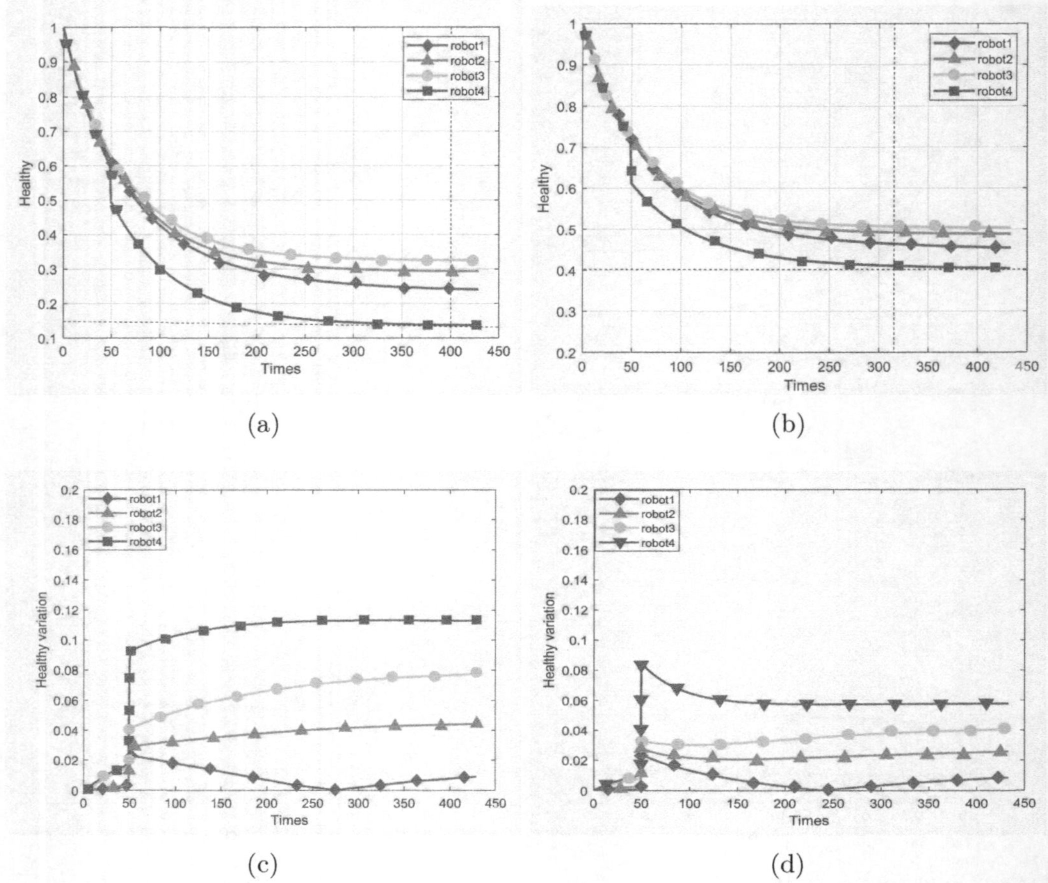


Figure 7.6: Comparison before and after adding HOM in the third scenario: (a) $h_i(t)$ of the robot under normal conditions; (b) $h_i(t)$ of the robot under HOM; (c) $e_i(t)$ under normal conditions; (d) $e_i(t)$ under HOM.

As can be seen from Fig. 7.6c, if there is no health optimization management(HOM), the differences in robot health diverge with time and are large compared with the average value, while in Fig. 7.6d, the differences are small.

Scenario 4: due to the abnormal fault of the motor of a single robot, the power consumption rate of the robot increases

Figs. 7.7a and 7.7b describe the results of scenario 4, in which each robot is initialized with the same health level. Fig. 7.7a depicts the health of each robot during formation control without health optimization management. When all robots reach the final configuration, the health of robot 4 decreases more than the rest, reaching the final configuration at $h_i(t) = 0.02$. In Fig. 7.7b, with health optimization management, the same robot extends its life and does not reach the final configuration until $h_i(t) = 0.3$, increasing its life by 28%.

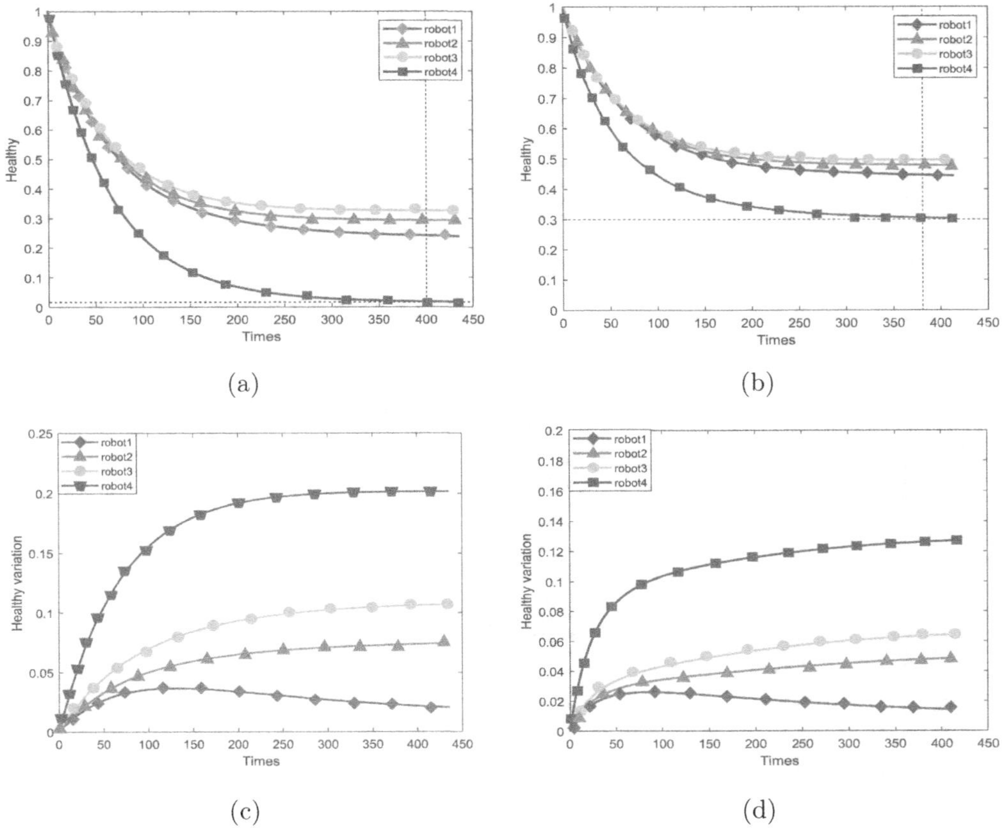


Figure 7.7: Comparison before and after adding HOM in the fourth scenario: (a) $h_i(t)$ of the robot under normal conditions; (b) $h_i(t)$ of the robot under HOM; (c) $e_i(t)$ under normal conditions; (d) $e_i(t)$ under HOM.

As seen from Fig. 7.6c, if there is no health optimization management, the differences in robot health will diverge greatly with the passage of time, while in Fig. 7.6d, the differences are small.

In the above four cases, the health optimization management greatly increases the lifetime of the weakest robot, and the adjustment of the existing framework is very small. Health optimization management is very important for multi-robot formations to ensure that the formation can operate for a long period, to reduce health loss, and to ensure the reliability of robots in the task.

It can be seen from Table 7.1 that by applying intelligent health balancing to each robot, the overall health value can be improved by about 20% to 30%, and the whole robot formation is more robust to health changes caused by many problems that may reasonably be expected. The system can better deal with accidental damage caused by collision, mechanical failure, or insufficient maintenance.

Table 7.1: Comparison of simulation results in four scenarios before and after adding the HOM algorithm

| | $h_i(t)$ (normal situation) | $h_i(t)$ (HOM) | Percentage increase in system health value |
|---|---|---|---|
| Scenario 1 | 0.26 | 0.48 | 22% |
| Scenario 2 | 0.04 | 0.38 | 34% |
| Scenario 3 | 0.15 | 0.40 | 25% |
| Scenario 4 | 0.02 | 0.30 | 28% |

Scenario 5: Formation simulation and results analysis after robot stops working

When $h_i(t) \leq 0$ due to other reasons, the robot is considered to be unable to work normally. In order not to affect the overall task of the formation, the insertion construction method was used to regroup multi-robots and re-divide the task area. Taking the square formation as an example, as shown in Figs. 7.8a and 7.8b, four robots form a square formation with the number of iterations $k = 120$. After that, when the robot in the lower right corner cannot work normally due to a fault, and $h_i(t) \leq 0$, as shown in Fig. 7.8e, the insertion construction method is used to make the other healthy robots adopt a new formation and reconstruct a new division of the new service area.

As shown in Fig. 7.8c, when the number of iterations reaches 280, the remaining three robots adopt a new formation in the area and realize a new division of the area. The red plus sign in the figure is the final position of the faulty robot. The position error and cost function diagrams for this process are shown in Figs. 7.8d and 7.8f. The simulation in Fig. 8.8 highlights the critical self-healing capability of the system. The loss of an agent does not lead to mission failure; instead, the remaining team members autonomously and smoothly reconfigure to cover the abandoned area, showcasing the inherent fault tolerance of the proposed distributed framework.

As shown above, when the insertion construction method is introduced into the CVT, it can not only reduce the impact on other robots when there is a faulty robot in the formation but can also compensate for the negative impact of the faulty robot as far as possible. This allows the robots in the formation to better respond to the changes in the environment, and the formation control can better adapt to the actual situation, improving the execution robustness of the robot formation.

### 7.3.2 Simulation Analysis under Non-uniform Density Function

In this section, the effectiveness of the proposed distributed robot coverage algorithm is verified through simulation. The simulation environment consists of a $10m \times 10m$ area to be covered, with a group of 10 ground mobile robots deployed, all sharing the same kinematic model, to cover points of higher density in the environment. Three common scenarios are considered in the subsequent sections. In the first scenario, the W-CVT algorithm is applied under a constant density function to cover the task

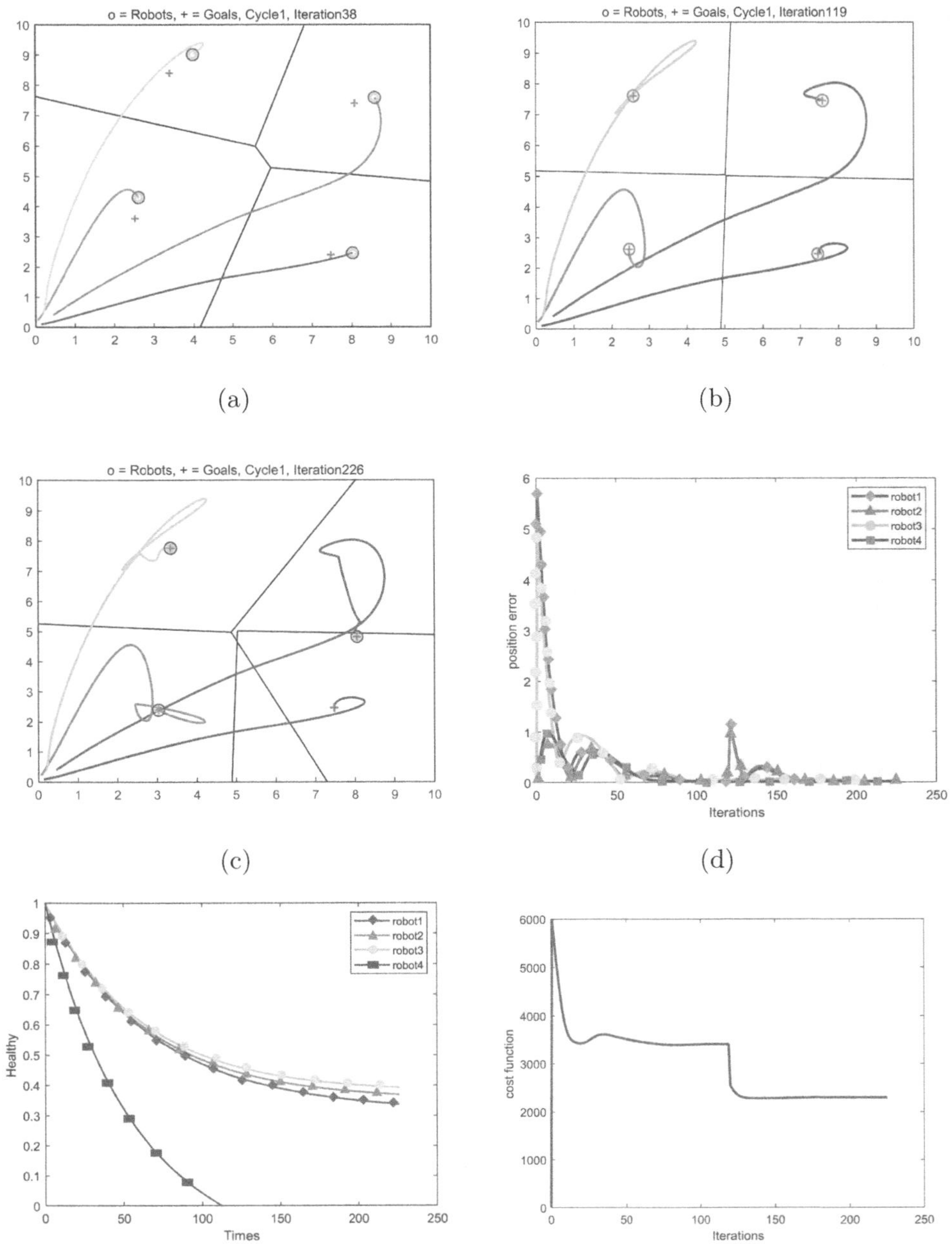


(a) (b)

(c) (d)

(e) (f)

Figure 7.8: Robot formation after the introduction of the insert construction method: (a) $k = 36$; (b) $k = 109$; (c) $k = 280$; (d) position error; (e) $h_i(t)$ of each robot; (f) cost function.

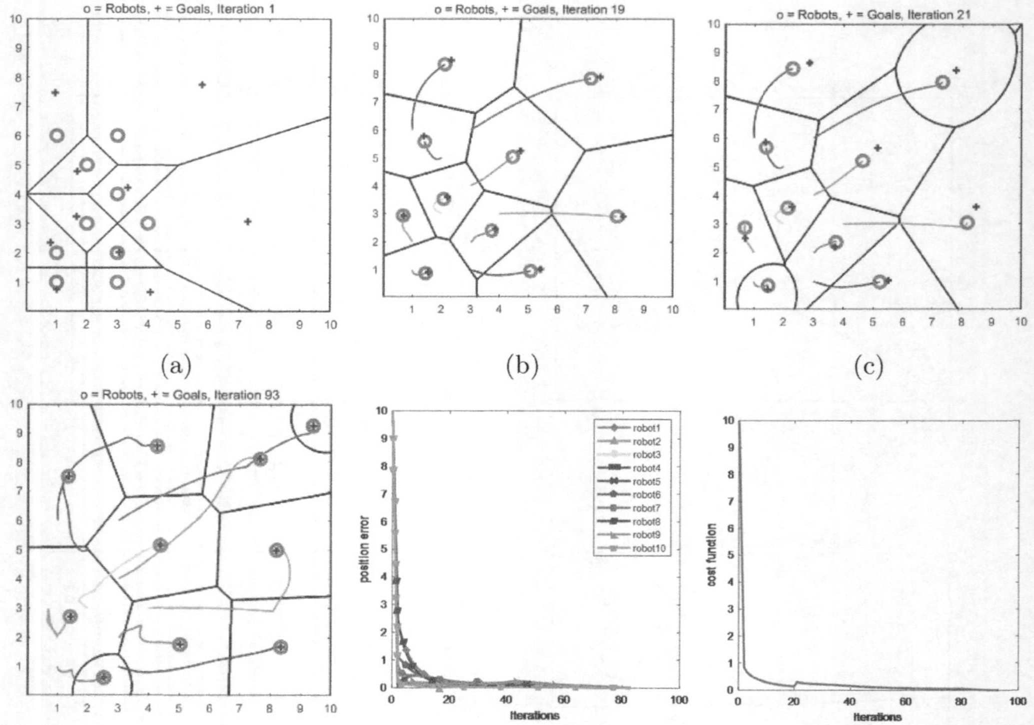


(a) (b) (c)

(d) (e) (f)

Figure 7.9: Multi-robot area coverage based on W-CVT with uniform density function: (a) $k = 1$; (b) $k = 19$; (c) $k = 21$; (d) $k = 93$; (e) position error; (f) cost function.

area, highlighting its ability to partition the area based on the health values of the robots, optimize their positions, and provide an optimal coverage solution even when robot health degrades.

Scenario 1: W-CVT area coverage algorithm under uniform density function

In the first scenario, it is assumed that before $k = 20$, all ground robots have no health degradation, i.e., $h_1 = h_2 = \cdots = h_{10} = 1$. At $k = 20$, two robots experience sudden health degradation, i.e., $d_1 = d_2 = d_3 = d_4 = d_5 = d_9 = d_{10} = 0$, and $d_6 = d_7 = 0.2$. All robots are controlled by the W-CVT algorithm to cover the task area, with the density function $p(q) = 1$. The 10 robots start moving from the lower-left corner of the task area, as shown in Fig. 7.9a. Among the 10 robots, those with $d = 0$ (healthy robots) are indicated by light gray circles, while those with $d = 0.2$ (robots with sudden health degradation) are indicated by dark gray circles. Figs. 7.9a and 7.9b show the movement trajectories of the 10 robots before $k = 20$. It can be seen that all robots are positioned at the edges of their Voronoi diagrams, which are on the perpendicular bisectors between robots, forming a symmetric distribution. After $k = 20$, the movement trajectories of all robots under the control of the W-CVT algorithm are shown in Figs. 7.9c and 7.9d. The final convergence to the optimal configuration is shown in Fig. 7.9d. It can be observed that the light gray robots (those without health degradation) are responsible for larger cells, while the dark gray robots (those with health degradation) are responsible for smaller cells. The cell partitioning is asymmetric, with the cell boundaries biased toward the side of the robots with health degradation. Fig. 7.9e shows the error curve between the current positions and centroid positions of the 10 robots throughout the process. It can be observed that at $k = 20$, the position errors of all robots undergo significant changes. This is because, under the control of the W-CVT algorithm, the cells responsible for each robot incorporate the impact of health values, leading to new cell partitions and thus new centroid positions. Compared to those before $k = 20$, the position errors show noticeable changes. Eventually, at $k = 93$, the position errors of all robots converge to 0, indicating that all robots have reached their desired positions under the coverage control algorithm. Fig. 7.9f displays the coverage cost function throughout the process. It can be seen that around $k = 20$, the cost function experiences some growth. This is due to the algorithm beginning to account for the health-degraded robots, optimizing their positions based on health values during Voronoi tessellation, and re-partitioning the area. After $k = 20$, the robots start converging to the optimal configuration. Finally, after $k = 93$, the cost function also converges to the new optimum.

Fig. 7.10a shows the health value $h_i(t)$ degradation curves of the 10 robots before the introduction of health optimization management in Scenario 1. It can be observed that before $k = 20$, the health values of the 10 robots show minimal degradation. At $k = 20$, the health values of robots 6 and 7 drop sharply under the given conditions. By $k = 110$, all 10 robots show some health degradation due to the coverage process. The health values of the 8 robots without sudden decay average around 0.43. Fig. 7.10b shows the health value $h_i(t)$ degradation curves of the 10 robots after the introduction of health optimization management. It can be seen that at $k = 93$, the health values of all 10 robots are reduced by the health optimization algorithm, resulting in a certain degree of health value reduction (left shift). The health values at the optimal configuration are improved by about 16%. From Fig. 7.10, it can be observed that robots with the same initial health values show differences in their final health values after reaching the optimal configuration. This is due to variations in the step size of each robot during each iteration and the number of iterations required to reach the final stable state, i.e., differences in movement time and control gains among the robots. These factors lead to differences in the final health values.

From the simulation, it can be observed that the W-CVT algorithm, as a spatial coverage solution for robots with health degradation, can partition the Voronoi tessellation based on the health values of the robots. Specifically, robots with greater health degradation are assigned smaller Voronoi tessellation, providing an optimal configuration solution.

Scenario 2: W-CVT area coverage algorithm under non-uniform density function

In the second scenario, it is assumed that before $k = 20$, all ground robots have no health degradation, i.e., $h_1 = h_2 = \cdots = h_{10} = 1$. At $k = 20$, four robots experience sudden health degradation, i.e., $d_1 = d_2 = d_3 = d_4 = d_5 = d_{10} = 0$, and $d_6 = d_7 = d_8 = d_9 = 0.2$. This simulates sudden failures encountered by robots during normal operation, leading to a sudden deterioration in their health conditions.

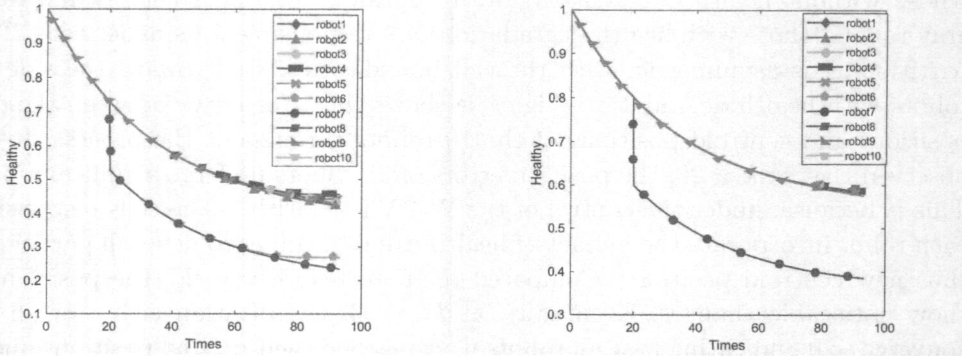


(a) (b)

Figure 7.10: Diagram of robot health loss in Scenario 1: (a) The $h_i(t)$ of the robot in Normal Conditions; (b) the $h_i(t)$ of the robot in Health Optimization Management.

All robots are controlled by the W-CVT algorithm to cover the task area. Fig. 7.11a shows the density function of the task area, which is given by (13), where $(x_c, y_c) = (7, 7)$. The deeper the black color, the higher the priority of the area. The 10 robots start moving from random positions within the task area, as shown in Fig. 7.11b. Among the 10 robots, those with $d = 0$ and $d = 0.2$ are represented by light gray and dark gray circles, respectively. Figs. 7.11b and 7.11c show the movement trajectories of the 10 robots before $k = 20$, where all robots are symmetrically distributed. After $k = 20$, the movement trajectories of all robots are shown in Figs. 7.11c and 7.11d. The final convergence to the optimal configuration is shown in Fig. 7.11d, where the light gray robots are responsible for larger Voronoi tessellation compared to the dark gray robots. From Fig. 7.11e, it can be observed that compared to before $k = 20$, there are significant changes in position errors. This is attributed to the W-CVT algorithm causing new Voronoi tessellation partitions and generating new cell centroids. Eventually, at $k = 110$, the position errors of all robots converge to 0, reaching the desired positions. Fig. 7.11f shows that around $k = 20$, the cost function increases due to considering robot health degradation. After $k = 20$, the robots start converging to the optimal configuration. Finally, after $k = 73$, the cost function also converges to the optimum. The process stops iterating and reaches the optimal coverage configuration when both the cost function and position errors approach zero at $k = 110$.

$$\sigma(x, y) = e^{-\delta\left(a(x-x_c)^2+b(y-y_c)^2\right)}. \tag{7.13}$$

Fig. 7.12a shows the health value consumption curve $h_i(t)$ of 10 robots without health optimization management in Scenario 2. It can be observed that before $k = 20$, the health value consumption of the 10 robots is relatively low. At $k = 20$, the health values of robots 6, 7, 8, and 9 drop sharply under the given conditions. By $k = 110$, the health values of all 10 robots have decreased due to the coverage process. The health values of the 6 robots that did not experience sharp decay average around 0.36,

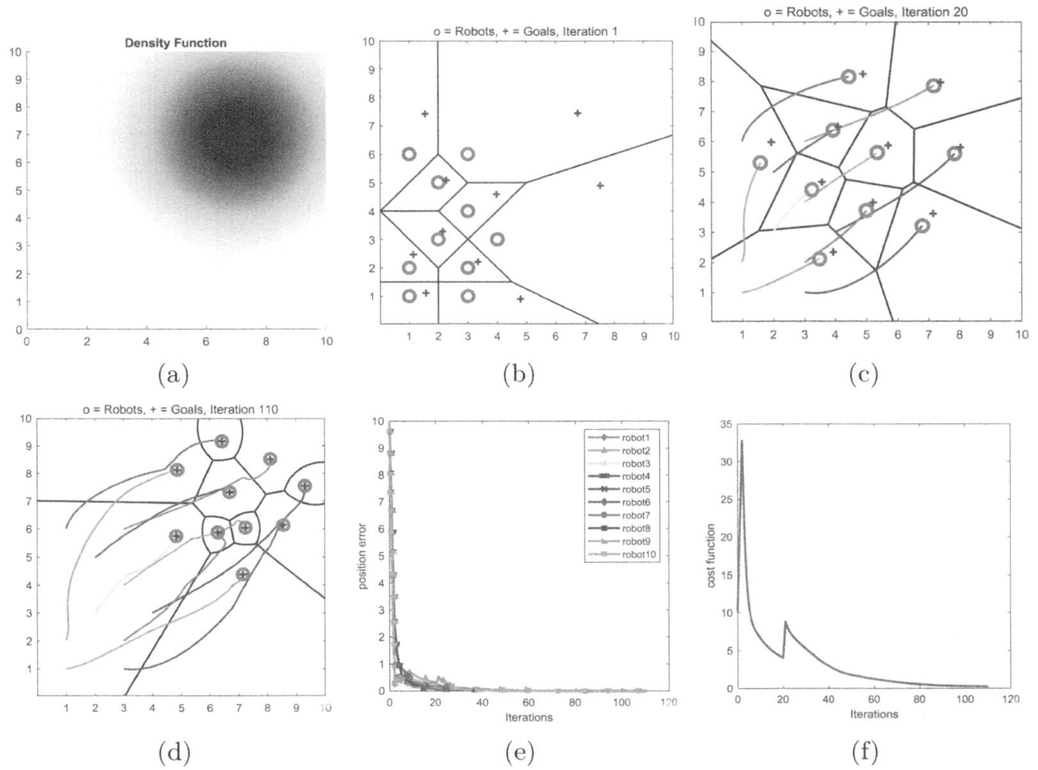


Figure 7.11: Multi-robot area coverage based on W-CVT with under non-uniform density function (Scenario 2): (a) density function; (b) $k = 1$; (c) $k = 20$; (d) $k = 110$; (e) position error; (f) cost function.

while the health values of the 4 robots that experienced sharp decay average around 0.18. Fig. 7.12b shows the health value consumption curve $h_i(t)$ of 10 robots with health optimization management in Scenario 2. It can be observed that at $k = 110$, the health value consumption of the 10 robots is reduced to some extent. The health values of the 6 robots that did not experience sharp decay average around 0.53, while the health values of the 4 robots that experienced sharp decay average around 0.34. This represents an improvement of approximately 16% in health values compared to the optimal configuration.

Scenario 3

In the third scenario, it is assumed that before $k = 20$, all ground robots have no health issue, i.e., $h_1 = h_2 = \cdots = h_{10} = 1$. After $k = 20$, four robots experience sudden health consumption, i.e., $d_1 = d_2 = d_3 = d_4 = d_5 = d_{10} = 0$, and $d_6 = d_7 = d_8 = d_9 = 0.2$. Fig. 7.13a shows the density function of the task area, which is given by (7.13), where $(x_c, y_c) = \{(2,8),(8,2)\}$. The initial positions of the 10 robots are shown in Fig. 7.13b, with colors and health consumption values consistent with the third scenario. Figs. 7.13b to 7.13c illustrate the movement trajectories of the 10 robots before $k = 20$. After $k = 20$, the movement trajectories of all robots are shown

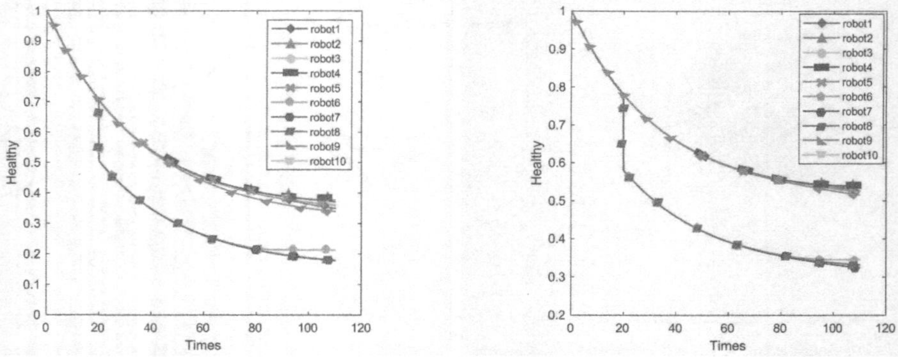


(a) (b)

Figure 7.12: Diagram of robot health loss in Scenario 2: (a) The $h_i(t)$ of the robot in normal conditions; (b) the $h_i(t)$ of the robot in Health Optimization Management.

in Figs. 7.13c to 7.13d, eventually converging to the optimal configuration depicted in Fig. 7.13d. From Fig. 7.13e, it can be observed that at $k = 20$, the position errors of all robots change significantly compared to before $k = 20$. Eventually, at $k = 64$, the position errors of all robots tend to 0, indicating that all robots have reached their desired positions under the control of the coverage algorithm. Fig. 7.13f shows that around $k = 20$, the cost function increases slightly due to the introduction of health values. After $k = 20$, the robots reconverge to the optimal configuration, and the cost function finally converges to the optimum after $k = 64$. Differences in control gains and stabilization iterations among robots lead to variations in the final health values of robots with the same initial health values.

Fig. 7.14a shows the health value depletion curves of 10 robots under Scenario 3 without health optimization management. Before $k = 20$, the health values of the 10 robots exhibit relatively low reduction. After $k = 20$, the health values of robots 6, 7, 8, and 9 abruptly decrease under the given conditions. By $k = 64$, the health values of the 6 robots without abrupt decay and the 4 robots with abrupt decay decrease to approximately 0.5 and 0.35, respectively. Fig. 7.14b shows the health value depletion curves of 10 robots under Scenario 3 with health optimization management. At $k = 64$, the health value depletion of the 10 robots is significantly reduced. The health values of the 6 robots without abrupt decay and the 4 robots with abrupt decay decrease to approximately 0.71 and 0.52, respectively. The health values have improved by about 19% compared to the optimal configuration.

In the three cases above, the distributed coverage control method based on weighted Voronoi diagrams effectively addresses the distributed coverage configuration of robots with health degradation. This method adjusts the size of Voronoi tessellation according to the health status of the robots. By incorporating the health status of robots into the position optimization problem of coverage control and minimizing the coverage cost function, the method ensures efficient coverage. The addition of health optimization significantly increases the lifespan of robots with the lowest health values while requiring minimal adjustments to the existing framework. Health

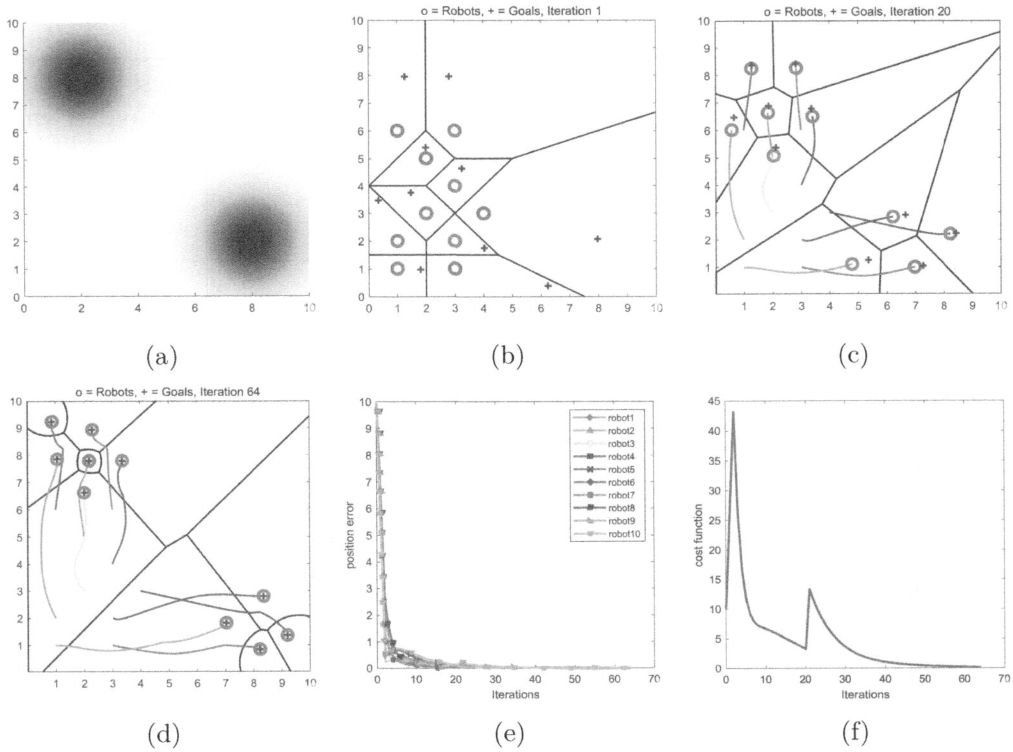


Figure 7.13: Multi-robot area coverage based on W-CVT with under non-uniform density function (Scenario 3): (a) Density function; (b) $k = 1$; (c) $k = 20$; (d) $k = 64$; (e) position error; (f) cost function.

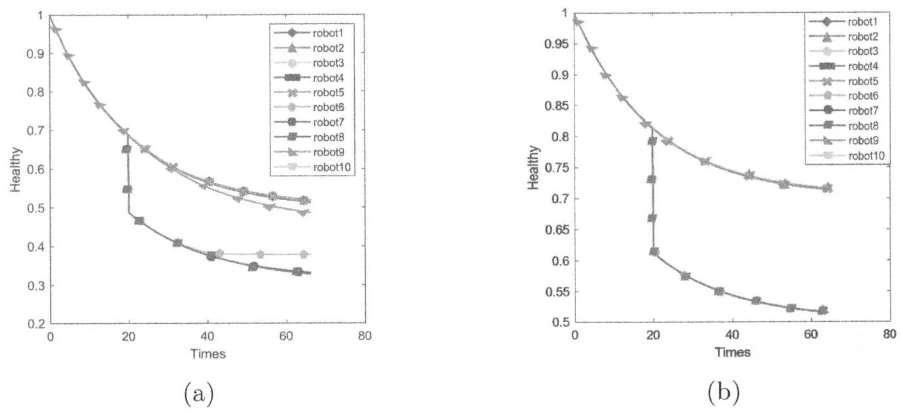


Figure 7.14: Diagram of robot health loss in Scenario 3: (a) The $h_i(t)$ of the robot in normal conditions; (b) the $h_i(t)$ of the robot in Health Optimization Management.

Table 7.2: Comparison of simulation results in three scenarios before and after adding the health optimization management algorithm

| | $h_i(t)$ (normal situation) | $h_i(t)$ (HOM) | Percentage increase in system health value |
|---|---|---|---|
| Scenario 1 | 0.43/0.23 | 0.59/0.39 | 16% |
| Scenario 2 | 0.36/0.18 | 0.53/0.34 | 16% |
| Scenario 3 | 0.50/0.35 | 0.71/0.52 | 19% |

optimization management ensures long-term coverage operation and prevents excessive wear and tear over extended periods. As shown in Table 7.2, applying health balancing to each robot improves the overall health value by approximately 17%. This makes multi-robot systems more robust against health variations caused by numerous unreasonable conditions. The system can better handle unexpected damages caused by collisions, mechanical failures, or insufficient maintenance.

## 7.4 Chapter Summary

This chapter introduces multi-robot formation control algorithms and coverage control algorithms based on health optimization management for multi-robot systems. By quantifying the health conditions of robots and constructing a health optimization manager, these methods enhance the endurance of unhealthy robots. This allows the system to better handle mechanical failures and unforeseen damages, thereby improving the reliability of robots in executing tasks across different scenarios. In coverage control, a distributed coverage control method based on weighted Voronoi diagrams is introduced. This method incorporates the health degradation of robots into the position optimization problem, providing optimal configurations for robots while minimizing the coverage cost function. In formation control, the endurance of unhealthy robots is maximized without altering the existing framework. This improves the stability and robustness of unhealthy robots, enabling them to better respond to environmental changes.

The methods presented in this chapter can be applied to robot systems of different dimensions and heterogeneous groups with varying performance and motion models. Future work could be on making more precise characterization of robots with different health conditions, further enhancing their performance and the control performance of the group.

## Bibliography

[1] Brett Bethke, Jonathan P How, and John Vian. Group health management of UAV teams with applications to persistent surveillance. In *2008 American Control Conference*, pages 3145–3150. IEEE, 2008.

[2] Nikhil Nigam. Dynamic replanning for multi-UAV persistent surveillance. In *AIAA Guidance, Navigation, and Control (GNC) Conference*, page 4887, 2013.

[3] Farid Sharifi, Youmin Zhang, and Amir G Aghdam. Coverage control in multi-vehicle systems subject to health degradation. In *2013 International Conference on Unmanned Aircraft Systems (ICUAS)*, pages 1119–1124. IEEE, 2013.

[4] Alyssa Pierson, Wilko Schwarting, Sertac Karaman, and Daniela Rus. Weighted buffered Voronoi cells for distributed semi-cooperative behavior. In *2020 IEEE International Conference on Robotics and Automation (ICRA)*, pages 5611–5617. IEEE, 2020.

[5] Mert Turanli and Hakan Temeltas. Multi-robot energy-efficient coverage control with Hopfield networks. *Stud. Inform. Control*, 29(2):179–188, 2020.

[6] Yangquan Chen, Zhongmin Wang, and Jinsong Liang. Automatic dynamic flocking in mobile actuator sensor networks by central Voronoi tessellations. In *IEEE International Conference Mechatronics and Automation, 2005*, volume 3, pages 1630–1635. IEEE, 2005.

[7] Subhrajit Bhattacharya, Robert Ghrist, and Vijay Kumar. Multi-robot coverage and exploration on Riemannian manifolds with boundaries. *The International Journal of Robotics Research*, 33(1):113–137, 2014.

[8] Beth Boardman, Troy Harden, and Sonia Martinez. Limited range spatial load balancing in non-convex environments using sampling-based motion planners. *Autonomous Robots*, 42(8):1731–1748, 2018.

[9] Enrique Teruel, Rosario Aragues, and Gonzalo López-Nicolás. A distributed robot swarm control for dynamic region coverage. *Robotics and Autonomous Systems*, 119:51–63, 2019.

[10] Brett Bethke, Jonathan P How, and John Vian. Group health management of UAV teams with applications to persistent surveillance. In *2008 American Control Conference*, pages 3145–3150. IEEE, 2008.

[11] YangQuan Chen, Zhongmin Wang, and Jinsong Liang. Optimal dynamic actuator location in distributed feedback control of a diffusion process. *International Journal of Sensor Networks*, 2(3-4):169–178, 2007.

[12] Juan Chen, Bo Zhuang, YangQuan Chen, and Baotong Cui. Diffusion control for a tempered anomalous diffusion system using fractional-order PI controllers. *ISA Transactions*, 82:94–106, 2018.

[13] Johan Hatleskog, Sorin Olaru, and Morten Hovd. Voronoi-based deployment of multi-agent systems. In *2018 IEEE Conference on Decision and Control (CDC)*, pages 5403–5408. IEEE, 2018.

[14] Vishnu G Nair and KR Guruprasad. GM-VPC: An algorithm for multi-robot coverage of known spaces using generalized Voronoi partition. *Robotica*, 38(5):845–860, 2020.

[15] Jorge Cortés. Coverage optimization and spatial load balancing by robotic sensor networks. *IEEE Transactions on Automatic Control*, 55(3):749–754, 2010.

[16] Justin M Bradley and Ella M Atkins. Optimization and control of cyber-physical vehicle systems. *Sensors*, 15(9):23020–23049, 2015.

[17] Alyssa Pierson, Wilko Schwarting, Sertac Karaman, and Daniela Rus. Weighted buffered Voronoi cells for distributed semi-cooperative behavior. In *2020 IEEE International Conference on Robotics and Automation (ICRA)*, pages 5611–5617. IEEE, 2020.

CHAPTER 8

# Multi-Robot Capture Based on Voronoi Tessellations

## 8.1 Introduction

THE capture task refers to the collaborative effort of multiple entities, utilizing their individual capabilities to constrain a target within a specific region or area [1]. This task is widely applied in fields such as military, security, and autonomous driving. In multi-robot systems, capture tasks are particularly significant because multiple robots can cooperate to encircle and capture the target, thus improving the efficiency and flexibility of task execution. Based on the availability of target information, capture problems can be divided into two main scenarios: the known target scenario and the unknown target one.

In the case of a known target, robots can obtain the precise location and state of the target in advance, making the capture task relatively simple to perform. However, in the case of an unknown target, robots must rely on local perception, collaborative search, and other strategies to locate and capture the target, which greatly increases the complexity of the task, involving dynamic sensing and multi-robot collaboration [2]. In this context, Voronoi partitioning, as an effective spatial division method, has been widely applied in multi-robot capture tasks [3]. It helps improve capture efficiency and optimize robot collaboration by appropriately dividing the regions among robots. This chapter focuses on the multi-robot capture problem in both known and unknown target scenarios, particularly exploring how to use the Voronoi partitioning method to enhance capture performance and address challenges in practical applications. The unique duality of Voronoi tessellations makes them particularly suitable for this dual-scenario problem: their geometric properties enable structured encirclement when the target is known, while their spatial decomposition capabilities provide a natural framework for efficient, distributed area search when the target is unknown.

DOI: 10.1201/9781003714569-8

### 8.1.1 Multi-Robot Capture Strategy with Known Target Location

For target capture with known target position, a robot system can typically utilize the target's known position, velocity, trajectory, and other information to plan and capture efficiently. The challenge in such tasks lies in how to design an effective capture strategy that ensures that the robots can surround the target in the shortest time and prevent the target from escaping. To achieve this, the robots need to make dynamic adjustments during the capture process, taking into account the potential movement patterns of the target, and adjust the formation and strategy accordingly Fundamentally, this challenge can be broken down into two coupled sub-problems: the optimal assignment of capture positions around the target and the coordinated, collision-free motion of robots toward these positions.

There have been extensive methods for solving the problem of capturing a target with a known location including reinforcement learning-based strategies [4, 5], game theory approaches [6–8], planar reach-avoid games [9, 10], consensus-based methods [11], leader-follower strategies [12], and the unique approach of parallel optimization based on time minimization [13]. However, these methods lack a unified strategy for both encirclement and formation, leading to increased computational complexity and difficulties in achieving the capture task. Inspired by the Voronoi tessellations characteristics discussed earlier, this chapter integrates Voronoi diagrams with multi-robot target capture tasks. Voronoi diagrams -based formation control offers flexibility and strong robustness, while Voronoi cells naturally create collision-free safe zones for multiple agents [14–16], providing inherent collision avoidance capabilities [17, 18]. This often results in phase-based strategies, where formation control and target approach are handled by separate, switched controllers, which can lack smoothness and robustness against dynamic targets.

For the case of multi-targets, Wang et al. [19] proposed an adaptive multi-target Voronoi capture algorithm. They utilized a minimum distance criterion to enable the multi-drone formation to lock onto specific targets, repeating the process until all targets are captured. Alyssa et al. [20] investigated the minimum area capture in a multi-target scenario, and imposed constraints on the drone velocities with simulations in both 2D and 3D environments. However, they did not provide a clear explanation of the formation capture strategy in 3D. Although the minimum area method noted above can achieve target capture, the research on this capture strategy remains limited to the 2D plane. Moreover, the enclosing circle is not complete, allowing the targets to escape in bounded 3D environments. According to the variation of the capture angles, we found that only considering the distance factor, without constraining the capture angles, the drones will experience noticeable vibrations in order to adjust their positions during practical operations. Therefore, it is necessary to investigate the omni-directional capture strategy based on 3D-Voronoi tessellations in complex environments. Moreover, a simple 2D enclosing formation is insufficient in 3D space, as it fails to constrain the target's movement along the vertical axis, creating a clear escape route.

Note that the aforementioned methods mostly assume that the locations of the targets are known, allowing the robots to execute the capture task directly. However, in practical applications, target locations are often unknown, requiring the robots to first perceive the presence and distribution of the targets before effectively carrying out the capture task execution.

### 8.1.2 Multi-Robot Capture Strategy with Unknown Target Location

When the target location is unknown, robots must first obtain comprehensive information about the number and the distribution of targets before effectively executing the encirclement task. In such cases, perception and information gathering become critical preliminary steps, forming the foundation for successful encirclement. Therefore, before encirclement, it is crucial to study how to perceive targets in an unknown environment. Currently, common target perception methods for encirclement include probabilistic inference methods [21,22], graph theory-based approaches [23,24], coverage search methods [25,26], and learning-based methods [27,28], each with its unique advantages, limitations, and application scenarios.

Probabilistic inference methods model the target's state using predictive techniques such as Bayesian inference to estimate its localization and trajectory. These methods offer advantages in highly uncertain environments. Zhang et al. [29] proposed a target localization approach using a set of mobile sensors that rely solely on local distance-based measurements. By maximizing the Fisher information about the target in sensor measurements, their method improved target localization estimation and enhanced the robustness of the localization model. Ma et al. [30] introduced a Gaussian maximum entropy-based target search algorithm, which reduces the computational complexity of target localization by approximating the function maximum with a normal distribution and computing mutual information. This approach enables target localization with limited noisy observations. However, probabilistic inference methods tend to have high computational complexity, and as the number of robots increases, the computational burden grows significantly, limiting their applicability in large-scale systems.

Graph theory-based methods model the environment as a graph structure and use graph search algorithms for target searching. Duberg et al. [31] proposed a map-update-based target search method, which updates and extends the graph-based dense planning structure with each map update, ensuring that robots always move toward the optimal target. Peng et al. [32] developed a target localization method based on graph neural networks, which aggregates information from neighboring robots in a multi-robot system to achieve global optimization in heterogeneous robot systems. This method demonstrates strong robustness and adaptability, effectively handling dynamic variations in the number of robots and tasks.

Coverage search methods primarily focus on efficiently covering the entire target area to ensure thorough scanning and searching. These methods are highly efficient and easy to implement. Zhao et al. [33] proposed an event-triggered mechanism for full-coverage target search, task allocation, and encirclement of dynamic targets. Their approach utilizes the Prometheus project as the fundamental control framework for multi-UAV systems and employs a Gazebo simulation platform with four camera-equipped drones to validate the feasibility of the proposed "event-trigger-event-processing" framework. Wang et al. [34] introduced a cooperative search and pursuit strategy based on a "Z-shaped" path search pattern, which distributes search resources evenly within the area to enhance search efficiency. They also incorporated an artificial potential field method to ensure collision avoidance while capturing targets. However, in practical applications, coverage search methods often require considering algorithm constraints specific to the application scenario and may involve high computational costs. In recent years, learning-based methods, particularly reinforcement learning and deep learning, have gained significant attention for solving multi-robot target perception and encirclement problems. Reinforcement learning enables robots to continuously interact with the environment and optimize strategies, making it well-suited for scenarios with uncertain target localizations and dynamic environment changes. Alagha et al. [35] proposed two multi-agent deep reinforcement learning models for target localization. The first model achieves fast and cost-effective localization using cooperative agents and a specific reward function, while the second model improves computational efficiency through sparse rewards and expert demonstrations. Experimental results demonstrated their effectiveness in terms of localization time, cost, and learning stability. Building on top of this, the same research team [36] further enhanced their approach using transfer learning, integrating deep learning models with deep reinforcement learning knowledge. This allowed for precise localization estimation when the target was unreachable, resulting in faster learning speeds and lower computational complexity. Despite the promising capabilities of learning-based methods in handling uncertainty and dynamic environments, several challenges remain. First, these methods require extensive environment interactions during training, leading to prolonged training times and high computational resource. Second, the convergence and stability of learning algorithms remain critical technical issues that need to be addressed.

This review highlights a clear dichotomy in multi-robot capture research: strategies for known targets often focus on geometric encirclement, while strategies for unknown targets are dominated by search and perception. A significant gap exists in developing a unified framework that can seamlessly integrate both phases. This chapter aims to bridge this gap by leveraging the dual nature of Voronoi tessellations. We will demonstrate how this single geometric tool can be adapted to formulate an omni-directional capture strategy (ODMV) for known targets, and a perception-driven search and capture strategy for unknown targets, thereby providing a cohesive and efficient solution for the complete multi-robot capture problem.

## 8.2 Multi-Robot Capture Strategy with Known Target location

### 8.2.1 Omni-Directional Minimum Area (ODMA) Capture Strategy

The principle of the minimum area strategy in a 2D plane is to divide a new Voronoi cell for the target based on the Voronoi tessellations and obtain the direction for each robot to minimize the area of the Voronoi cell where the target is located. In a 2D

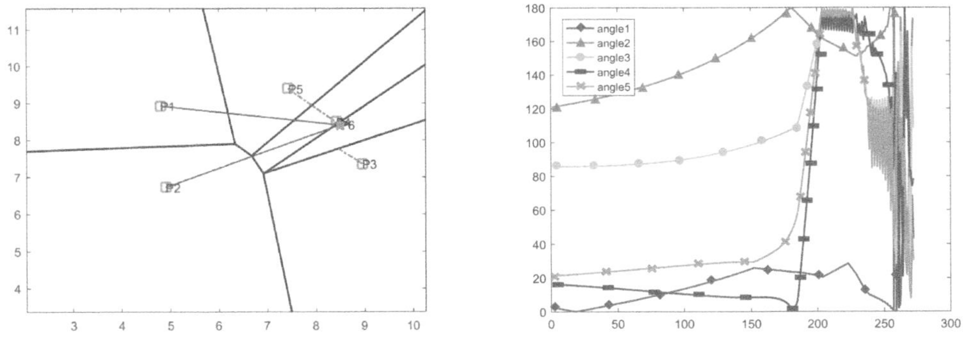


(a) (b)

Figure 8.1: The capture result of minimum area strategy: (a) The final distribution of drones and target; (b) the variability of the angle between two robots.

environment, if we define the Voronoi cell containing the target to be $V_k$, its area $A_k$ is given by:

$$V_k = \left\{ q \in V_k \mid \|q - x_k\| < \|q - x_{p_i}\|, i = 1, 2, 3, \dots N \right\}, \tag{8.1}$$

$$A_k = \int_{V_K} dq. \tag{8.2}$$

The direction of motion for the capture robot that results in the fastest area reduction is along the negative gradient. This direction can be expressed as:

$$u_p = -\frac{\frac{\partial A_k}{\partial x_p}}{\left\| \frac{\partial A_k}{\partial x_p} \right\|} = \frac{C_i - x_p}{\|x_p - C_i\|'}, \tag{8.3}$$

where $C_i$ is the centroid of the shared boundary $l$, and $x_p$ is the position of the capture robot. However, this pure minimum area strategy does not restrict the capture angle of drones. As shown in Fig. 8.1, the green symbols P1-P5 represent the capture drones, and P6 is the target. The red dashed lines is the distance from the adjacent capture to the edge of the target Voronoi cell, and the blue lines is the distance from the non-adjacent capture robot to the target. The angles between any two robots and P6 are denoted as angle 1 to angle 5. Fig. 8.1a is the final distribution of capture robots under the minimum area strategy. It can be observed that due to the convergence distance being satisfied between P4 and P6, the overall formation completed the capture task early. Fig. 8.1b shows the changes of angle 1–angle 5. We can see that after the 200th iteration, significant oscillations of the angles can be observed. This is caused by the absence of a constraint on the motion direction of robots. It not only fails to achieve effective capture but also affects the stable motion of robots.

In response to this situation, this chapter improves on the basis of minimum area by restricting the change rate of the Voronoi cell boundary at the target. When the change rate of the final boundary is equal, it can present a better capturing effect. This is the principle of the **omni-directional minimum area (ODMA)** strategy, with its control law defined as:

$$u_p = \frac{C_i - x_p}{\|x_p - C_i\|} \cdot \frac{\partial L}{\partial x_p}. \tag{8.4}$$

Here, the term $\frac{\partial L}{\partial x_p}$ acts as a scalar modulation factor related to the rate of change of the boundary or angular separation, serving to adjust the robot's speed to enforce a uniform, omni-directional encirclement.

The effectiveness of this improved strategy is shown in Fig. 8.2. Fig 8.2a shows the final distribution of capture robots under omni-directional minimum area strategy. It can be seen that the Voronoi cell formed by each robot and target is approximately pentagonal. From the angle changes in Fig. 8.2b, the angles tend to be equal, specifically converging to 72 degrees. This indicates that the ODMA is effective in achieving a successful capture of target.

### 8.2.2 Variable Step Wolf Pack Algorithm (WPA)

After determining the capture direction of ODMA, we introduce the variable step WPA as the motion method of drones. The details of WPA can be found in [37], and the fixed step of traditional WPA is improved to a variable-step version:

$$\text{step} = \frac{d_{\max}}{d_{\min}} \left| x_p - x_k \right|, \tag{8.5}$$

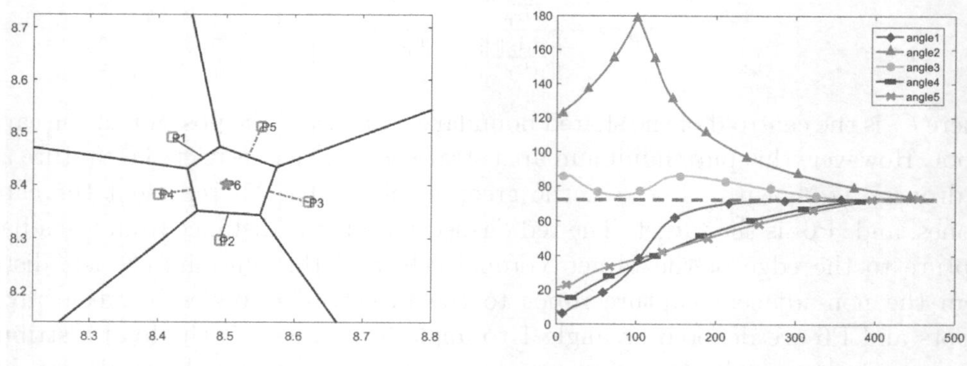


(a) (b)

Figure 8.2: The capture result of omni-directional minimum area strategy: (a) The final distribution of drones and target; (b) the variability of the angle between two robots.

where $|x_p - x_k|$ is the distance between drones and target, and $d_{max}$ and $d_{min}$ represent the maximum and minimum distance between drones and target in the capture process, respectively. The ratio of these distances is used to prevent the drones from having excessively large or small steps. As the drones approach the target, the ratio will gradually approach 1, enabling a smooth transition of step.

### 8.2.3 Minimum Volume Capture Strategy

In the case where the target position is known, Huang et al. [20] studied the minimum area strategy in a 2D plane. Its principle is to divide a new Voronoi cell for the target based on the Voronoi tessellations and obtain the direction for each robot to minimize the area of the Voronoi cell where the target is located. Under the control of the corresponding strategy, the area of the target Voronoi cell gradually decreases until the target is captured. Based on the research in [20], this section applies the minimum area capture strategy to the multi-drone formation capture in 3D space. Due to the difference between 3D and 2D spaces, the minimum area capture strategy becomes the minimum volume capture strategy in 3D space. The Voronoi cell where the target is located can be represented as:

$$V_k^3 = \left\{ q \in \Omega^3 \mid \sqrt{\sum_{N=x,y,z} (q_i - x_k)^2 \leq \sum_{N=x,y,z} (q_j - x_p)^2},\ i, j = 1, 2, \ldots, N, \forall j \neq i \right\}. \tag{8.6}$$

The volume of $V_k$ is:

$$V_{V_k} = \int_{V_k} dq, \quad q \in V_k. \tag{8.7}$$

Taking its derivative and combining (8.1), the change rate of its volume can be expressed as:

$$\dot{V}_{V_k} = \sum_{j \in N_k} \int_{A_j} \left[ \frac{(x_p - q)\, \dot{x}_p}{\|x_p - x_k\|} - \frac{(x_k - q)\, \dot{x}_k}{\|x_p - x_k\|} \right] dq, \tag{8.8}$$

where $N_k$ is the surface number of $V_k$, $A_j$ is the $j$th surface of $V_k$, and $x_k$ and $x_p$ are the coordinates of the target and the capture drones, with common factors:

$$\dot{V}_{V_k} = \sum_{j=1} \int_{A_j} dq \left[ \frac{\left(x_p - \int_{A_j} q dq / \int_{A_j} dq\right)}{\|x_p - x_k\|} \right] - \sum_{j=1} \int_{A_j} dq \left[ \frac{\left(x_k - \int_{A_j} q dq / \int_{A_j} dq\right)}{\|x_p - x_k\|} \right]. \tag{8.9}$$

According to the physical meaning of integration, the common factor in (8.9) can be expressed as:

$$A_k^j = \int_{A_j} dq, \quad C_k^j = \frac{\int_{A_j} q\, dq}{\int_{A_j} dq}, \tag{8.10}$$

where $A_k^j$ is the area of the $j$th surface, and $C_k^j$ is the centroid of $A_j$. Then (8.9) can be simplified as:

$$\dot{V}_{V_k} = \sum_{j=1} A_k^j \left[ \frac{(x_p - C_k)}{\|x_p - x_k\|} \right] \dot{x}_p - \sum_{j=1} A_k^j \left[ \frac{(x_k - C_k)}{\|x_p - x_k\|} \dot{x}_k \right]. \tag{8.11}$$

During the approach of the target by multiple robots, the changes in the 3D-Voronoi diagram of the target are directly related to the drones near the target. Therefore, the rate of change of the Voronoi cells can also be expressed as:

$$\dot{V}_{V_k} = \frac{\partial V_k}{\partial x_k} \dot{x}_k + \sum_{j=1}^{N_n} \frac{\partial V_k}{\partial x_{p_j}} \dot{x}_{p_j} + \sum_{i=1}^{N-N_n} \frac{\partial V_k}{\partial x_{p_i}} \dot{x}_{p_i}, \tag{8.12}$$

where $N_n$ is the number of the 3D-Voronoi cells adjacent to $x_k$. Combined with (8.11), it can be obtained:

$$\sum_{j=1}^{N_n} \frac{\partial V_k}{\partial x_{p_j}} \dot{x}_{p_j} + \sum_{i=1}^{N-N_n} \frac{\partial V_k}{\partial x_{p_i}} \dot{x}_{p_i} = A_k^j \sum_{j=1}^{N_k} \frac{(x_{p_j} - C_k)}{\|x_{p_j} - C_k\|}. \tag{8.13}$$

In the study presented in this chapter, the movement of non-adjacent robots is not directly related to the changes in the target 3D-Voronoi cell, so (8.13) can be written as:

$$\sum_{j=1}^{N_n} \frac{\partial V_k}{\partial x_{p_j}} \dot{x}_{p_j} = A_k^j \sum_{j=1}^{N_k} \frac{(x_{p_j} - C_k)}{\|x_{p_j} - C_k\|}. \tag{8.14}$$

Using the negative gradient direction with the fastest volume reduction, the motion direction of the adjacent robot $x_p$ can be obtained:

$$u_p = -\frac{\partial V_k / \partial x_p}{\|\partial V_k / \partial x_p\|} = \frac{C_k^j - x_p}{\|x_p - C_k^j\|}, \tag{8.15}$$

where $C_k^j$ is the centroid of the surface adjacent to the target. After combining the minimum volume capture strategy with 3D-Voronoi tessellations, the motion direction of adjacent robots can be obtained.

However, because the motion of adjacent robots are not limited, the minimum volume strategy cannot guarantee the capture of the target, which means that in some dynamic scenarios, it will increase the space for the escape of dynamic target. Therefore, improvement of the minimum volume method is needed.

### 8.2.4 Omni-Directional Minimum Volume (ODMV)

As established in the previous section, the unconstrained minimum volume strategy can lead to non-uniform encirclements. To address this, we introduce the **Omni-Directional Minimum Volume (ODMV)** strategy, which augments the minimum volume principle with an explicit constraint to enforce a stable, symmetric capture formation.

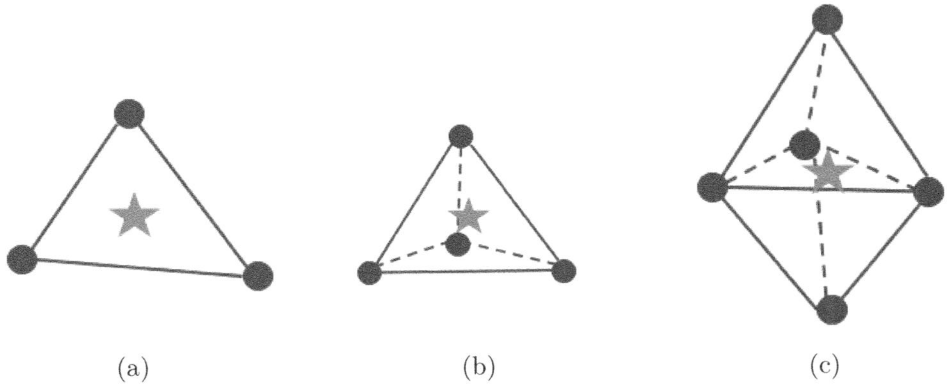

Figure 8.3: Visual representation of omni-directional capture of target under a differing number of robots: (a) $n = 3$; (b) $n = 4$; (c) $n = 5$.

The ODMV strategy refers to constraint of the motion directions of the robots during the capture process. This allows the drone formation to achieve an omni-directional capture of the target, resembling a regular polygon. The visual representation of the capture is depicted in Fig. 8.3. The green stars is the known position of the target.

**Theorem 1.** *To capture the target in all directions by the minimum volume strategy, the same change rate on each surface of the target 3D-Voronoi cell must be maintained.*

**Proof.** According to the nature of regular polygon, each surface is equal:

$$A_1 = A_2. \tag{8.16}$$

Derive motion rate from Equation (7.16):

$$\frac{\partial A_1}{\partial x_{p_1}} = \frac{\partial A_2}{\partial x_{p_2}}, \tag{8.17}$$

which means the motion rate of each surface is the same. By the volume calculation of the Voronoi cell, it can be obtained that under the minimum volume strategy, the volume change of the target 3D-Voronoi cell is:

$$\dot{V}_{V_k} = \frac{A}{x_p - x_k} \left[ -\left(x_p - C_i\right) - \left(x_k - C_i\right) \dot{x}_k \right]. \tag{8.18}$$

Because the volume change of the target 3D-Voronoi cell is directly related to adjacent robots, (8.18) can be changed to:

$$\dot{V}_V \Big|_{p_{i,i=1,2,\dots N}} = \frac{A}{x_p - x_k} \left[ -\left(x_p - C_i\right) - \left(x_k - C_i\right) \dot{x}_k \right]. \tag{8.19}$$

The second derivative of (8.19) is:

$$\ddot{V}_V\Big|_{p_{i,i=1,2,\ldots,N}} = \frac{A}{x_p - x_k}\left[-\dot{x}_{p_i} - \dot{x}_k^2 - (x_k - C_i)\,\ddot{x}_k\right], \tag{8.20}$$

when $i$ takes different values, the corresponding results will not be the same. Thus, if you want to have the same rate of change of the volume in each direction, it is necessary to ensure that:

$$\dot{x}_{p_1} = \dot{x}_{p_2} \Rightarrow \frac{\partial A_{p_1}}{\partial x_{p_1}} = \frac{\partial A_{p_2}}{\partial x_{p_2}}. \tag{8.21}$$

The motion speed of the robot adjacent to the target tends to be equal, consistent with the conclusion drawn by (8.4). Therefore, the motion rate of adjacent robots becomes:

$$u_p = -\frac{\partial A}{\partial x_p} \cdot \frac{\partial V_k}{\partial x_p} \Big/ \left\|\frac{\partial V_k}{\partial x_p}\right\| = \frac{\partial A}{\partial x_p} \cdot \frac{C_i - x_p}{\|x_p - C_i\|}. \tag{8.22}$$

This is the final control law for the ODMV strategy. It elegantly combines two components: the unit vector $\frac{C_i - x_p}{\|x_p - C_i\|}$ provides the base direction of motion (toward the shared face's centroid, to minimize volume), while the scalar term $\frac{\partial A}{\partial x_p}$ acts as a dynamic gain that modulates the robot's speed to enforce the omni-directional constraint from Theorem 1.

**Theorem 2.** *To ensure that the robots achieve uniform capture in terms of angles, it is necessary to maintain equal distance between the robots and the target.*

**Proof.** Let the angle between any two adjacent robots $p_1$ and $p_2$ be denoted as $\theta$. Then the angle can be expressed using the cosine rule as follows:

$$\theta = \arccos \frac{L_1^2 + L_2^2 - L_3^2}{2L_1L_2}, \tag{8.23}$$

where $L_1$ and $L_2$ represent the distance between $p_1$, $p_2$ and the target, and L3 is the distance between two adjacent drones. In achieving the omni-directional capture of the target, the distance between each drone and the target is the same, denoted as $L_1 = L_2$; the equation can be simplified as follows:

$$\theta = \arccos \frac{2L_1^2 - L_3^2}{2L_1L_2}. \tag{8.24}$$

The derivative of the angle $\theta$ with respect to time is given by:

$$\frac{d\theta}{dt} = \frac{-1}{\sqrt{1 - X^2}}\frac{\partial X}{\partial t}, \tag{8.25}$$

where $X$ can be denoted by:

$$X = \frac{2L_1^2 - L_3^2}{2L_1}, \tag{8.26}$$

$$\frac{\partial X}{\partial t} = \frac{L_1 L_3 \frac{\partial L_3}{\partial t} - L_3^2 \frac{\partial L_1}{\partial t}}{L_3^3}. \tag{8.27}$$

Based on (8.27), it can be seen that when the distance between adjacent robots and the target approach $L_1 \to L_3$, the angle variation between any two adjacent robots tends to be the same, which means that $\frac{\partial X}{\partial t} = 0$.

Therefore, the drones are capable of achieving the omni-directional capture of the known target, with the angle between any two drones and the target tending to be equal.

### 8.2.5 Convergence Proof of Minimum Volume Strategy

To prove that the target 3D-Voronoi cell no longer increases under ODMV, (7.10) is substituted into (8.18), as follows:

$$\dot{V}_{V_k} = \frac{A}{x_p - x_k} \left[ (x_p - C_i)\, \dot{x}_p - (x_k - C_i)\, \dot{x}_k \right]. \tag{8.28}$$

According to (8.22), it can be simplified as follows:

$$\dot{V}_{V_k} = \frac{A}{x_p - x_k} \left[ -\frac{\partial A}{\partial x_p} (x_p - C_i) - (x_k - C_i)\, \dot{x}_k \right]. \tag{8.29}$$

Setting $\dot{V}_{V_k} = 0$, one gets

$$-\frac{\partial A}{\partial x_p} (x_p - C_i) - (x_k - C_i)\, \dot{x}_k = 0, \tag{8.30}$$

$$\dot{x}_k = \frac{(C_i - x_p)}{C_i - x_k} \cdot \frac{\partial A}{\partial x_p}. \tag{8.31}$$

We can see that $\dot{V}_{V_k}$ is always no greater than 0, and only when the target moves toward the centroid of the shared surface will it be taken as 0. Therefore, under the influence of the motion direction $u_p$, the volume of the target 3D-Voronoi will never increase.

Assuming the target is moving such that the second term is negligible or non-positive, and assuming that as a robot moves closer, the shared face area increases ($\partial A/\partial x_p > 0$), then the first term $-\frac{\partial A}{\partial x_p}(x_p - C_i)$ contributes negatively to $\dot{V}_{V_k}$. This implies that the control action of the robots always works to shrink the volume.

To prove that the distance between the robot and the target is always reduced when $\dot{V}_{V_k} = 0$, define $D$ as:

$$D = x_p^2 - x_k^2. \tag{8.32}$$

The derivative of $D$ is:

$$\dot{D} = 2\,(x_p - x_k)\,(\dot{x}_p - \dot{x}_k). \tag{8.33}$$

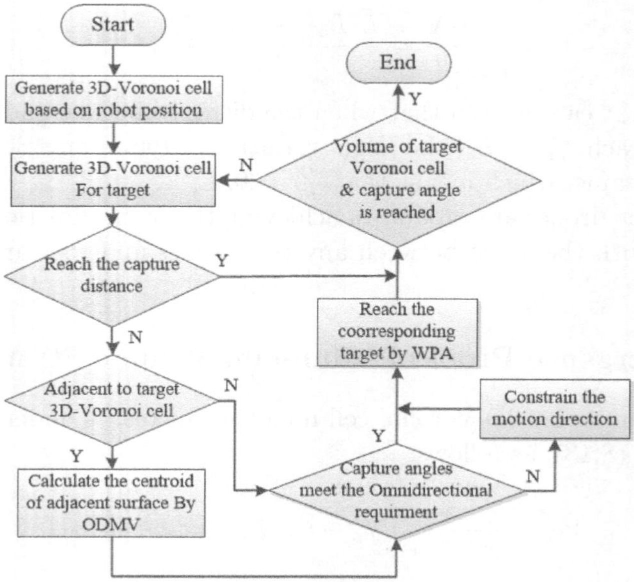


Figure 8.4: The overall logical framework of ODMV algorithm.

Substituting (8.31) to (8.33):

$$\dot{D} = 2\left(x_p - x_k\right)\left(\frac{\left(C_i - x_p\right) - \left(C_i - x_k\right)}{\left(C_i - x_k\right)\left(C_i - x_p\right)}\right)\frac{\partial A}{\partial x_p}. \tag{8.34}$$

Due to the nature of 3D-Voronoi tessellations, the distance from $C_i$ to the target is equal to the distance from the capture robot, $C_i - x_k = C_i - x_p$, so (8.34) can be written as:

$$\dot{D} = -2\frac{\left(x_k - x_p\right)^2}{\left(C_i - x_p\right)^2}\frac{\partial A}{\partial x_p}, \tag{8.35}$$

which represents that the distance between the robot and the known target is always decreasing. Fig. 8.4 represents the overall logical framework of the algorithm. The pseudocode is shown in Algorithm 9.

### 8.2.6 Simulations Analysis

To validate the performance of the proposed ODMV algorithm, a series of simulation experiments were conducted in MATLAB 2016b. The experiments are designed to systematically evaluate the algorithm's effectiveness, robustness, and scalability across four increasingly challenging scenarios: a single static target, multiple static targets, a single dynamic target, and a static target in an environment with obstacles.

This section simulates the above algorithm on the Windows system using MATLAB 2016b as the simulation platform, in a 3D environment of 10 m × 10 m × 10 m. The positions of the robot and the target are represented by black and green squares, respectively. The red line represents the captured movement trajectory. In order to

**Algorithm 9** The pseudocode of the ODMV algorithm.

***Initialization***
Initialize drones' position $\{x_p\}_{i=1}^{n}$; number $n$; task area $Q \subseteq \mathbb{R}^3$; target $\{x_e\}$
***Procedure***
Generate 3D-Voronoi cells $\{V_i^3\}_{i=1}^{n+1}$ and calculate the centroids $\{z_i\}_{i=1}^{n+1}$
Ensure the position of the head-wolf
**while** Drones and target are in capture distance **and** capture angle meets the standard **do**
    Drones move to the head-wolf drone
    **if** The current 3D-Voronoi cell is adjacent to the target 3D-Voronoi cell **then**
        Calculate the centroid of the surface adjacent to the target point $Z_i$
        Adjacent robot to centroid $Z_i$
        Calculate the change rate for each surface and make corrections to drones
        **if** The capture angle of drones satisfies the omni-directional capture **then**
            The motion direction provided by the ODMV is the moving direction of drones
        **else**
            A constraint is imposed on the motion direction based on the variation of the capture angles, resulting in a new motion direction
        **end if**
    **else**
        Move directly to the target
    **end if**
**end while**

minimize the volume of the target 3D Voronoi unit, we set the interception distance to 0.1m and used the robot's capture angle of the target as the effectiveness indicator of the capture process. Simulation analysis was conducted in single target, multi-target, moving target, and environments with obstacles.

### Formation Capture Under Single Target

After improving the minimum volume strategy, this section effectively constrains the motion of robots, allowing more robots to participate in the capture process and enabling the formation to effectively capture targets. The capture process is shown in Fig. 8.5.

To visually compare the capture angle during the capture process, the capture angle is introduced below:

$$\delta = \frac{\alpha}{2\pi}, \tag{8.36}$$

where the capture angle is $\alpha$. The changes in the capture angle of the drone formation to the target are shown in Fig. 8.6.

Fig. 8.6 shows the change in the capture angle of the multi-robot formation. The capture angles of the X-Y, X-Z, and Y-Z planes are used to reflect the encirclement of the formation against the target in a 3D environment.

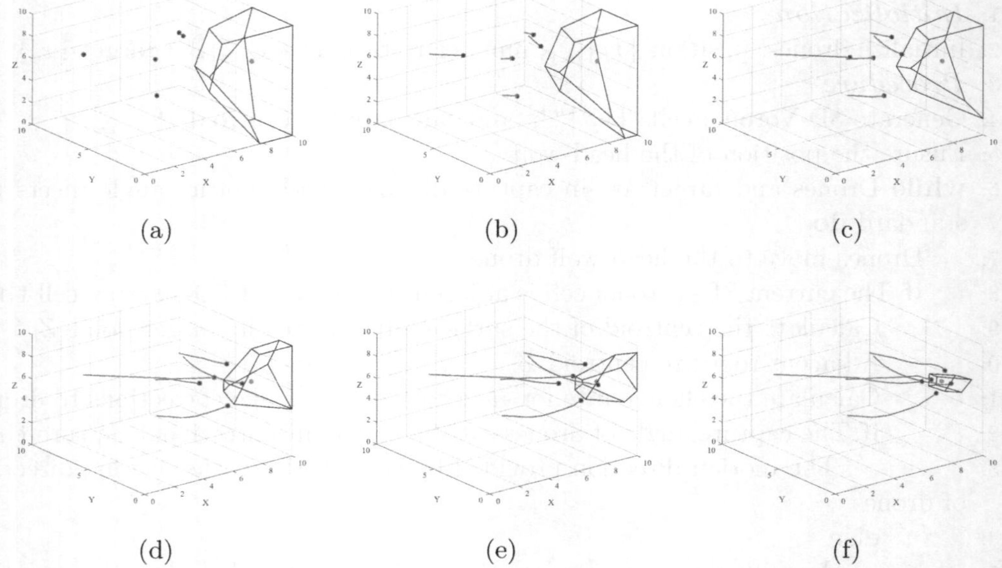


Figure 8.5: 3D-Voronoi capture based on ODMV strategy: (a) $k = 0$; (b) $k = 5$; (c) $k = 10$; (d) $k = 20$; (e) $k = 30$; (f) $k = 45$.

**Formation Capture Under Multiple Target**

The ODMV is also appropriate for multi-target capture. In many practical scenarios, the robot formation needs to capture more than one target. Therefore, in this section, simulations are conducted with two targets, and the number of capture robots is increased from 5 to 10. The robot formation determines the capture target based on the distance. The capture process is shown in Fig. 8.7.

Fig. 8.7 shows that the capture strategy based on the ODMV can accomplish the capture process in a two target mission. This observation will also be applicable for three or more targets.

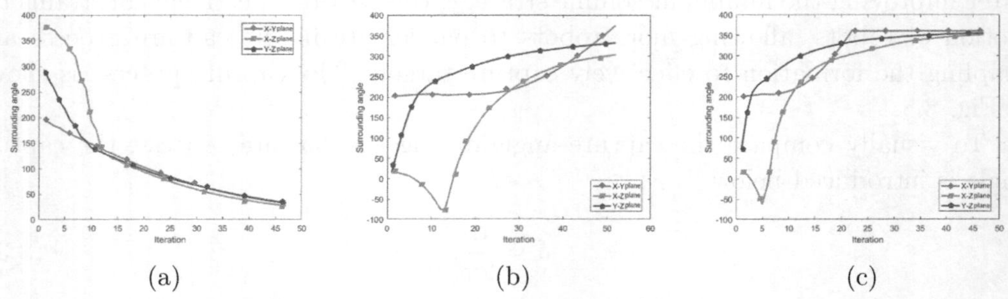


Figure 8.6: Capture angle change before and after adding minimum volume strategy: (a) Variable-step-size WPA formation capture; (b) minimum volume formation capture; (c) ODMV formation capture.

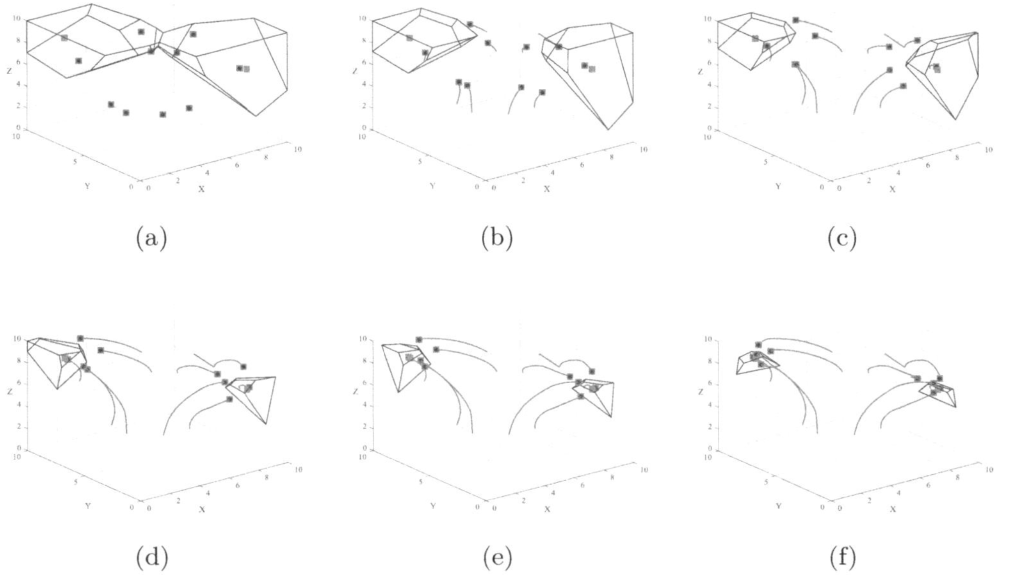

Figure 8.7: Capture of multi-robot under two target: (a) $k = 0$; (b) $k = 10$; (c) $k = 17$; (d) $k = 26$; (e) $k = 38$; (f) $k = 45$.

Fig. 8.8 shows the capture angle variation of the target in the upper left corner during the capture process. The final angles in Figs. 8.8a and 8.8b do not satisfy the effective capture. In Fig. 8.8b, due to the absence of constraints on the movement.

**Formation Capture Under Dynamic Target**

Under the dynamic target environment, the motion trajectory of the target is represented by a blue curve, and the robot's motion trajectory is represented by a red curve. We can see that under the ODMV strategy, multiple robots are eventually

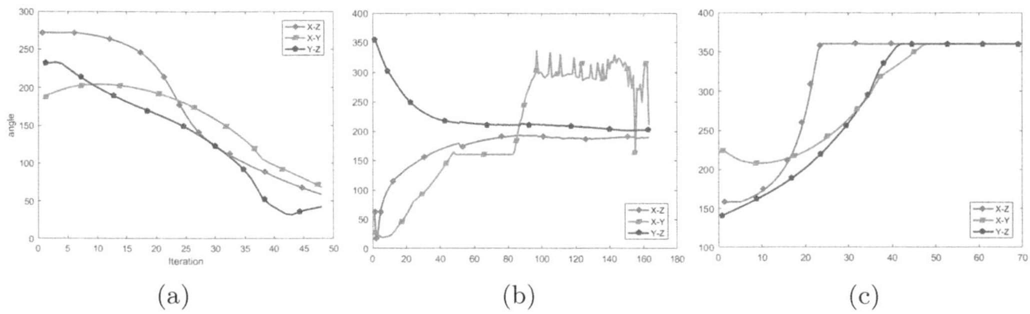


Figure 8.8: The change of capture angles under two targets: (a) Variable-step-size WPA formation capture; (b) minimum Volume formation capture; (c) ODMV formation capture.

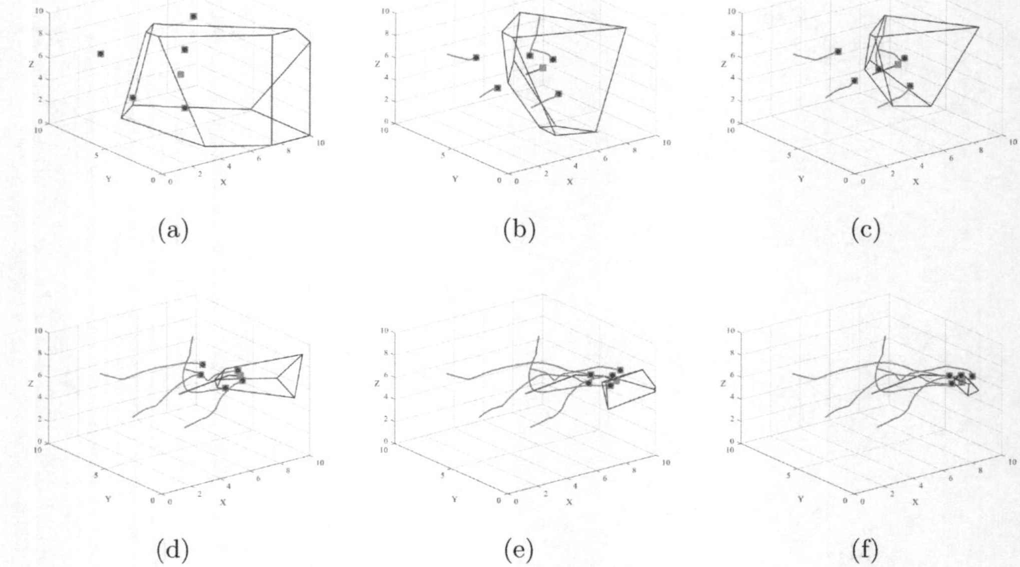


Figure 8.9: Capture of multi-robot under dynamic target: (a) $k$ = 0; (b) $k$ = 9; (c) $k$ = 17; (d) $k$ = 25; (e) $k$ = 34; (f) $k$ = 47.

able to complete the omni-directional capture of this dynamic target. As can be seen in Fig. 8.9, the ODMV can still complete the capture task of this target under the case of the dynamic target.

As can be seen in Fig. 8.9, the ODMV can still complete the capture of this target under the case of a dynamic target. Similar to Section 7.2.4.2, we conducted a comparative experiment on the capture angles in WPA, minimum volume, and the ODMV, as shown in Fig. 8.10.

Fig. 8.10 illustrates the variation of capture angles under a dynamic target. From Figs. 8.10a and 8.10b, it can be observed that both the WPA and the minimum

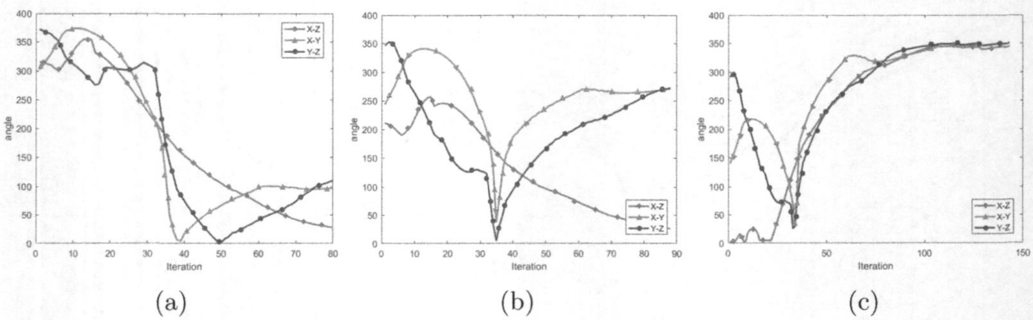


Figure 8.10: The change of capture angles under dynamic target: (a) Variable step size WPA formation capture; (b) minimum volume formation capture; (c) ODMV formation capture.

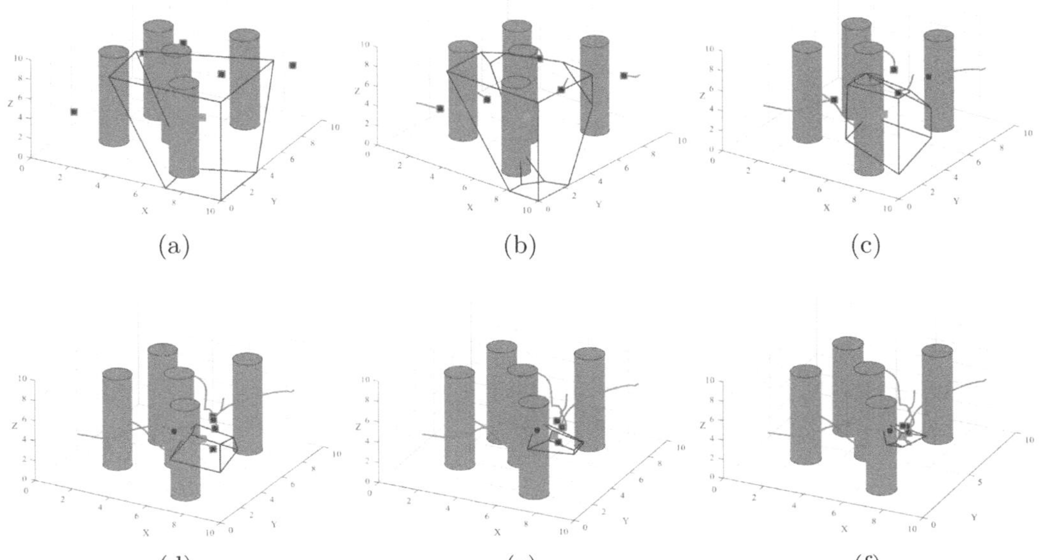

Figure 8.11: Capture of multi-robot under obstacles: (a) $k = 0$; (b) $k = 9$; (c) $k = 18$; (d) $k = 24$; (e) $k = 33$; (f) $k = 52$.

volume strategy fail to achieve effective encirclement of the target. In Fig. 8.10c, the capture angles finally reach to 360, and the omni-directional capture is accomplished.

**Formation Capture Under Obstacle Environment**

In the obstacle environment, the process of formation capture is shown in Fig. 8.11.In order to facilitate the observation of the capture process, we adjusted the perspective of the simulation picture and the transparency of the surface of the obstacle: In the simulation process of Fig. 8.11, we use the artificial potential field (APF) method for obstacle avoidance between drones and obstacle; The variation of capture angles are shown in Fig. 8.12.

## 8.3 Multi-Robot Capture Strategy with Unknown Target location

When conducting a pursuit of a target with an unknown position, the robot must first sense its environment. However, the uncertainty and dynamic changes of the environment impose higher demands on the robot's sensing capabilities. To improve the accuracy of the robot's target perception in an unknown environment and effectively process large volumes of data, this section adopts Variational Sparse Gaussian Process Regression (VS-GPR) as the core algorithm. Gaussian Process Regression (GPR) is a probabilistic, non-parametric method that provides reliable predictions and uncertainty estimates for unknown regions. Variational Sparse Gaussian Process Regression enhances the application ability in large-scale data-driven environmental

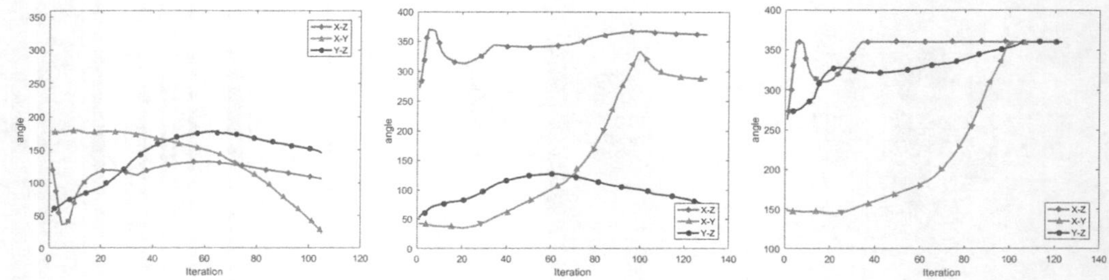


(a) (b) (c)

Figure 8.12: The change of capture angles under obstacles: (a) Variable step size WPA formation capture; (b) minimum volume formation capture; (c) ODMV formation capture.

perception tasks by using variational inference to approximate the posterior distribution of the Gaussian process.

### 8.3.1 Variational Sparse Gaussian Process Regression

To perceive and model the unknown environment, we require a method that can build a continuous representation of the target distribution from sparse sensor measurements. Gaussian Process Regression (GPR) is an ideal non-parametric tool for this, as it provides not only a mean prediction but also a variance, which quantifies the model's uncertainty. However, standard GPR is computationally intensive, scaling poorly with the number of data points. To enable real-time application in a multi-robot system, this section employs a computationally efficient approximation known as Variational Sparse Gaussian Process Regression (VS-GPR).

In recent years, Sparse Gaussian Process Regression (SGPR) has developed rapidly. The core idea is to approximate the full Gaussian process model by using sparse inducing points. Given the real-time requirements of environmental perception, which demands fast computation speeds and high prediction accuracy, this section employs the VS-GPR method to train the model and use it to guide the robot's motion.

In essence, "Sparse" refers to the use of a small set of representative points (called inducing points) to approximate the full dataset, which drastically reduces computational cost. "Variational" refers to the mathematical technique used to optimize the properties of these inducing points to best explain the observed data.

Variational Free Energy (VFE) is an efficient variational inference technique that approximates the posterior distribution through variational inference. By minimizing the Kullback-Leibler (KL) divergence between the variational distribution and the true posterior distribution, the model parameters are optimized. The variational distribution is adjusted to reduce the model's overfitting to noise and possible outliers in the training data, rather than directly fitting the data. This largely mitigates the overfitting problem in prior approximations, and the approximate posterior distribution converges toward the true posterior. First, the posterior distribution is parameterized as a set of inducing points, and optimization is performed by minimizing the KL divergence between the variational and true posterior distributions. In this process, by adjusting the positions of the inducing points, the variational distribution gradually approximates the true posterior, thus reducing computational complexity. The following provides a detailed introduction to the theory of VS-GPR [38].

Set the inducing points as $\mathbf{Z} = \{z_i\}_{i=1}^{M}$, with corresponding function values $\boldsymbol{u} = \{u_i\}_{i=1}^{N} = \{\boldsymbol{f}(z_i)\}_{i=1}^{N}$. Assuming that these inducing points are given, after training the model, the prediction points and training samples are independent, can be expressed as:

$$p_r\left(\boldsymbol{f}_* \mid y\right) = \int p_r\left(\boldsymbol{f}_* \mid \boldsymbol{u}\right) p_r(\boldsymbol{u} \mid y) d\boldsymbol{u}, \tag{8.37}$$

where $p_r(\boldsymbol{f}_*|y)$ is the conditional probability of the function $\boldsymbol{f}$ given the observed data $y$, and $p_r(\boldsymbol{u}|y)$ is the posterior distribution of the inducing function values given the observed data $y$. The score represents the integration over the entire domain for the value of $\boldsymbol{u}$.

Taking the natural logarithm of both sides of equation (8.38) and applying Jensen's inequality, the lower bound of the log marginal likelihood can be computed:

$$\begin{aligned} &\log p_r(y) = \log \int p_r(y|\boldsymbol{f}, \boldsymbol{u}) p_r(\boldsymbol{f}, \boldsymbol{u}) d\boldsymbol{f} d\boldsymbol{u} = \\ &\log \int p_r(y|\boldsymbol{f}, \boldsymbol{u}) p_r(\boldsymbol{f}, \boldsymbol{u}) \frac{q(\boldsymbol{f}, \boldsymbol{u})}{q\boldsymbol{f}, \boldsymbol{u})} d\boldsymbol{f} d\boldsymbol{u} \geq \\ &\int q(\boldsymbol{f}, \boldsymbol{u}) \log \frac{p_r(y|\boldsymbol{f}, \boldsymbol{u}) p_r(\boldsymbol{f}, \boldsymbol{u})}{q(\boldsymbol{f}, \boldsymbol{u})} d\boldsymbol{f} d\boldsymbol{u}, \end{aligned} \tag{8.38}$$

where $q(\boldsymbol{f}, \boldsymbol{u})$ is referred to as the variational distribution. When $q(\boldsymbol{f}, \boldsymbol{u}) = p_r(\boldsymbol{f}, \boldsymbol{u}|y)$, the equality holds.

Assume that the form of $q(\boldsymbol{f}, \boldsymbol{u})$ as:

$$q(\boldsymbol{f}, \boldsymbol{u}) = p_r(\boldsymbol{f} \mid \boldsymbol{u}) \phi(\boldsymbol{u}). \tag{8.39}$$

When $q(\boldsymbol{f}, \boldsymbol{u})$ approximates $p_r(\boldsymbol{f}, \boldsymbol{u}|y)$, $\phi(\boldsymbol{u}) \approx p_r(\boldsymbol{u}|y)$, and simultaneously, $p_r(\boldsymbol{f}|u) \approx p_r(\boldsymbol{f}|\boldsymbol{u}, y)$.

By eliminating the numerator in the fraction of the expression $p_r(\boldsymbol{f}|\boldsymbol{u})$, can be obtained:

$$\begin{aligned} &\log p_r(y) \geq \int p_r(\boldsymbol{f} \mid \boldsymbol{u}) \phi(\boldsymbol{u}) \\ &\log \frac{p_r(y \mid \boldsymbol{f}) p_r(\boldsymbol{u})}{\phi(\boldsymbol{u})} d\boldsymbol{f} d\boldsymbol{u} \triangleq F_V(Z, \phi) \end{aligned}, \tag{8.40}$$

where $F_V(Z, \varphi)$ is the objective function. The position of the inducing points $\mathbf{Z}$ and the transformation distribution $\varphi(\boldsymbol{u})$ are optimized to maximize $F_V(Z, \varphi)$. When the equality holds, $\log p_r(y) = F_V(Z, \varphi)$, $q(\boldsymbol{f}, \boldsymbol{u}) = p_r(\boldsymbol{f}, \boldsymbol{u}|y)$, $\varphi(u) = p_r(\boldsymbol{u}|y)$, $p_r(\boldsymbol{f}|\boldsymbol{u}, y) = p_r(\boldsymbol{f}|\boldsymbol{u})$.

Further calculate the analytical solution for the optimal $\varphi(\boldsymbol{u})$:

$$\phi(\boldsymbol{u}) = \mathcal{N}\left(\sigma^{-2} K_Z \Sigma K_{ZX} y, K_Z \Sigma K_Z\right), \tag{8.41}$$

where $\Sigma$ is a matrix used in the variational Gaussian process to compute the parameters of the variational distribution, $\Sigma = \left[K_Z + \sigma^{-2} K_{ZX} K_{XZ}\right]^{-1}$. Substituting the optimal $\varphi(\boldsymbol{u})$ into $F_V(Z, \varphi)$ gives:

$$F_v(Z) = \log\left[\mathcal{N}\left(0, \boldsymbol{\sigma}^2 Q_X + I_N\right)\right] - \frac{1}{2\boldsymbol{\sigma}^2} \operatorname{Tr}(\tilde{K}), \tag{8.42}$$

where $K_Z$ is the kernel matrix of the **Z** space, which measures the similarity between data points in the **Z** space. $K_{ZX}$ is the cross-kernel matrix between the $X$ and $Z$ spaces, reflecting the similarity between data points in $X$ space (the original data space) and inducing points in **Z** space. $K_{XZ}$ is the transpose of $K_{ZX}$. $Q_X = K_{XZ} K_Z^{-1} K_{ZX}$, $\tilde{K} = K_X - K_{XZ} K_Z^{-1} K_{ZX}$.

After optimization is complete, predictions can be made using $\varphi(\boldsymbol{u})$:

$$p_r\left(\boldsymbol{f}_* \mid y\right) = \mathcal{N}\left(\boldsymbol{m}_*, \boldsymbol{\sigma}_*^2\right). \tag{8.43}$$

The analytical solution for the mean and variance:

$$\boldsymbol{m}_* = \boldsymbol{\sigma}^{-2} k_{*Z} \Sigma K_{ZX} y, \tag{8.44}$$

$$\sigma_*^2 = k_* - k_{*Z} K_Z^{-1} k_{Z^*} + k_{*Z} \Sigma k_{Z^*}, \tag{8.45}$$

where $k_{*Z} = k(x_*, Z)$ and $k_{Z*} = k(Z, x_*)$.

The VFE method introduces sparsity in the model. It only retains information from a subset of data points, enabling efficient modeling and prediction for large-scale datasets. The training time complexity is $O(NM^2)$, and the prediction time complexity is $O(M^2)$.

### 8.3.2 Environmental Awareness Strategy

With the VS-GPR model providing both mean predictions and uncertainty estimates, the next challenge is to design a control strategy that intelligently guides the robots to explore the environment. This leads to the classic exploration-exploitation dilemma: should the robots move toward areas with a high predicted target density (exploitation) to refine the location of known targets, or should they move toward areas of high uncertainty (exploration) to discover new targets and improve the global model? A purely exploitative strategy risks missing undiscovered targets, while a purely explorative one is inefficient.

This section proposes an environmental perception method based on a centroid Voronoi partition as the basic framework. It utilizes the robot's movement in the environment to acquire prior information, and then updates the variational sparse Gaussian process regression model to obtain a predicted global density function model.

Finally, the robot is guided to locate the pollution source in the unknown environment.

First, the following assumptions are made:

**Assumption 1.** There are $n$ robots performing environmental sensing tasks in the region $\boldsymbol{Q} \subseteq \boldsymbol{R}^3$, where the region density function is initially unknown.

**Assumption 2.** The communication between $n$ robots is maintained at all times, with occasional communication delays.

The multi-robot control under the traditional CVT algorithm is constrained by the density function, where robots have a higher interest in areas with higher density. In perception tasks in unknown environments, robots need to cover more areas to find the location of the target. In adaptive sampling, the information entropy method performs well. Entropy can quantify the degree of uncertainty; a higher entropy value indicates greater uncertainty in the model about the environment, requiring more information to reduce this uncertainty. Therefore, by using the entropy of the model's error variance as a weight, the robots are encouraged to explore areas with lower information content (i.e., higher error variance) and higher model prediction uncertainty, rather than staying in areas that have already been explored or have more information. This strategy helps the robots explore from a global perspective, avoiding early concentration in a local area, and thus prevents them from getting stuck in a local optimum. The entropy is defined in information theory as [39]:

$$H(\sigma) = \sigma^2 \ln \sigma^2, \tag{8.46}$$

where $\sigma$ represents the posterior variance of the predictive model.

The cost function of the CVT algorithm can be reformulated as:

$$J_e = \sum_{i=1}^{n} \int_{\mathcal{V}_i} [\varepsilon(\boldsymbol{q}) + H(\sigma)] \left\| \boldsymbol{q} - \mathbf{p}_i \right\|^2 d\boldsymbol{q}, \tag{8.47}$$

where $\varepsilon(\boldsymbol{q})$ represents the predicted concentration value of the points in the Voronoi region.

The introduction of information entropy creates a trade-off for the robot between exploration and exploitation. When the robot identifies unexplored areas, it increases the weight assigned to those regions. Therefore, the incorporation of information entropy helps the CVT algorithm achieve perceptual coverage more quickly while also reducing model errors.

$$\begin{aligned} M_{\mathcal{V}_i} &= \int_{\mathcal{V}_i} \rho(\boldsymbol{q}) d\boldsymbol{q} + \int_{\mathcal{V}_i} \sigma^2 \ln \sigma^2 d\boldsymbol{q} \\ \boldsymbol{C}_{\mathcal{V}_i} &= \frac{1}{M_\rho} \int_{\mathcal{V}_i} \boldsymbol{q}\rho(\boldsymbol{q}) d\boldsymbol{q} + \frac{1}{M_{H(\sigma)}} \int_{\mathcal{V}_i} \boldsymbol{q}\sigma^2 \ln \sigma^2 d\boldsymbol{q} \end{aligned}, \tag{8.48}$$

where $M_\rho = \int_{\mathcal{V}_i} \rho(\boldsymbol{q}) d\boldsymbol{q}$, $M_H(\sigma) = \int_{\mathcal{V}_i} \sigma^2 \ln \sigma^2 d\boldsymbol{q}$.

Therefore, the centroid of the Voronoi region is composed of both the density centroid and the uncertainty component.

The position control law of the robot can be updated as:

$$\dot{\boldsymbol{p}}_i = -k_p \left(\boldsymbol{p}_i - \boldsymbol{C}_{\mathcal{V}_i}\right), \tag{8.49}$$

where $k_p$ represents a constant used to adjust the robot's speed to prevent it from moving too rapidly.

To cheek the stability of the overall closed-loop control system, a Lyapunov function is chosen as:

$$V = \sum_{i=1}^{n} \int_{\mathcal{V}_i} \frac{k_1}{2} \left\|\boldsymbol{q} - \boldsymbol{p}_i\right\|^2 f(\boldsymbol{q}) d\boldsymbol{q}, \tag{8.50}$$

where $f(\boldsymbol{q}) = \varepsilon(\boldsymbol{q}) + H(\sigma)$, $k_1$ is a constant.

Taking its derivative:

$$\dot{V} = \sum_{i=1}^{n} \int_{\mathcal{V}_i} k_1 \left(\boldsymbol{q} - \boldsymbol{p}_i\right)^T f(\boldsymbol{q}) d\boldsymbol{q} \dot{\boldsymbol{p}}_i. \tag{8.51}$$

Substituting $\dot{\boldsymbol{p}}_i$ into the controller, then $\dot{V}$ becomes:

$$\begin{aligned} \dot{V} =& \sum_{i=1}^{n} \int_{V_i} k_1 \left(\boldsymbol{q} - \boldsymbol{p}_i\right)^T f(\boldsymbol{q}) d\boldsymbol{q} \left[k_p \left(\boldsymbol{C}_{V_i} - \boldsymbol{p}_i\right)\right] = \\ & \sum_{i=1}^{n} -k_1 M_{\mathcal{V}_i} \left(\boldsymbol{C}_{V_i} - \boldsymbol{P}_i\right)^T k_p \left(\boldsymbol{C}_{V_i} - \boldsymbol{p}_i\right) = \\ & - k_1 k_p \sum_{i=1}^{n} M_{\mathcal{V}_i} \left\|\boldsymbol{C}_{\mathcal{V}_i} - \boldsymbol{p}_i\right\|^2 < 0 \end{aligned} \tag{8.52}$$

Given that $k_1$ and $k_p$ are positive values, and $M_{\mathcal{V}_i}$ is positive, then $\dot{V} \leq 0$.

Since $\dot{V} \leq 0$, according to LaSalle's Invariance Principle, it can be concluded that the system will converge to the largest invariant set in $\mathbf{Q}$.

### 8.3.3 Uniform Capture

After perceiving the target's position, the robots need to move to appropriate capture points to enclose the target and complete the capture task. However, the specific capture points for the robots are unknown and need to be dynamically assigned based on the distance between the robots and the pollution source. To address this issue, this section adopts the principle of minimizing the average capture distance and uses a particle swarm optimization algorithm to assign suitable capture points to the robots, ensuring an all-around enclosure of the pollution source. Suppose $n$ robots are involved in the capture task, then the number of capture points is $n$. These uniformly distributed capture points are subject to the following constraints:

$$\begin{cases} \alpha = \frac{2\pi}{n} \\ (x - x_c^2) + (y - y_c^2) = R^2 \\ \left\|x_p^i - x_o^j\right\| \leq r_d \end{cases}, \tag{8.53}$$

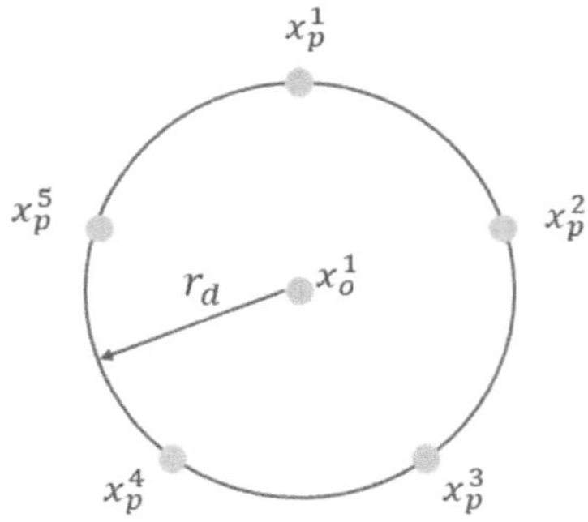


Figure 8.13: Uniform capture strategy for a source or target.

where $(x_c, y_c)$ represents the center position of the target, $\alpha$ is the angular difference between two capture points, and $x$, $y$ are the coordinates in the 2D space. After the target is uniformly captured, the $n$ capture robots are located on the same circumference, with $R$ representing the radius of the circle formed by the $n$ capture robots.

As shown in Fig. 8.13, the capture point distribution for five robots capturing a single target is illustrated. The green points represent the positions of the capture robots, the blue points represent the estimated target position through variational sparse Gaussian process regression, and the black circles represent the minimum capture range of the robots. The objective function to be optimized by the particle swarm algorithm is defined as:

$$J_{dx} = \min \sum_{i=1}^{n} \sum_{j=1}^{s} \int_{\varrho} \left\| \boldsymbol{x}_p^i - \boldsymbol{x}_o^j \right\|^2 dx. \tag{8.54}$$

This objective function is applicable to both single and multiple targets scenarios, where $i$ represents the number of robots and $j$ represents the number of targets being captured. The particle swarm optimization algorithm is used to determine which capture point belongs to which robot. This objective function ensures that the total travel distance for all capture robots is minimized.

The PSO algorithm searches for the optimal solution by simulating the movement of particles in the solution space. Each particle updates its position by adjusting its velocity based on its own historical experience and the information from the entire swarm. Ultimately, the PSO algorithm gradually converges to the optimal solution, achieving efficient capture point allocation. The steps for using the PSO algorithm to allocate capture points are as follows:

**Step 1**. Determine the capture point positions, the initial number of robots, and the task allocation constraints.

**Step 2**. Randomly initialize a swarm of particles, where each particle represents a task allocation scheme. Each particle contains the mapping relationship between the capture robots and the capture points.

Table 8.1: Simulation parameter settings

| Parameter | Parameter values |
|---|---|
| Task switching threshold | 4.5 |
| Coordinates of obstacles $(m)$ | (5.0, 1.5) (1.2, 3.5) (4.0, 5.0) |
| Obstacle size | $0.5m \times 0.5m \times 3m$ |
| Maximum speed of UAV | $0.3m/s$ |
| Robot-target capture distance | $0.5m$ |

**Step 3**. Calculate the objective function (fitness function) value for each particle based on the task allocation scheme it represents.

**Step 4**. Update each particle's velocity and position according to the principles of the particle swarm optimization algorithm. The velocity update considers the particle's own historical best position, the global best position, and random factors. The position update is then adjusted based on the new velocity.

**Step 5**. Repeat Steps 3 and 4 until the stopping condition is met. Typically, the stopping condition can be either the maximum number of iterations or the fitness reaching a predefined threshold.

**Step 6**. After the iterations are complete, select the task allocation scheme represented by the particle with the optimal fitness as the final result.

The pseudocode for the multi-robot environmental perception and capture algorithm is shown in Algorithm 10:

### 8.3.4 Simulations Analysis

This section mainly focuses on the simulation verification and analysis of the algorithm proposed in the previous text. Simulations are conducted in single-target, multi-target, and dynamic target scenarios to verify the feasibility of the algorithm. First, the simulation environment and parameters are set. The environment is configured to be $7m \times 7m \times 7m$ meters, with the number of capture robots set to 5 and 6, respectively, and 3 fixed obstacles are included to test the algorithm's obstacle avoidance performance. The number of capture points is equal to the number of robots. To facilitate the observation of the capture robots' movement trajectories, the Voronoi regions are not displayed in this simulation. Table 8.1 provides the experimental parameter settings, including task switching thresholds, the size and coordinates of obstacles, the speed of the capture robots, and other related parameters.

**Formation Capture Under Sigle Target**

In the single-target capture task, the target is located at (5m, 4.5m, 4m). During the simulation, five capture robots first sense the information of the unknown environment and predict the localization of the target point to be captured, ultimately achieving a uniform capture of the target. Each capture robot in the task must avoid obstacles in its path. The simulation results of the VS-GPR algorithm are shown in Fig. 8.14, where the curves in Figs. 8.14a, 8.14b, and 8.14c represent the movement

**Algorithm 10** The multi-robot environmental perception and capture algorithm.

***Initialization***
The initial position information of $n$ robots $\{p_i\}_{i=1,2,\ldots,n}$, task area $\boldsymbol{Q} \subseteq \boldsymbol{R}^3$, task switching threshold $T$, maximum number of iterations $t$, minimum distance $r_d$ between capture robots and target points, and maximum speed $v_{\max}$ of the capture robots.
***Procedure***
**while** Maximum number of iterations for formation $t$ **do**
  **for** $i = 1, \ldots, n$ **do**
    Obtain the position information of robot $p_i$
    Calculate the distance between each robot and any point in the area: $\|\boldsymbol{q} - \boldsymbol{p}_i\|$
    Construct the constrained Voronoi region $v_i$ for each robot based on the SVM method
    **if** The density value of the capture point localization in the robot detection is lower than the task switching threshold $T$ **then**
      Obtain environmental prior information of the robot's localization.
      Compute the variational sparse Gaussian process regression model to obtain the posterior distribution of the predicted field.
      Calculate the change rate for each surface and make corrections to the robot.
      Drive the robot to the centroid localization $\boldsymbol{p}_i = \boldsymbol{C}_{v_i}$
    **else**
      Calculate the distance between different capture robots and the predicted target: $\left\|\boldsymbol{x}_p^i(t) - \boldsymbol{x}_0^j(t)\right\|_2$
      Optimize the objective function based on the PSO algorithm: $J_{ds}$
      Assign the optimal capture points $x_0$ to different robots
      **if** $\left\|\boldsymbol{x}_p^i(t) - \boldsymbol{x}_0^j(t)\right\|_2 < r_d$ **then**
        The capture task is complete
      **else**
        Drive the robot to the capture point localization: $x_p^J = x_o^j$
      **end if**
    **end if**
  **end for**
  Combine historical information to obtain the explored point set $X = \{\boldsymbol{x}_n\}_{n=1}^N$
**end while**

trajectories of different capture robots with different colors. The yellow sphere represents the target localization, and the three black cylinders represent obstacles in the environment.

From the analysis of the simulation results, it can be concluded that the SVM-based separating hyperplane method used in this section ensures that all five robots can avoid obstacles during the formation process, guaranteeing the safety of the for-

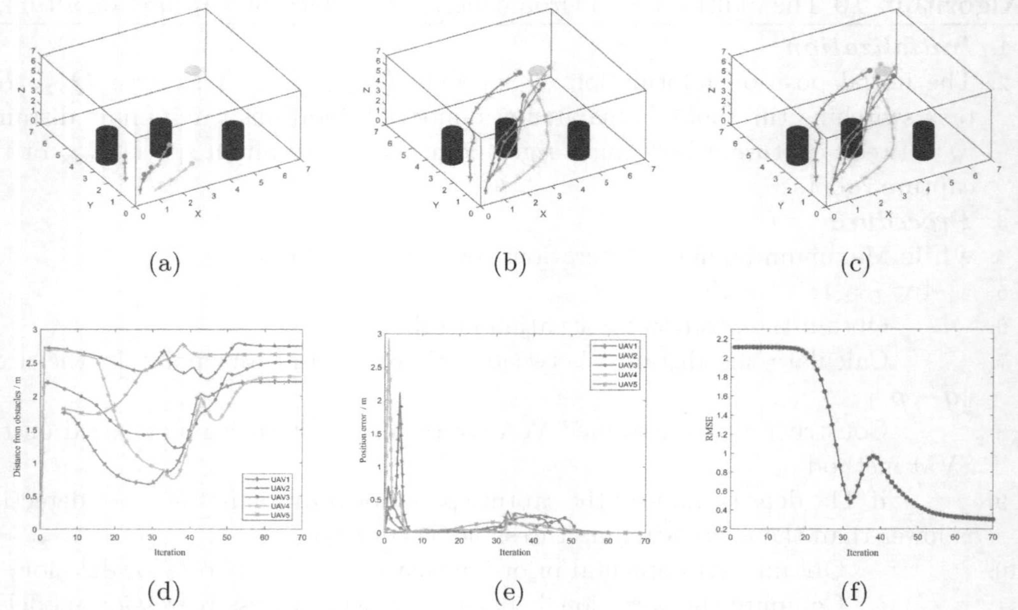


Figure 8.14: Five robots capture single target by VS-GPR: (a) The initial capture process ($k = 15$); (b) during the capture process ($k = 35$); (c) complete uniform capture ($k = 70$); (d) distance between robot and obstacle; (e) position error from the target point; (f) root mean squared error of the model.

mation. As shown in Fig. 8.14e, the position error between the robots and the target localization continuously decreases with the increase in the number of iterations, ultimately converging. It is important to note that the target point is set as the Voronoi centroid during the environmental perception phase, and as the capture point during the formation and capture phase. Fig. 8.14f shows that as the five robots move within the task area, the perceived environmental model becomes increasingly accurate.

**Formation Capture Under Multiple Target**

Considering that there may be multiple targets in more complex scenarios, simulations were conducted with 6 robots capturing 2 to 5 targets in a bounded environment. In this simulation, the robots need to update the prediction field model by moving within the environment, accurately predicting the locations of the targets, and then each robot performs the capture of the respective target.

The positions of the multiple targets are set at $(2m, 5m, 3.5m)$, $(5m, 3m, 5.5m)$, $(4m, 5m, 3.5m)$, $(3m, 4m, 3.5m)$, and $(5m, 1.5m, 5m)$. The number of targets is adjusted based on the simulation requirements. Here, simulation results for 2 and 4 targets are presented. The capture process is shown in Figs. 8.15 and 8.16. The results demonstrate that the algorithm can successfully achieve effective capture of multiple unknown targets by multi-robot, while ensuring the safety of the robots. Figs. 8.15(f) and 8.16(f) indicate that the model established by VS-GPR has high

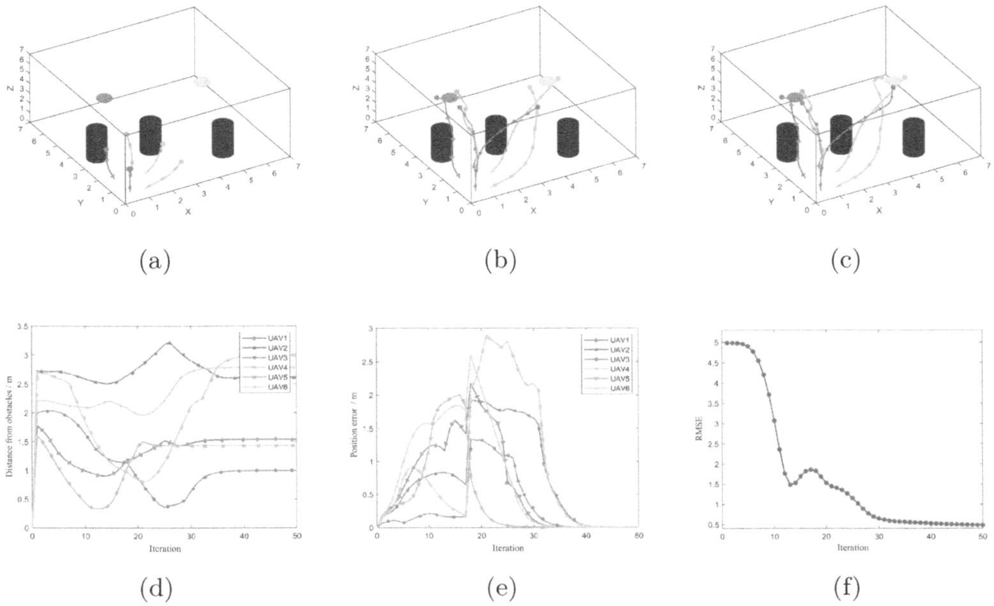


Figure 8.15: Six robots capture two targets by VS-GPR: (a) The initial capture process ($k = 15$); (b) during the capture process ($k = 38$); (c) complete uniform capture ($k = 57$); (d) distance between robot and obstacle; (e) position error from the target point; (f) root mean squares error of the model.

accuracy. Figs. 8.15c shows that when there are 4 targets in the scene, each target is assigned to 1, 1, 1, and 3 capture robots, respectively. The comparison between the VS-GPR method and the GPR algorithm for multi-robot pursuit of multiple target is shown in Table 8.2.

The simulation results demonstrate that the total iteration time of the VS-GPR algorithm increased by 10.46% and 4.32%, respectively, when containing 2 to 4 targets. However, no significant time increase was observed when scaling up to 5 targets, which is attributed to constraints in spatial scope and robot population limitations. The average single-iteration time of VS-GPR was measured at 2.05 seconds, compared to 5.57 seconds for the conventional GPR-based containment approach. These findings confirm that the proposed multi-robot containment strategy achieves reduced iteration time while maintaining high model prediction accuracy, ultimately enabling

Table 8.2: Comparison of time consumption for multi-robot capture multi-targets

| Target number | GPR time (s) | VS-GPR time (s) |
|---|---|---|
| 2 | 278.63 | 102.67 |
| 3 | 319.37 | 113.41 |
| 4 | 345.56 | 118.32 |
| 5 | 352.17 | 118.84 |

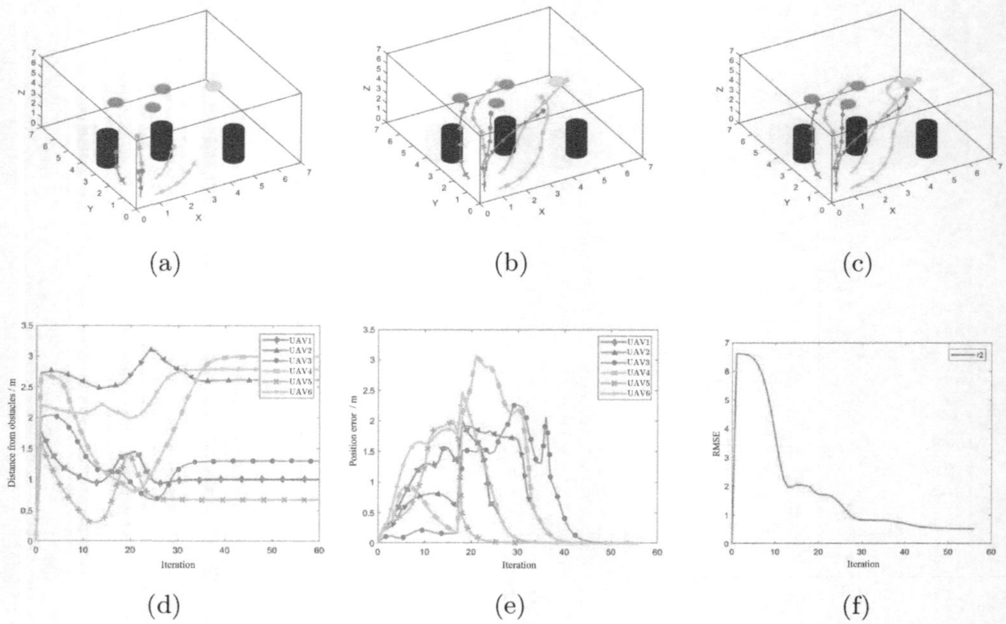


Figure 8.16: Six robots capture four targets by VS-GPR: (a) The initial capture process (k = 15); (b) during the capture process (k = 38); (c) complete uniform capture (k = 57); (d) distance between robot and obstacle; (e) position error from the target point; (f) root mean square error of the model.

successful target containment. Notably, to explicitly visualize each robot's trajectory during containment, the simulation employed frequent iterations with smaller UAV speeds, resulting in smoother motion paths at the expense of prolonged total iteration duration relative to practical experimental conditions.

**Formation Capture Under Dynamic Target**

In the dynamic target simulation environment, 5 capture robots are set to capture 1 dynamic target. Throughout the simulation, we assume that the target does not adopt an evasion strategy and only moves in a straight line. The initial and final position coordinates of the dynamic target are (10, 45, 50) and (50, 45, 50), respectively. The movement path of the target is represented by a dashed line. The simulation results are shown in Fig. 8.17.

This simulation demonstrates the framework's robustness against a dynamic target. The VS-GPR model continuously updates based on new sensor readings, allowing the system to track the moving target. Consequently, the assigned capture points are also updated dynamically. As shown in Fig. 8.17, the team of robots successfully adjusts its formation to follow and eventually encircle the target at its final position, validating the algorithm's applicability in non-static environments.

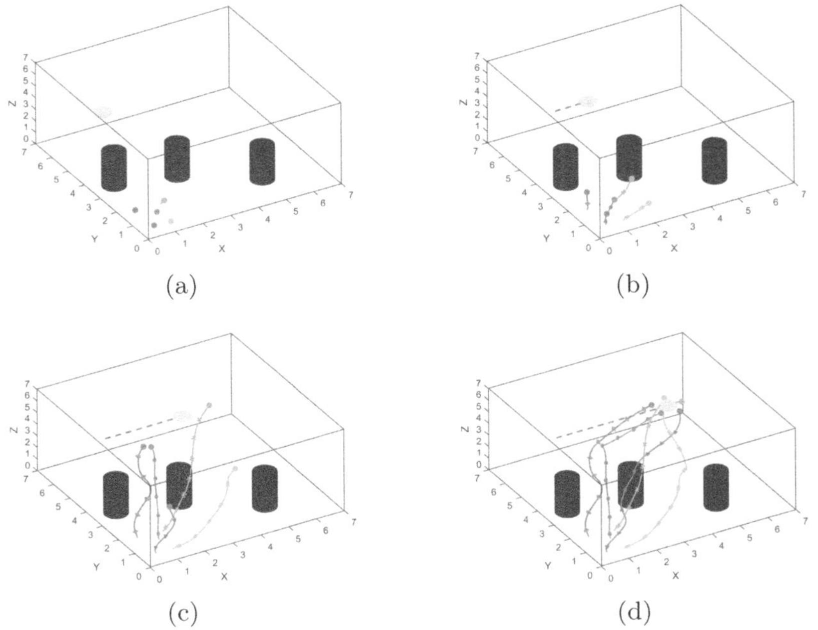


Figure 8.17: Six robots capture dynamic target by VS-GPR: (a) The initial capture process ($k = 1$); (b) during the capture process ($k = 15$); (c) during the capture process ($k = 42$); (d) complete uniform capture ($k = 73$).

## 8.4 Chapter Summary

The aim of this chapter is to study multi-robot capture methods based on Voronoi diagrams, which are mainly divided into capture with known target positions and multi-robot capture with unknown target positions. In the case where the target position is known, an omni-directional minimum volume 3D-Voronoi diagram algorithm for cooperative target capture by multi-robot is presented. By extending the Voronoi diagrams into a three-dimensional space, this method ensures collision avoidance during the capture process and constrains the capture angles, thereby improving the effectiveness of the capture. Additionally, the Variable Step Wolf Pack (VSWP) algorithm is introduced for motion planning, enabling robots to dynamically adjust their capture strategies in complex environments, and making the approach applicable to multi-target capture.

In the case where the target position is unknown, a distributed perception and capture algorithm based on variational sparse Gaussian process regression is presented. Considering that the traditional Gaussian process regression has high computational complexity and is unsuitable for handling large datasets, this chapter first takes into account the impact of obstacles and introduces a centroid-Voronoi partitioning algorithm with separating hyperplanes for dynamic task region planning by robots. Second, multi-robot movement exploration within the task area is utilized to acquire environmental information. The posterior distribution of the unknown environment is approximated using the variational free energy method, achieving efficient perception of the unknown environment. Finally, a particle swarm optimization algorithm is employed to dynamically assign capture points to the capture robots, ensuring that multiple robots achieve a comprehensive and evenly distributed capture of the target.

In future research, dynamic task environments will be further considered, where the target may change its position or strategy during movement. On this basis, the introduction of an escape strategy for the target could be explored, making the capture process more aligned with real-world application scenarios.

## Bibliography

[1] Wei Li, Wenhao Yan, Huaguang Shi, Si Li, and Yi Zhou. Multi-robot cooperative pursuit-evasion control: A deep reinforcement learning approach based on prioritized experience replay. In *Proceedings of the 2024 8th International Conference on Control Engineering and Artificial Intelligence*, pages 120–127, 2024.

[2] Yuanliang Xue, Guodong Jin, Tao Shen, Lining Tan, Nian Wang, Jing Gao, and Lianfeng Wang. Consistent representation mining for multi-drone single object tracking. *IEEE Transactions on Circuits and Systems for Video Technology*, 2024.

[3] Meng Zhou, Zihao Wang, Jing Wang, and Zhengcai Cao. Multi-robot collaborative hunting in cluttered environments with obstacle-avoiding Voronoi cells. *IEEE/CAA Journal of Automatica Sinica*, 11(7):1643–1655, 2024.

[4] Wenbo Du, Tong Guo, Jun Chen, Biyue Li, Guangxiang Zhu, and Xianbin Cao. Cooperative pursuit of unauthorized UAVs in urban airspace via multi-agent reinforcement learning. *Transportation Research Part C: Emerging Technologies*, 128:103122, 2021.

[5] Jiawei Xia, Yasong Luo, Zhikun Liu, Yalun Zhang, Haoran Shi, and Zhong Liu. Cooperative multi-target hunting by unmanned surface vehicles based on multi-agent reinforcement learning. *Defence Technology*, 29:80–94, 2023.

[6] John L Salmon, Landon C Willey, David Casbeer, Eloy Garcia, and Alexander Von Moll. Single pursuer and two cooperative evaders in the border defense differential game. *Journal of Aerospace Information Systems*, 17(5):229–239, 2020.

[7] Emre Koyuncu and Gokhan Inalhan. Exploiting delayed and imperfect information for generating approximate UAV target interception strategy. *Journal of Intelligent & Robotic Systems*, 69:313–329, 2013.

[8] Nick-Marios T Kokolakis and Kyriakos G Vamvoudakis. Bounded rational Dubins vehicle coordination for target tracking using reinforcement learning. *Automatica*, 149:110732, 2023.

[9] Zhiyuan Sun, Hanbing Sun, Ping Li, and Jin Zou. Cooperative strategy for pursuit-evasion problem in the presence of static and dynamic obstacles. *Ocean Engineering*, 279:114476, 2023.

[10] Mikhail Khrenov, Phillip Rivera-Ortiz, and Yancy Diaz-Mercado. Geometric feasibility for defense manifold maintenance in planar reach-avoid games against a fast evader. *IFAC-PapersOnLine*, 54(20):807–813, 2021.

[11] Weihao Li, Kaiyu Qin, Mengji Shi, Jingliang Shao, and Boxian Lin. Dynamic target enclosing control scheme for multi-agent systems via a signed graph-based approach. *IEEE/CAA Journal of Automatica Sinica*, 10(2):560–562, 2023.

[12] Jinliang Shao, Lei Shi, Yuhua Cheng, and Tieshan Li. Asynchronous tracking control of leader–follower multiagent systems with input uncertainties over switching signed digraphs. *IEEE Transactions on Cybernetics*, 52(7):6379–6390, 2021.

[13] Renato F dos Santos, Ragesh K Ramachandran, Marcos AM Vieira, and Gaurav S Sukhatme. Parallel multi-speed pursuit-evasion game algorithms. *Robotics and Autonomous Systems*, 163:104382, 2023.

[14] Kunal Shah and Mac Schwager. Grape: Geometric risk-aware pursuit-evasion. *Robotics and Autonomous Systems*, 121:103246, 2019.

[15] Wei Sun and Panagiotis Tsiotras. Sequential pursuit of multiple targets under external disturbances via Zermelo–Voronoi diagrams. *Automatica*, 81:253–260, 2017.

[16] Marco Casini and Andrea Garulli. On the advantage of centralized strategies in the three-pursuer single-evader game. *Systems & Control Letters*, 160:105122, 2022.

[17] Xi Li, Guoyuan Qi, and Limin Zhang. Time-varying formation dynamics modeling and constrained trajectory optimization of multi-quadrotor UAVs. *Nonlinear Dynamics*, 106(4):3265–3284, 2021.

[18] Bailing Tian, Pinpin Li, Hanchen Lu, Qun Zong, and Lei He. Distributed pursuit of an evader with collision and obstacle avoidance. *IEEE Transactions on Cybernetics*, 52(12):13512–13520, 2021.

[19] Yan Wang, Guohao He, Yuedong Ma, Guojie Kong, and Jianwei Gong. Research on multi-robots self-organizing cooperative pursuit algorithm based on Voronoi graph. In *2020 39th Chinese Control Conference (CCC)*, pages 3840–3844. IEEE, 2020.

[20] Alyssa Pierson, Zijian Wang, and Mac Schwager. Intercepting rogue robots: An algorithm for capturing multiple evaders with multiple pursuers. *IEEE Robotics and Automation Letters*, 2(2):530–537, 2016.

[21] Ben Tu, Axel Gandy, Nikolas Kantas, and Behrang Shafei. Joint entropy search for multi-objective Bayesian optimization. *Advances in Neural Information Processing Systems*, 35:9922–9938, 2022.

[22] Yang Xu, Ronghao Zheng, Senlin Zhang, Meiqin Liu, and Shoudong Huang. CARE: Confidence-rich autonomous robot exploration using Bayesian kernel inference and optimization. *IEEE Robotics and Automation Letters*, 8(10):6755–6762, 2023.

[23] Mariliza Tzes, Nikolaos Bousias, Evangelos Chatzipantazis, and George J Pappas. Graph neural networks for multi-robot active information acquisition. In *2023 IEEE International Conference on Robotics and Automation (ICRA)*, pages 3497–3503. IEEE, 2023.

[24] Dongming Zhou, Zhengbin Pang, and Wei Li. Enhanced causal reasoning and graph networks for multi-agent path finding. In *2024 International Joint Conference on Neural Networks (IJCNN)*, pages 1–8. IEEE, 2024.

[25] Yasser Bouzid, Yasmina Bestaoui, and Houria Siguerdidjane. Quadrotor-UAV optimal coverage path planning in cluttered environment with a limited onboard energy. In *2017 IEEE/RSJ International Conference on Intelligent Robots and Systems (IROS)*, pages 979–984. IEEE, 2017.

[26] Sung-Won Cho, Jin-Hyoung Park, Hyun-Ji Park, and Seongmin Kim. Multi-UAV coverage path planning based on hexagonal grid decomposition in maritime search and rescue. *Mathematics*, 10(1):83, 2021.

[27] Meng Zhou, Xinheng Wang, Chang Wang, and Jing Wang. Multi-robot cooperative target search based on distributed reinforcement learning method in 3D dynamic environments. *Drones and Autonomous Vehicles*, 1(4):10012, 2024.

[28] Xiaofeng Li, Jie Ren, and Yunbo Li. Multi-mode filter target tracking method for mobile robot using multi-agent reinforcement learning. *Engineering Applications of Artificial Intelligence*, 127:107398, 2024.

[29] Tianpeng Zhang, Victor Qin, Yujie Tang, and Na Li. Distributed information-based source seeking. *IEEE Transactions on Robotics*, 39(6):4749–4767, 2023.

[30] Haitong Ma, Tianpeng Zhang, Yixuan Wu, Flavio P Calmon, and Na Li. Gaussian max-value entropy search for multi-agent Bayesian optimization. In *2023 IEEE/RSJ International Conference on Intelligent Robots and Systems (IROS)*, pages 10028–10035. IEEE, 2023.

[31] Daniel Duberg and Patric Jensfelt. Ufoexplorer: Fast and scalable sampling-based exploration with a graph-based planning structure. *IEEE Robotics and Automation Letters*, 7(2):2487–2494, 2022.

[32] Juntong Peng, Hrishikesh Viswanath, and Aniket Bera. Graph-based decentralized task allocation for multi-robot target localization. *IEEE Robotics and Automation Letters*, 9(11):10676–10683, 2024.

[33] Yan Gao, Chenggang Bai, Lei Zhang, and Quan Quan. Multi-UAV cooperative target encirclement within an annular virtual tube. *Aerospace Science and Technology*, 128:107800, 2022.

[34] Hongmin Wang, Jiaqiang Tian, Lingyun Wei, and Yufeng Zhuang. Multi-unmanned aerial vehicles cooperative searching and tracking strategy for multiple moving targets. *Control Theory & Applications/Kongzhi Lilun Yu Yinyong*, 38(7), 2021.

[35] Ahmed Alagha, Rabeb Mizouni, Jamal Bentahar, Hadi Otrok, and Shakti Singh. Multiagent deep reinforcement learning with demonstration cloning for target localization. *IEEE Internet of Things Journal*, 10(15):13556–13570, 2023.

[36] Ahmed Alagha, Rabeb Mizouni, Shakti Singh, Jamal Bentahar, and Hadi Otrok. Adaptive target localization under uncertainty using multi-agent deep reinforcement learning with knowledge transfer. *Internet of Things*, 29:101447, 2025.

[37] Hu-Sheng Wu, Fengming Zhang, and Lushan Wu. New swarm intelligence algorithm – Wolf Pack Algorithm. *Systems Engineering and Electronics*, 35(11):2430–2438, 2013.

[38] Linda SL Tan, Victor MH Ong, David J Nott, and Ajay Jasra. Variational inference for sparse spectrum Gaussian process regression. *Statistics and Computing*, 26:1243–1261, 2016.

[39] Alfred Wehrl. General properties of entropy. *Reviews of Modern Physics*, 50(2):221, 1978.

CHAPTER 9

# Reinforcement Learning and Multi-Robot Capture

## 9.1 Introduction

With the increasing application of multi robot systems in complex tasks, traditional capture control methods are facing more and more challenges. Especially in dynamic and uncertain environments, how to efficiently coordinate cooperation and task allocation among multiple robots has become a key issue in research. Traditional rule-based control methods are difficult to adapt to more complex environments, while reinforcement learning (RL), especially deep reinforcement learning (DRL), provides new ideas for multi robot collaborative decision-making with its excellent self-learning and adaptation capabilities. Reinforcement learning is a learning method based on reward and punishment mechanisms. In multi robot encirclement tasks, multi-agent deep reinforcement learning (MADRL) combines the powerful feature extraction ability of deep neural networks with the decision optimization ability of reinforcement learning, and can solve decision problems in high-dimensional state and action spaces. In multi robot capture tasks, adopting the MADRL framework can help each robot automatically learn how to perform capture tasks in complex environments, especially in situations where multi robot collaboration is required, enabling efficient cooperation. Each robot observes local environmental conditions and takes actions, collaborates with other robots and interacts with the environment, ultimately learning the optimal trapping strategy.

As a representative problem in multi-agent system research, multi-robot cooperative capture has significant application value in fields such as dynamic environment monitoring, disaster rescue, and group security. Its core objective is to enable multiple robots to collaboratively achieve rapid capture of a dynamic escaping target while avoiding environmental obstacles and inter-agent collisions [1]. However, traditional rule-based control methods are difficult to adapt to complex and dynamic environments, highlighting the urgent need for more intelligent collaborative decision-making mechanisms. Against this background, reinforcement learning (RL), especially deep reinforcement learning (DRL), has shown remarkable advantages. By means of an end-to-end interactive learning mechanism, DRL provides a new paradigm for

DOI: 10.1201/9781003714569-9

multi-robot cooperative decision-making [2–4]. Nevertheless, the partial observability of reinforcement learning makes it challenging for agents to effectively capture global key information, and the sparse reward mechanism often causes policy learning to fall into local optima [5, 6]. These limitations severely restrict the practical performance of DRL in multi-robot systems.

To enhance cooperative efficiency, researchers have explored integrating structured methods with deep reinforcement learning (DRL). Voronoi partitioning, as a spatial division strategy, has been widely applied in multi-robot tasks to allocate independent working regions for each robot. The combination of Voronoi diagrams and reinforcement learning provides a structured and intelligent cooperative solution for multi-robot systems. Through geometric constraints, Voronoi diagrams can effectively reduce collision risks among robots while reinforcement learning continuously optimizes decision-making [7]. However, standard Voronoi diagrams do not consider the physical size of robots and the requirements of dynamic obstacle avoidance. To address this limitation, the Buffered Voronoi Cell (BVC) introduces a safety radius to optimize region partitioning. This approach offers a more robust underlying framework for multi-robot cooperation, further improving system safety and collaboration.

At the algorithmic level, as the complexity of reinforcement learning problems increases, solely relying on value-based methods faces the curse of dimensionality in high-dimensional state spaces, while pure policy-based methods suffer from slow convergence. Therefore, the Actor–Critic approach, which combines the advantages of both policy and value functions, has become an effective solution for addressing high-dimensional state space problems [8]. Based on this framework, improved algorithms have been developed to enhance the efficiency of multi-UAV capture of maneuvering targets and exhibit strong generalization capability in complex dynamic environments [9]. Other studies have proposed reinforcement learning–based path planning algorithms to address the dependence on environmental information and the lack of real-time performance in traditional planning tasks [10]. These approaches demonstrate that the Actor–Critic framework has significant advantages in multi-agent cooperative tasks; however, in specific multi-robot target capture scenarios, it remains necessary to integrate reinforcement learning with other control strategies to further improve performance and efficiency.

Although the Actor–Critic framework provides an effective foundation for multi-agent cooperation, its application to multi-robot target capture still requires integration with other control methods to improve capture performance and efficiency. Several studies have proposed reinforcement learning–based control approaches guided by potential field models [11], reactive planners for autonomous robotic exploration in complex environments [12], and control models combining deep neural networks with deterministic policy gradient algorithms [13]. Although these approaches have achieved remarkable results in target capture tasks, deep reinforcement learning still faces challenges in practical applications, such as information overload from the environment and limited policy interpretability. These limitations have motivated researchers to explore more cognitively inspired computational mechanisms—namely, the attention mechanism. By simulating the selective attention characteristics of the human visual system, the attention mechanism dynamically allocates computational resources to key information nodes [14], providing a new perspective for overcoming the inherent limitations of DRL. Incorporating attention mechanisms into DRL can enhance the model's capability to process environmental states, particularly under partial observability and complex conditions, thereby improving decision quality and information aggregation [15].

In multi-agent scenarios, the integration of attention mechanisms can exhibit stronger cooperative effects. Various studies have proposed Actor–Critic algorithms integrated with attention mechanisms to improve learning efficiency in multi-agent systems [16], algorithms that enhance attention modeling within centralized Critics to effectively capture teammates' policies [17], and hybrid attention mechanisms combined with multi-agent learning frameworks to improve information extraction in complex environments [18]. Other research has focused on integrating global and local attention mechanisms for cooperative UAV control [19], introducing self-attention mechanisms into Critic networks to address continuous control problems in cooperative settings [20], and combining hybrid attention models with experience replay to improve sample efficiency in multi-robot decision-making [21]. Furthermore, hierarchical attention Actor–Critic algorithms have been developed to reduce the dimensionality of both action and state spaces [22].

## 9.2 Theoretical Foundation of Reinforcement Learning Capture

The core of the multi robot encirclement task is to enable the pursuer group to form efficient and collaborative decision-making strategies through autonomous learning, in order to successfully encircle escapees in complex dynamic environments. Traditional rule-based control methods are difficult to adapt to the uncertainty of the environment and the complex interactions between intelligent agents. And reinforcement learning, as well as its combination with deep learning, provides a new paradigm for solving this sequential decision-making problem. It enables robots to self-optimize from trial and error through continuous interaction with the environment, ultimately mastering advanced strategies for collaborative capture without explicit programming. This chapter will systematically elaborate on the core theoretical basis that supports reinforcement learning capture strategies. First, starting from the most basic reinforcement learning framework, it will analyze its core components and operational mechanisms.

### 9.2.1 Basic Analysis of Reinforcement Learning Framework

The reinforcement learning framework mainly consists of several core components: agents, environment, state, action, and reward. An intelligent agent is like an autonomous decision-maker, capable of perceiving the state of the environment and selecting corresponding actions based on its own strategies. The environment is the external world in which the intelligent agent is located, which receives the agent's actions and returns new states and rewards according to certain rules. Action is the behavior that an intelligent agent can take in its current state, while state is a description of the environment that contains the information needed for the agent to make decisions. And the reward is the feedback of the environment on the agent's actions, which is a scalar value. The agent's goal is to learn strategies that can maximize cumulative rewards through continuous interaction with the environment.

Reinforcement learning can be broadly categorized into two paradigms—Value-Based and Policy-Based methods—depending on the optimization objective. Both approaches tackle sequential decision-making problems from distinct perspectives: value estimation and policy search. Value-Based methods (such as Q-Learning and DQN) iteratively update the state–action value function $Q(s, a)$ or the state value function $V(s)$, thereby deriving the optimal policy indirectly. Their advantages lie in theoretical convergence guarantees (e.g., the Bellman optimality equation) and high sample efficiency, making them especially suitable for discrete action spaces (e.g., game control and discrete decision tasks). However, these methods rely heavily on the accuracy of value function estimation and face computational bottlenecks in finding the maximal action–value in continuous action spaces. Policy-Based methods (such as REINFORCE and PPO [23,24]), on the other hand, directly parameterize the policy function $\pi_\theta(a \mid s)$ and optimize the parameters $\theta$ via gradient ascent to maximize the expected cumulative reward. These methods naturally support continuous action outputs (e.g., robotic motion control) and encourage stochastic exploration through entropy regularization. Nevertheless, they suffer from high variance, slow convergence, and susceptibility to local optima.

The value function plays an important role in reinforcement learning. It is used to evaluate the predictive value of the long-term reward that an agent can obtain by taking an action in a certain state. The value function provides an agent with a way to measure the advantages and disadvantages of different states and actions. Specifically, the value function can take two forms: The state value function $V^\pi(s)$ represents the expected cumulative reward that the agent can obtain when following a policy $\pi$ starting from state $s$. It measures the long-term return after executing policy $\pi$ beginning from a particular state. he action value function $Q^\pi(s, a)$, on the other hand, represents the expected cumulative reward that the agent can obtain by taking action $a$ in state $s$, and then following policy $\pi$. The main difference between the two lies in that the state value function is based solely on the state, while the action value function considers both the state and the specific action taken within that state. In reinforcement learning, strategy is the rules to bridge the gap between the perception (state) and behavior (operation) of agents. It is a component of active learning and optimization, which determines how the agent selects actions in each state to achieve the goal of the agent. According to the different decision-making methods, strategies can be divided into two categories: deterministic strategies and stochastic strategies. The core characteristics of deterministic strategy are its uniqueness and certainty. It establishes a direct and clear mapping from the state space to the action space, and directly specifies the specific actions that the agent should take in a given state. The core feature of randomness strategy is its probability and exploratory. Instead of outputting a definite action, it outputs the probability distribution of all possible actions in a given state.

The method based on value function and the method based on strategy have their own advantages and limitations. In order to achieve the best performance in complex decision-making tasks, a fusion framework that can learn from each other is very important. Actor critical framework is the mainstream paradigm emerging under this demand. It realizes the coordination and decoupling of strategy optimization and value evaluation through the exquisite dual network structure. The Actor (policy network) takes the environment state as the input and outputs the specific action, which is responsible for directly parameterizing and executing the policy; The Critic (value network) is used to evaluate the value of state action and provide a low variance gradient signal for the policy gradient update of the policy actor. The key to the efficient collaboration of actor critical framework is the Temporal Difference (TD) Error, which perfectly replaces the Monte Carlo estimation, solves its high variance problem, and realizes efficient and stable strategy learning while retaining the flexibility of strategy optimization. For example, the MADDPG algorithm used in this chapter is the excellent expansion of actor critical framework in multi-agent scenarios, which perfectly embodies its collaborative concept of "centralized training and decentralized execution." The centralized critical uses global information to evaluate the collaborative value, and the decentralized actor performs actions based on local observations, which realizes the efficiency and stability of multi-agent collaborative decision-making.

### 9.2.2 Bellman Equation and Reinforcement Learning

Behrman equation is the theoretical cornerstone of reinforcement learning. It decomposes the seemingly complex long-term decision-making problem into a recursive calculation problem based on the current immediate reward and future state value. In the framework of Markov decision process (MDP), Behrman equation provides a core mathematical tool and theoretical guarantee for solving the optimal strategy.

decision process provides a formal and standardized mathematical model for sequential decision problems. The core assumption is Markov, that is, the future state of the system depends only on the current state and the selected action, and has nothing to do with the historical state trajectory. This feature significantly simplifies the complexity of the problem, enables the recursive computer system advocated by Behrman equation to be realized, and then helps the agent to find the optimal strategy in the case of dynamic uncertainty of the environment. The core idea of Bellman equation is that it recursively expresses the value evaluation method of the optimal strategy. In MDP, the optimal value function (or state action value function) in any state can be recursively calculated by the optimal value of the current immediate reward and subsequent states. This recursive relationship profoundly embodies the principle of time consistency, that is, the optimal decision must take into account both the immediate benefits of the current action and the long-term impact on the future state.

As mentioned above, in reinforcement learning, Bellman equation constitutes the theoretical cornerstone and core iteration mechanism of value function and strategy optimization. It decomposes the long-term cumulative return into the expectation of immediate reward and future return after discount by recursion, which provides a mathematical guarantee for the convergence of the algorithm. Its two core forms, state value function and action value function, jointly define the goal of solving the optimal strategy. The bellman optimal equation can be expressed in two forms: the state value function and the action value function (Q function), specifically as follows.

The Bellman optimality equation in the form of the state value function describes the maximum cumulative reward that can be obtained by following the optimal policy starting from a certain state $s$. It indicates that the optimal value of state $s$, denoted as $V^*(s)$, is equal to the sum of the immediate reward $R(s, a)$ obtained after selecting an optimal action $a$ from the current state, and the weighted expected value of executing the optimal policy from the next state $s'$, represented as $V^*(s')$. The recursive relationship can be expressed as:

$$V^*(s) = \max_a \sum_{s'} P(s' \mid s, a) \left[R(s, a) + \gamma V^*(s')\right]. \tag{9.1}$$

Here, $R(s, a)$ denotes the immediate reward obtained after taking action $a$ in state $s$. $P(s' \mid s, a)$ represents the probability of transitioning from state $s$ to state $s'$ after executing action $a$. $\gamma$ is the discount factor, which determines the weighting of future rewards.

The Bellman optimality equation in the form of the action value function (Q-function) describes the maximum cumulative reward that can be obtained after taking an action $a$ in a given state $s$. This equation recursively computes the current immediate reward $R(s, a)$ and the value of the optimal action in the subsequent state, thus obtaining the optimal state–action value. It can be expressed as:

$$Q^*(s, a) = \sum_{s'} P(s' \mid s, a) \left[R(s, a) + \gamma \max_{a'} Q^*(s', a')\right]. \tag{9.2}$$

This equation indicates that the optimal value of taking action $a$ in state $s$ is the expected immediate reward $R(s, a)$ plus the discounted expected value of the next state $s'$. In other words, the value of each optimal state–action pair $Q^*(s, a)$ can be computed recursively through the current reward $R(s, a)$ and the discounted maximum $Q$-value of the subsequent state $s'$ after taking the optimal action $a'$. Solving a Markov Decision Process essentially means finding $V^*$ and $Q^*$. The convergence property of the Bellman equation provides the theoretical guarantee that reinforcement learning algorithms can achieve the optimal policy. Through value iteration or policy iteration methods, the Bellman equation ensures that, under certain conditions, the agent will eventually converge to the optimal policy. It is worth noting that, in multi-agent coordination or complex tasks such as cooperative environments, the standard Bellman equation needs to be extended to joint action–value functions.

## 9.3 Modeling and Environment Design of Reinforcement Learning Capture

Markov decision process (MDP) is the core mathematical framework to describe the interaction between agent and environment in reinforcement learning. It regards the interaction between the agent and the environment as a discrete time series. At each time step, the agent selects an action according to the current state, and the environment transfers to a new state according to the agent's action and current state, and feeds back an instant reward to the agent. Robots affect the environment by perceiving the current state and executing actions, and the execution of actions will cause the environment to transfer to a new state with a certain probability. At the same time, the environment feeds back the reward value to the robot according to the reward function. In order to obtain greater rewards, robots continue to explore the optimal strategy. The sequential decision problem under uncertainty can be modeled by Markov decision process.

An important characteristic of MDP is Markov property, that is, the future state only depends on the current state and action, and has nothing to do with the previous historical state. This feature greatly simplifies the complexity of the problem, so that we can use some mathematical methods and algorithms to solve the optimal strategy. Through MDP, we can abstract complex practical problems into a mathematical model, which provides a solid foundation for the design and application of reinforcement learning algorithm.

### 9.3.1 Partially Observable Markov Decision Process

In most multi-robot tasks, each robot cannot fully observe the entire environment; instead, it can only perceive partial environmental information. This leads to the Partially Observable Markov Decision Process (POMDP), which is a fundamental mathematical framework for handling decision-making problems under partial observability. A POMDP is typically represented by a seven-element tuple $S, O, A, T, R, Z, \gamma$ where $S$ denotes the set of environment states, $O$ represents the set of observation states for the robots $O = \{o_1, o_2, \ldots, o_n\}$, and $A = \{A_1, A_2, \ldots, A_n\}$ denotes the joint action space of all robots. $T(s_t, a_t, s_{t+1})$ is the state transition function, $R$ is the joint cumulative reward function, $Z$ is the observation function, and $\gamma$ is the discount factor. In the multi robot Markov decision process, each robot selects an action to interact with the environment according to its decision strategy, and then the joint actions of all robots affect the environment and get a joint reward feedback. The total cumulative reward $R$ can be expressed as:

$$R = \sum_{i=1}^{n} \sum_{t} \gamma^t r_i^t, \tag{9.3}$$

where $t$ is the number of steps in the partially observable Markov decision process, $r_i^t$ denotes the reward obtained by agent $i$ at time step $t$, and $\gamma^t$ represents the discount factor applied to future rewards.

The multi robot cooperative capture problem is described as a partially observable Markov decision process, each robot selects actions according to local observations. The joint actions will affect the environment and generate joint rewards. The goal of the robot is to maximize its own cumulative discount reward by learning the optimal strategy, while the overall goal of the multi robot system is to find the optimal

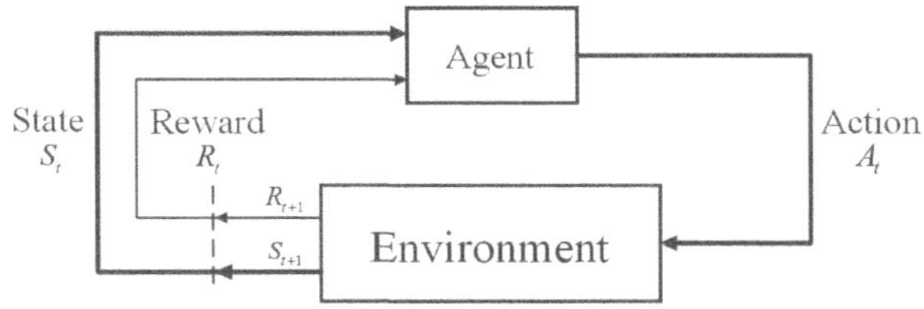


Figure 9.1: Schematic diagram of interaction between agent and environment.

joint strategy and maximize the cumulative joint reward, so as to complete the task of collaborative capture. As shown in Fig. 9.1, the interaction between agent and environment in Markov process is illustrated.

### 9.3.2 Modeling of Multi-Robot Cooperative Capture Environment

In the multi robot cooperative hunting environment, it is assumed that there are $n$ pursuers who jointly complete the task of capture $m$ escapees. There are $k$ random static obstacles in the environment, and the initial positions are generated randomly. The pursuer needs to surround the escapee in his encirclement circle to complete the encirclement task. Next, the state space, action space and reward function observed by the robot are described:

Observation state space: the observation space refers to all the information observed by the robot during the interaction with the environment. The observation state space of robot $i$ includes its own position, position relative to other robots, position relative to obstacles and position relative to targets. Obviously, with the increase of the number of robots in the environment, the dimension of observation state will also increase, which makes the processing of input state more difficult, and also poses a challenge to find the best search strategy. Specifically, the observation of robot $i$ can be represented as:

$$O_i = \begin{cases} p_i \\ p_{\text{robot}_1}, p_{\text{robot}_2}, \cdots, p_{\text{robot}_{n-1}} \\ p_{l_1}, \cdots, p_{l_k} \\ p_{o_1}, \cdots, p_{o_m}, \end{cases} \tag{9.4}$$

where $p_i$ denotes the position (or state) of robot $i$, $p_{\text{robot}_1}, p_{\text{robot}_2}, \cdots, p_{\text{robot}_{n-1}}$ represent the positions of other robots, $p_{l_1}, \ldots, p_{l_k}$ correspond to landmark information and $p_{o_1}, \ldots, p_{o_m}$ denote observations of surrounding obstacles.

Robot action space: in the robot's action space, the action space is defined as the acceleration of the robot in the $x$ and $y$ directions, denoted as $acc_x$ and $acc_y$. The initial velocities of both the pursuer and the evader are set to zero, and the maximum speed and acceleration of the evader are slightly higher than those of the pursuer.

Reward space: for the reward space, the pursuer receives a positive reward $r_{succ}$ after successfully capturing the evader. At the same time, a distance-based reward

$r_{dist}$ is added during the capture process to encourage the pursuer to approach the evader. This helps avoid situations where sparse rewards make it difficult for the robot to learn an optimal policy. The original design of the reward function is as follows:

The pursuer gets a positive reward after successfully capturing the fugitive, which is the target of the pursuer. Therefore, the target reward is defined as $r_{succ}$ as shown in formula (9.5).

$$r_{succ} = \begin{cases} k, & \text{if success} \\ 0, & \text{otherwise,} \end{cases} \tag{9.5}$$

where $k$ is a positive constant representing the reward value given when the pursuer successfully captures the evader.

The maximum process reward is denoted as $max\_reward$, and $min\_dist$ represents the minimum distance between the pursuer and the evader. Let $d_i = \|adv\_pos_i - good\_pos\|$ denote the Euclidean distance between the $i^{th}$ pursuer and the evader, where $adv\_pos_i$ and $good\_pos$ are the positions of the pursuer and the evader, respectively. The distance-based reward for the $i^{th}$ pursuer, $r_{dist}(a_i)$, is defined as follows:

$$r_{dist}(a_i) = \begin{cases} max\_reward \cdot \dfrac{min\_dist}{d_i}, & \text{if } d_i > min\_dist \\ 0, & \text{if } d_i \leq min\_dist, \end{cases} \tag{9.6}$$

In addition, robots should avoid collisions during movement. Therefore, a collision penalty $r_{collision}$ is introduced, as shown in formula (9.7). Here, $a_i$ represents the $i^{th}$ pursuer, $d_i$ is the Euclidean distance between pursuer $a_i$ and the evader, and $d_{ij} = \|adv\_pos_i - adv\_pos_j\|$ denotes the distance between two pursuers.

$$r_{collision}(a_i) = -0.2\left[\mathbb{1}(d_i \leq 0.07) + \sum_{j\neq i} \mathbb{1}(d_{ij} \leq 0.07)\right], \tag{9.7}$$

where $\mathbb{1}(\cdot)$ denotes the Indicator Function, which is defined as:

$$\mathbb{1}(\text{condition}) = \begin{cases} 1, & \text{if the condition is true} \\ 0, & \text{otherwise.} \end{cases}$$

In this context, $\mathbb{1}(d_i \leq 0.07)$ represents whether the $i^{th}$ pursuer is too close to the evader, and $\mathbb{1}(d_{ij} \leq 0.07)$ indicates whether two pursuers $i$ and $j$ are too close to each other. This penalty encourages robots to maintain safe distances and avoid collisions during cooperative pursuit.

After summing all individual rewards, the total reward for the $i^{th}$ pursuer can be expressed as follows, where $\varepsilon$ represents the total number of collision events:

$$r_{total}(a_i) = \sum_{i=1}^{n} \left(r^i_{succ} + r^i_{dist} + \varepsilon r^i_{collision}\right). \tag{9.8}$$

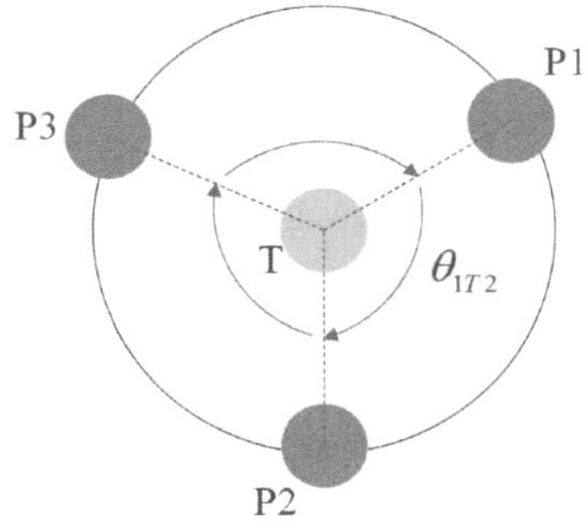


Figure 9.2: Schematic diagram of ideal hunting effect.

To achieve a uniform encirclement effect, it is assumed that the ideal encirclement consists of $n$ pursuers evenly distributed around the evader. Let the angle formed by pursuer $i$, its clockwise neighboring pursuer $j$, and the evader $T$ be denoted as $\theta_{ij}$, which represents the relative angular separation between two adjacent pursuers. For example, when three pursuers are evenly distributed clockwise around the evader $T$, the ideal encirclement configuration is illustrated in Fig. 9.2.

To ensure that the three surrounding robots are as evenly distributed around the evader as possible, let $\theta_{ij} = 120°$. During the encirclement process, a certain angular deviation is allowed, which can be set to 10°. Thus, the ideal encirclement angle can be defined as $110° \leq \theta_{iTj} \leq 130°$. Therefore, the constraint condition can be expressed as$|\theta_{iTj} - 120°| \leq 10° \quad (i, j \in N; i \neq j)$. An angle reward function $r_{\text{angle}_i}$ is designed and added to the total reward function $R$, where $r_{\text{angle}_i}$ is defined as in formula 10.9:

$$r_{\text{angle}_i} = \begin{cases} -(\theta_{iTj} - 110°)^2, & 0 \leq \theta_{iTj} \leq 110° \\ -(\theta_{iTj} - 110°)(\theta_{iTj} - 130°) + h \times \left(e^{(\theta_{3T1} - \theta_{2T1})} - 1\right), & 110° < \theta_{iTj} \leq 130° \\ l \times \left(e^{(\theta_{3T1} - \theta_{2T1})} - 1\right) + e^{(\theta_{iTj} - \theta_{1Tj})} - 1, & \theta_{iTj} > 130° \end{cases} \tag{9.9}$$

where $\theta_{iTj}$ represents the angle between the evader and two adjacent robots $i$ and $j$. $h$ and $l$ are weighting coefficients ranging from 0 to 1, which adjust the reward strengths across different angular regions.

$$r_{\text{circle}} = 10, \quad 110° \leq \theta_{ij} \leq 130°. \tag{9.10}$$

When the encircling robots form a nearly ideal surrounding formation, an additional positive reward $r_{\text{circle}}$ can be provided to enhance the coordination among robots during the capture stage, thereby improving the probability of a successful encirclement and better guiding the robots' encircling behavior. The reward $r_{\text{circle}}$ is defined as shown in formula (9.10).

## 9.4 Multi-Agent Actor-Critical Algorithm for Capture Tasks

In the multi-agent environment, the task of reinforcement learning is not only to optimize the behavior of a single agent but also to consider the interaction and cooperation between agents. The actor critical framework mentioned above provides a method combining value function optimization and strategy optimization to solve this problem efficiently. Specifically, actor is responsible for executing strategies and optimizing action selection, critical evaluates the value of state action pairs, and provides feedback signals to actor network. However, in Multi-Agent Reinforcement Learning, the nonstationarity of the environment, the interaction of strategies among agents and credit allocation make it a great challenge to directly extend the single agent actor critical method to multi-agent scenarios.

In multi-agent reinforcement learning, first, the Behrman equation needs to be extended to the form of joint action value function to effectively deal with the interaction and collaborative decision-making of multiple agents. The traditional Behrman equation usually only considers the state and action of a single agent, but in a multi-agent system, the behavior between agents is interdependent. Therefore, the introduction of joint action value function has become the core tool for strategy optimization in multi-agent systems. In a multi-agent environment, the joint action-value function $Q^\pi(s_1, a_1, \ldots, s_N, a_N)$ evaluates the value of the joint state and joint action of all agents. It can be expressed by the following equation:

$$Q^\pi(s_1, a_1, \ldots, s_N, a_N) = \mathbb{E}\left[\sum_{t=0}^{\infty} \gamma^t r_t \mid s_1^0 = s_1, a_1^0 = a_1, \ldots, s_N^0 = s_N, a_N^0 = a_N\right], \tag{9.11}$$

where $s_i$ and $a_i$ represent the state and action of the $i$-th agent, respectively. The term $r_t$ denotes the immediate reward obtained after all agents perform the joint action $(a_1, a_2, \ldots, a_N)$ at time step $t$. The Bellman equation of the joint action value function can be expressed as follows:

$$\begin{aligned} Q^\pi(s_1, a_1, \ldots, s_N, a_N) &= R(s_1, a_1, \ldots, s_N, a_N) \\ +\gamma \sum_{s_1', a_1', \ldots, s_N', a_N'} & P(s_1', a_1', \ldots, s_N', a_N' \mid s_1, a_1, \ldots, s_N, a_N) Q^\pi(s_1', a_1', \ldots, s_N', a_N'), \end{aligned} \tag{9.12}$$

where $P(s_1', a_1', \ldots, s_N', a_N' \mid s_1, a_1, \ldots, s_N, a_N)$ represents the probability of transitioning from the current states and actions $(s_1, s_2, \ldots, s_N)$ and $(a_1, a_2, \ldots, a_N)$ to the next states and actions $(s_1', s_2', \ldots, s_N')$ and $(a_1', a_2', \ldots, a_N')$. The term $Q^\pi(s_1', a_1', \ldots, s_N', a_N')$ denotes the value of joint action of the next state–action pair. The equation recursively calculates the joint value of each state action pair, helping agents in a multi-agent system to discover the optimal policy.

This section mainly uses the multi-agent deep deterministic policy gradient (MADDPG) algorithm to discuss the application of marl in the multi robot capture scene. MADDPG algorithm is specially designed to solve the continuous action space problem in multi-agent environment by extending the depth deterministic strategy gradient (DDPG) method. The actor critical framework is adopted, and the centralized training decentralized execution (CTDE) architecture is used to optimize the strategy and value evaluation of multi-agent. It has been proved that it can effectively deal with the problem of multi-agent cooperation and competition. In the previous article, we have described the expansion of the joint action value function and its Bellman equation, which provides a theoretical framework of recursive optimization strategy for multi-agent systems. However, how to effectively solve the joint-action value function in the actual algorithm is the core difficulty of multi-agent deep reinforcement learning. MADDPG algorithm shows the concrete implementation of this theory.

In the multi-robot reinforcement learning capture process, the Actor network of robot $i$ takes the observed state $o_i^t$ at time step $t$ as input and generates a specific capture action. All participating agents learn to determine the optimal joint deterministic policy and output the corresponding join-action. After interacting with the environment, each pursuer receives an immediate reward $r_i^t$ and updates its state, while the experience is stored in the replay buffer. The input of the Critic network consists of the environmental state and the joint actions, and its output is the state–action value function, which evaluates the actions generated by the Actor network given the corresponding states. Each robot's Actor network updates its policy by maximizing the expected value of the Critic function through the policy gradient method, as shown below:

$$\nabla_{\theta_i} J(\pi_i) = \mathbb{E}_{s,a}\left[\nabla_{a_i} Q_i^{\pi}(s_1, a_1, \ldots, s_N, a_N)\nabla_{\theta_i}\pi_i(o_i)\right], \tag{9.13}$$

where $\theta_i$ denotes the parameters of the Actor network for robot $i$, $Q_i(s_1, a_1, \ldots, s_N, a_N)$ represents the joint action-value function provided by the Critic network, and $\pi_i(o_i)$ is the robot policy $i$ under its local observation $o_i$. Through this mechanism, the Actor network utilizes the evaluation of the critic network of the joint action-value function for optimization, allowing each agent's policy to gradually converge to the optimal policy.

In MADDPG, the primary task of the Critic network is to evaluate the value of the current state–action pairs. The Critic network computes the value of each state–action pair through the joint action value function and updates this value using the Bellman equation. According to the Bellman expansion, the target $Q$ value of the Critic network can be calculated as follows:

$$y = r_i + \gamma Q_i'(s_1', a_1', \ldots, s_N', a_N'; \theta_i'), \tag{9.14}$$

where $s$ and $a$ denote the current state and action, and $s'$ represents the next state at the following time step. $Q_i'(s_1', a_1', \ldots, s_N', a_N'; \theta_i')$ is the joint action value estimated by the target network and is computed using the parameters $\theta_i'$ of the target network. The target value $y$ represents the temporal difference error of the Critic network.

By minimizing the TD error, the Critic network can update its evaluation of the state–action value.

To enable the Critic network to gradually approach the optimal joint action-value function, the mean squared error is used as the loss function to update the parameters $\theta_i$ of the Critic network. The loss function $L(\theta_i)$ is defined as follows:

$$L(\theta_i) = \mathbb{E}_{s,a,r,s'}\left[\left(Q_i(s_1, a_1, \ldots, s_N, a_N; \theta_i) - y\right)^2\right], \tag{9.15}$$

where $Q_i(s_1, a_1, \ldots, s_N, a_N; \theta_i)$ denotes the state–action value estimated by Critic network, and $y$ is the target value. By minimizing this loss function, the Critic network can iteratively optimize and more accurately estimate the joint action value function.

## 9.5 Combination of Voronoi Diagram and Reinforcement Learning

In the research of modern multi-agent systems, how to achieve efficient spatial collaboration and autonomous decision-making among agents is a core challenge. Traditional methods often separate path planning, task allocation and behavior control, which is difficult to adapt to the real-time requirements in complex dynamic environment. Voronoi diagram, as a classical geometric tool for spatial segmentation, and reinforcement learning, is a powerful sequential decision optimization framework. The deep integration of the two provides a new paradigm for solving the challenges of complex tasks.

Voronoi tessellation through its inherent nearest-neighbor principle dynamically partitions a distinct "area of responsibility" for each agent, thereby naturally decomposing complex global collaboration tasks into multiple parallel, structured subproblems. This not only provides the multi-agent system with provable safety guarantees for collision avoidance but also significantly reduces the search space and learning difficulty of cooperative strategies. However, although Voronoi tessellations offer a spatial partition based on the current positions of the agents, which evolves dynamically with their movement and provides a real-time changing spatial structure, they are inherently a passive geometric description tool. They merely reflect the current local spatial structure and cannot actively perform goal-oriented decision optimization. This is precisely where reinforcement learning demonstrates its distinct advantage. The integration of the two can establish complementary benefits between "spatial constraints" and "policy learning," enabling agents to perform more efficient zonal perception and policy optimization within the spatial domain. Within the reinforcement learning framework, Voronoi tessellations can be utilized for state representation, action constraints, or reward function design, thereby incorporating spatial topological features into the decision-making process. This guides the agents to learn how to optimize their spatial positions while ensuring safety, in order to collaboratively accomplish complex tasks such as coverage and capture. The following section introduces the integrated framework of these two methodologies:

Multi-Module Voronoi–RL Integration Framework: To address the integration of Voronoi diagrams and reinforcement learning (RL), a three-layer fusion architecture is constructed, comprising information fusion, reward coordination, and constraint control.

(1) Information Fusion Module:

The Voronoi topological structure is embedded into the agents' observation states, achieving a structured integration of geometric information and the decision-making network. By extending the state space and encoding spatial features, this layer provides spatial cognitive support for policy learning.

Traditional reinforcement learning usually makes decisions only based on local states, which is difficult to effectively describe the spatial topology and neighborhood interaction structure between agents, thus limiting its global coordination ability in multi-agent cooperation and spatial tasks. By dividing the environment into dynamically updated regions, Voronoi tessellation provides each agent with explicit neighborhood structure and spatial constraints, enabling it to integrate spatial distribution features in the decision-making process, and realize structured perception of the environment and more efficient collaborative optimization. In the traditional MADDPG framework, the observation state $o_i^t$ of agent $i$ mainly includes basic information such as position and velocity. After incorporating Voronoi diagrams, it becomes necessary to expand the state space to include spatial topological features as additional observations. For example, each agent's Voronoi cell area, boundary length, and number of neighboring cells can be encoded as geometric features, forming a high-dimensional state representation to be fed into the decision-making network. Compared with traditional observations of position and velocity, this extended observation provides the policy network with structured information about spatial partitioning. The extended observation state $o_i^{V,t}$ can be expressed as:

$$o_i^{V,t} = [o_i^t, V_i^{\text{area}}, V_i^{\text{boundary}}, N_i^{\text{adj}}, T_i^{\text{relation}}], \tag{9.16}$$

where $V_i^{\text{area}}$ represents the Voronoi cell area of agent $i$, reflecting the size of its responsibility region; $V_i^{\text{boundary}}$ denotes The total boundary length of the Voronoi cell corresponding to agent $i$ is used to characterize the geometric complexity of its controlled region and the degree of local spatial constraints; $N_i^{\text{adj}}$ indicates the number and relative positions of neighboring agents, representing the local topological structure; and $T_i^{\text{relation}}$ specifies whether the evader is located within the Voronoi region of the current pursuer agent. The extended observation state also provides supplementary information for subsequent reward design and action constraints, offering valuable prior knowledge for spatial coordination.

(2) Reward Coordination Module:

Reward functions are designed based on Voronoi geometric features to guide agents in learning spatially efficient cooperative strategies. Through reward shaping, geometric optimization objectives are transformed into learnable signals, which improve policy convergence efficiency and enhance global coordination.

In order to improve the cooperation efficiency of multi robot system in task space, this paper constructs a coverage balance reward based on Voronoi partition. The reward function uses the geometric characteristics of Voronoi elements to measure the spatial distribution uniformity of robots in the task area, so as to guide each robot to dynamically adjust its position and maintain a reasonable spatial separation and coverage ratio. In this way, it can not only optimize the global coverage effect of the group but also reduce the redundancy and collision risk between individuals, and achieve more efficient task execution and collaboration stability. In addition, the coverage balance reward not only ensures uniform distribution but also helps to improve the detection efficiency and the success rate of rounding up, providing more stable environmental feedback for subsequent strategy learning. The coverage balance reward $R^{balance}$ can be defined as:

$$R^{balance} = \alpha_{balance} \frac{\min_i V_i^{area}}{\max_i V_i^{area}}, \tag{9.17}$$

where $\alpha_{balance}$ denotes the maximum value of the coverage balance reward, which is set to 5. When all Voronoi cells have nearly equal areas, the ratio approaches 1, and the reward reaches its maximum. Conversely, when the area difference among Voronoi regions is large, the ratio approaches 0, resulting in a smaller reward. This definition directly reflects the uniformity of the Voronoi cell distribution.

(3) Constraint Control Module:

Buffered Voronoi regions are utilized to constrain the agents' action spaces, ensuring that the generated actions satisfy spatial allocation and safety principles. By applying action filtering and feasibility verification, this layer achieves a balance between hard constraints and soft optimization.

In the multi robot cooperative capture task, avoiding the collision between individuals is the key factor to ensure the task efficiency and system stability. This section improves the action selection mechanism in MADDPG method, that is, changing the action selection function $take_action$ to ensure that the robot will not leave the buffered Voronoi area after executing the action. At each time step, the buffered Voronoi diagram is used to generate the feasible action space for each robot, and the original discrete action space is mapped to the candidate action set in the Voronoi region.

Action Filtering Mechanism: During the cooperative encirclement process in multi-robot systems, each robot needs to select its next action based on the current state. However, since the action space is typically discrete (e.g., "move forward," "move backward," "turn left," "turn right"), each action produces a different effect. To avoid boundary-crossing behaviors, this work introduces an action filtering mechanism. Before executing an action, the robot predicts the resulting target position to ensure that the chosen action does not cause the robot to cross the Voronoi boundary. Each robot determines its candidate action $a_{ik}$ based on its current state $p_i$. The target position after executing the candidate action is predicted as follows. Assuming the candidate action $a_{ik}$ of robot $i$ is represented by its velocity and direction $(v_{ik}, \theta_{ik})$, the corresponding target position can be expressed as:

$$p_i^{\text{next}}(a_{ik}) = p_i + v_i \cdot \Delta t \cdot (\cos(\theta_i), \sin(\theta_i)), \tag{9.18}$$

where $\Delta t$ is the time step, and $p_i^{\text{next}}(a_{ik})$ denotes the position of robot $i$ after executing action $a_i$. Once the target position is obtained, it can be used in formula (9.19) to determine whether the selected action crosses the boundary. This equation serves as the filtering condition for boundary-violation actions.

$$p_i^{\text{next}}(a_{ik}) \in B_i \ \& \ d(p_i^{\text{next}}(a_{ik}), p_j) > 2\omega, \tag{9.19}$$

where $B_i$ represents the buffered Voronoi region of robot $i$, and $d(p_i^{\text{next}}(a_{ik}), p_j) = \|p_i^{\text{next}}(a_{ik}) - p_j\|$ denotes the Euclidean distance between the position of robot $i$ after executing the action and the position of another robot $j$. The safe action set $A_i^{\text{safe}}$ contains all feasible actions that ensure robot $i$ remains within its buffered Voronoi region boundary and avoids collisions with other robots after executing the action. It is defined as follows:

$$A_i^{\text{safe}} = \{a_{ik} \in A_i \mid d(p_i^{\text{next}}(a_{ik}), p_j) > 2\omega, \ \forall j \neq i, \ p_i^{\text{next}}(a_{ik}) \in B_i\}. \tag{9.20}$$

After obtaining the safe action set $A_i^{\text{safe}}$, the optimal action can be selected from the set according to the current policy. Under the framework of action selection and policy optimization, the output of the policy network is a probability distribution over the actions in the safe action set. To achieve differentiable optimization, the Gumbel–Softmax mechanism is introduced to approximate the discrete action distribution, allowing the gradient to pass through the discrete space and supporting backpropagation training of the policy network. After the output layer (*logits*) of the policy network, an action filtering operation is applied to eliminate potential boundary-crossing actions, ensuring that the execution point of each selected action remains within the buffered Voronoi region. Only the feasible actions filtered by the action filtering mechanism can be selected, ensuring that the pursuers maintain a collision-free and spatially consistent formation throughout the training process.

The three modules form a bottom-up closed-loop structure that extends from information perception to decision execution. This architecture not only maintains the adaptability and self-learning capability of reinforcement learning but also integrates the geometric optimization advantages of the Voronoi diagram to achieve dynamic spatial coordination. Through continuous interaction between perception, decision-making, and control, the system can effectively adapt to environmental changes, dynamically adjust the distribution of agents, and optimize cooperative strategies in real time. Consequently, the proposed framework enables more stable and efficient cooperative control in complex, dynamic spatial environments, ensuring both task performance and system robustness.

## 9.6 Attention-Enhanced MADDPG Capture Algorithm

Although maddpg algorithm can ensure the stability of the training environment, critical will comprehensively consider the state and action of other robots when evaluating robot behavior. However, with the increase of the number of participants in the environment, the input dimension of critical increases, which leads to the challenges of dimension explosion and information redundancy in information processing.

Critical is difficult to extract the part that is really useful for robot decision-making from a large amount of environmental information, which may lead to inaccurate action evaluation. At this time, how to effectively extract the information that is crucial to decision-making has become a key challenge. In order to solve this problem and enhance the information processing ability of robot in complex dynamic environment, this section proposes to integrate self-attention mechanism into Multi-Agent Reinforcement Learning. By introducing the attention mechanism into the critical network of each robot, the robot can dynamically and selectively focus on the most critical factors for decision-making according to the current state and global information.

### 9.6.1 Design of Self Attention Module

The attention mechanism is a weighting-based technique used to selectively focus on the most important parts of the input data. In reinforcement learning, the attention mechanism can effectively help agents extract key features from complex environmental information, thereby making more informed decisions. The basic idea is to measure the correlation between each input and the current policy, and assign different weights to the information based on this correlation. The attention function can be described as mapping a query and a set of key–value pairs to an output, where the Query $\mathbf{Q}$, Keys $\mathbf{K}$, and Values $\mathbf{V}$ are all vectors. The implementation of the attention mechanism proceeds as follows. Assume the input vector is $\mathbf{A}$. First, the query, key, and value spaces are constructed to perform the attention weight computation by projecting the input vector $\mathbf{A}$ into these spaces. The projections yield new vectors $\mathbf{Q}$, $\mathbf{K}$, and $\mathbf{V}$. Then, the similarity between $\mathbf{Q}$, $\mathbf{K}$, and $\mathbf{V}$ is computed using a dot-product operation, and the resulting attention weights reflect the relative importance of each piece of information to the current task. The mapping process is defined as:

$$\mathbf{Q} = \mathbf{A} \cdot \mathbf{W}^k = [q^1, q^2, \ldots, q^m], \tag{9.21}$$

$$\mathbf{K} = \mathbf{A} \cdot \mathbf{W}^k = [k^1, k^2, \ldots, k^m], \tag{9.22}$$

$$\mathbf{V} = \mathbf{A} \cdot \mathbf{W}^v = [v^1, v^2, \ldots, v^m]. \tag{9.23}$$

In a multi-agent system, each robot makes decisions based on its own partial observations and the global information exchanged with other agents. To enhance the Critic network's capability of processing global information, a self-attention module is designed in this study. The core idea of this module is to transform the input state and action information into query ($\mathbf{Q}$), key ($\mathbf{K}$), and value ($\mathbf{V}$) vectors, and compute attention scores among these vectors to dynamically select the parts most relevant to the current decision-making process.

In a multi-agent system, each robot makes decisions based on its own partial observations and the global information exchanged with other agents. To enhance the Critic network's capability of processing global information, a self-attention module is designed in this study. The core idea of this module is to transform the input state and action information into query ($\mathbf{Q}$), key ($\mathbf{K}$), and value ($\mathbf{V}$) vectors, and compute attention scores among these vectors to dynamically select the parts most relevant to the current decision-making process. The computation of $\mathbf{Q}$, $\mathbf{K}$, and $\mathbf{V}$ is shown in formula (9.21), (9.22), and (9.23). In the implementation, the input of the attention module consists of the states and actions of all agents. After being processed by the attention layer, each robot can dynamically adjust its attention weights based on the information obtained. During decision-making, the agents can not only focus on their own local observations but also consider the behaviors of other agents and the overall environmental state, thus making more efficient and cooperative decisions in dynamic environments.

The key advantage of the attention mechanism lies in its ability to dynamically adjust the focus of attention according to different task stages. At different phases of the encirclement process, agents may need to focus on different factors. In the initial searching stage, the pursuer uses its current state (such as position and velocity) as the query vector $\mathbf{Q}$, which is matched with the state information of surrounding agents and the evader as the key vector $\mathbf{K}$, to determine which regions or neighboring agents should receive more attention. As the pursuer gradually approaches the target, in the approaching stage it focuses on its relative distance and direction to the evader, using these as queries $\mathbf{Q}$, while the corresponding keys $\mathbf{K}$ represent the relative position, velocity, and possible acceleration of the evader. During the final capture stage, agents concentrate on the relative positions of other agents and obstacles, as well as collision avoidance strategies, in order to efficiently coordinate and complete the encirclement task. At each decision step, the agent selects an appropriate query $\mathbf{Q}$ based on its current state and matches it with the surrounding environment or the state information of other agents represented by $\mathbf{K}$, to compute the attention weights.

Specifically, after obtaining the required query and key vectors $\mathbf{Q}$ and $\mathbf{K}$, the attention weights are calculated using the dot-product operation, as shown in formula (9.24):

$$Attention\ Score = \frac{\mathbf{Q}_i \cdot \mathbf{K}_j^{\mathrm{T}}}{\sqrt{d_k}}, \tag{9.24}$$

where $\mathbf{Q}_i$ denotes the query vector of pursuer $i$, representing its current state, and $\mathbf{K}_j$ denotes the key vector of pursuer or evader $j$, representing the corresponding state information. $d_k$ indicates the dimensionality of the key vector, which is used to normalize the dot product and prevent excessively large values.

Next, the attention scores are normalized using the *softmax* function to obtain the attention weights corresponding to each key vector $\mathbf{K}$. The purpose of normalization is to ensure that the sum of all attention weights equals one, so that the influence of each agent or obstacle related to the current pursuer's decision is proportionally distributed. The normalization process is defined as follows:

$$\alpha_{ij} = \frac{\exp\left(\frac{\mathbf{Q}_i \cdot \mathbf{K}_j^{\mathrm{T}}}{\sqrt{d_k}}\right)}{\sum_k \exp\left(\frac{\mathbf{Q}_i \cdot \mathbf{K}_k^{\mathrm{T}}}{\sqrt{d_k}}\right)}, \tag{9.25}$$

where $\alpha_{ij}$ represents the attention weight assigned by pursuer $i$ to pursuer or evader $j$, and $\mathbf{Q}_i \cdot \mathbf{K}_j^{\mathrm{T}}$ denotes their similarity measure. The denominator $\sum_k \exp\left(\frac{\mathbf{Q}_i \cdot \mathbf{K}_k^{\mathrm{T}}}{\sqrt{d_k}}\right)$

represents the normalization of the similarity scores between the pursuer and all other agents or evaders. Finally, the attention weights are used to compute the weighted sum of the value vectors, yielding the final attention output. Through this process, each agent can dynamically adjust its policy representation based on the relative importance of the information. The formula is as follows, where $\mathbf{V}_j$ is the value vector of agent $j$:

$$\mathbf{V}_i^{\text{attn}} = \sum_j \alpha_{ij} \mathbf{V}_j, \tag{9.26}$$

The above formulas (9.24), (9.25) and (9.26) describe the calculation process of self-attention mechanism to help agents dynamically adjust their focus of attention according to the changes of environmental information in different stages of hunting. By calculating the similarity between Query and Key, the agent can determine which information is the most important for the current decision, and selectively focus on these information through weighted summation, effectively optimize the hunting process, and improve the collaborative efficiency of the agent in the complex dynamic environment.

### 9.6.2 MADDPG Algorithm with Integrated Attention Mechanism

In order to make the robot give different attention weights to the states and actions of its adjacent robots in the process of reinforcement learning training, the self-attention model is embedded into the Multi-Agent Reinforcement Learning algorithm. The state of the robot reflects the current information, while the action predicts the future information. Therefore, defining different attention levels for States and actions allows the extraction of the most critical current and future information. By modifying the critical network in MADDPG, the self-attention mechanism is introduced to improve the performance of the system in complex multi-agent environment. MADDPG adopts the mode of centralized training decentralized execution, so that each agent can share the global information in the training phase, so as to better cooperate and learn the optimal strategy. After introducing the attention mechanism, each agent's critical network can not only evaluate the value of the current state action pair, but also dynamically adjust the attention to the behavior of other agents according to the current state.

Specifically, the input of the critical network includes the global state and the action information of all agents. After being processed by the self-attention mechanism, a more accurate state action value function is output. This value function will be used to guide the actor network of each agent to adjust its strategy to optimize the execution efficiency and stability of the capture task. The attention mechanism ensures that each agent can use the global information more effectively in the training process, which improves the convergence speed and learning quality of the strategy. The critical network structure after adding the attention mechanism is shown in Fig. 9.3. The attention module is used to identify the state action pair with the highest relevance to the current decision, and output the weighted fusion of global information representation.

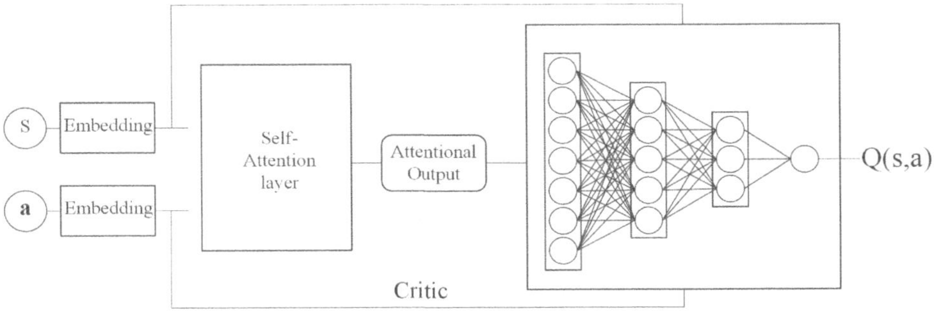


Figure 9.3: Critic framework integrated with self-attention mechanism.

The pseudo code of attention enhanced MADDPG capture algorithm is shown in algorithm 11. Based on the gradient framework of multi-agent depth deterministic strategy, the algorithm combines the buffering Voronoi region constraint and self-attention mechanism to enhance the cooperation and the efficiency of the capture among multi-agent. In each training round, each agent weights and fuses the multi-source state information through the self-attention module according to the current global state and neighbor state, so as to highlight the key interaction characteristics and realize the dynamic strategy adjustment. Then, the agent selects actions under the condition of following the constraint of buffering Voronoi area, interacts with the environment and obtains new status and reward signals, so as to complete a policy update.

### 9.6.3 Simulation Results and Analysis

This section uses the `simple_tag_v3` environment provided by PettingZoo to build a multi robot cooperative capture simulation system. Taking three pursuers capture a evader as an example, this environment can effectively simulate the cooperation and confrontation behavior between agents. The environment state space contains the complete state information of all agents, and the action space supports continuous action selection. In the experimental setup, the maximum number of steps per training episode, `episode_length`, is fixed at 50, and the agent position and obstacle distribution are initialized randomly at the beginning of each round. The training adopts the improved MADDPG algorithm framework, in which each pursuer agent learns the strategy through the Actor-Critic network. The algorithm uses action constraints based on Gumbel-softmax function [25] to ensure that the agent's actions are always kept within the buffered Voronoi area and the Critic network design with enhanced attention, so as to improve the global cooperation ability. To verify the effectiveness of the algorithm, this section compares the MADDPG capture method under buffered Voronoi with the benchmark algorithm. The training environment contains three pursuers, one evader and two static obstacles, which are represented by red, green and black spheres respectively. The basic settings of the training environment are the same. Specific simulation settings and training parameters are shown in as shown in Tables 9.1 and 9.2.

**Algorithm 11** Attention-Enhanced Multi-Agent Deep Deterministic Policy Gradient

**for** episode = 1 to 100000 **do**
Reset environment for each agent $i$ and obtain initial observation $o_i$
Initialize the global state $s$
Initialize cumulative update step $T_{\text{update}} \leftarrow 0$
Compute the buffered Voronoi region for each agent based on the initial state
**for** $t = 1$ to max-decision-step **do**
  **for** each agent **do**
    Obtain the current observation $o_i$ and all agents' joint state
    Apply the self-attention mechanism to all agent states and compute $\mathbf{Q}, \mathbf{K}, \mathbf{V}$
    Select action $a_i$ from the policy distribution
    Use boundary constraints to ensure the action remains inside its buffered Voronoi region
    Execute joint action $\mathbf{a} = [a_1, a_2, \ldots, a_n]$ and obtain rewards $\mathbf{r} = [r_1, r_2, \ldots, r_n]$ and new state $s'$
    Store $(s, \mathbf{a}, \mathbf{r}, s')$ into the replay buffer
  **end for**
  **if** replay buffer size > minimal size **then**
    Sample a mini-batch from the replay buffer
    **for** agent $i = 1$ to 4 **do**
      Set $y_i = r_i + \gamma Q'(s', a')$, and let $a'_i = \pi_i(s')$
      Update critic by minimizing loss $L(\theta_i) = \mathbb{E}_{s,a,r,s'}[(Q_i(s, a_1, \ldots, a_n; \theta_i) - y_i)^2]$
    **end for**
    Compute TD error $|\delta_i|$ for all agents, update sample importance, and reorder samples
    Update the priority $P(e)$ of samples in the replay buffer
  **end if**
  **for** each agent **do**
    Update the actor using policy gradients
    Soft update the target network parameters
  **end for**
  Update cumulative step counter $T_{\text{update}} \leftarrow T_{\text{update}} + 1$
**end for**
**end for**

In the baseline MADDPG training environment without any constraints, all pursuers' behaviors in each training episode depend entirely on their current states and action selections. The capture performance during training is shown in Fig. 9.4, and the pursuers' reward curves are shown in Fig. 9.5. During training, the absence of proper spatial constraints causes frequent collisions among robots, which increases negative rewards. Consequently, the reward curves often exhibit large oscillations, slowing down the policy learning process and reducing capture efficiency. Within a

Table 9.1: Parameters of the two-dimensional capture simulation environment

| Parameter | Value | Parameter | Value |
|---|---|---|---|
| Environment size | $1m\times 1m$ | Maximum speed of the evader | 1.3m/s |
| Radius of evader | 0.025m | Maximum speed of the pursuer | 1.0m/s |
| Radius of pursuer | 0.045m | Radius of obstacle | 0.1m |
| Acceleration of evader | $4m/s^2$ | Number of evaders | 1 |
| Acceleration of pursuer | $3m/s^2$ | Number of pursuers | 3 |

total of 50 training steps, the pursuers fail to form an effective encirclement by Step 24, with much of the time spent on path correction rather than cooperative advancement. The capture task is finally completed at Step 46. Moreover, since the baseline algorithm lacks structured constraints and global information filtering mechanisms, the pursuers tend to greedily chase the target while neglecting teamwork and coordination. This behavior often leads the strategy into shortsighted actions and local optima, resulting in initially high cumulative rewards that ultimately converge to a lower level.

The MADDPG framework combined with buffered Voronoi partitions divides the pursuers' operational space into safe regions, effectively reducing collisions among them. However, due to these spatial constraints, the initial capture behavior tends to be rigid—especially when the robots have not yet fully adapted to the regional partitioning. As a result, they may struggle to quickly learn optimal capture strategies, leading to less effective early-stage capture. As training progresses, the introduction of Voronoi constraints promotes better cooperation among the pursuers, significantly reducing collision conflicts. The pursuers can then focus more on target pursuit and collaborative coordination. Compared with the baseline algorithm, the

Table 9.2: Core training parameter settings of MADDPG

| Parameter | Value | Description |
|---|---|---|
| `actor_lr` | 1e-2 | Learning rate of the Actor network |
| `critic_lr` | 1e-2 | Learning rate of the Critic network |
| $\gamma$ | 0.95 | Discount factor (for future rewards) |
| `tau` | 1e-2 | Soft update parameter |
| `episode_length` | 50 | Steps per episode |
| `num_episodes` | 100000 | Total number of episodes |
| `batch_size` | 256 | Number of samples per batch from replay buffer |
| `buffer_size` | 100000 | Capacity of the replay buffer |
| `update_interval` | 100 | Steps between each network update |
| `hidden_dim` | 128 | Dimension of hidden layers in Actor/Critic networks |

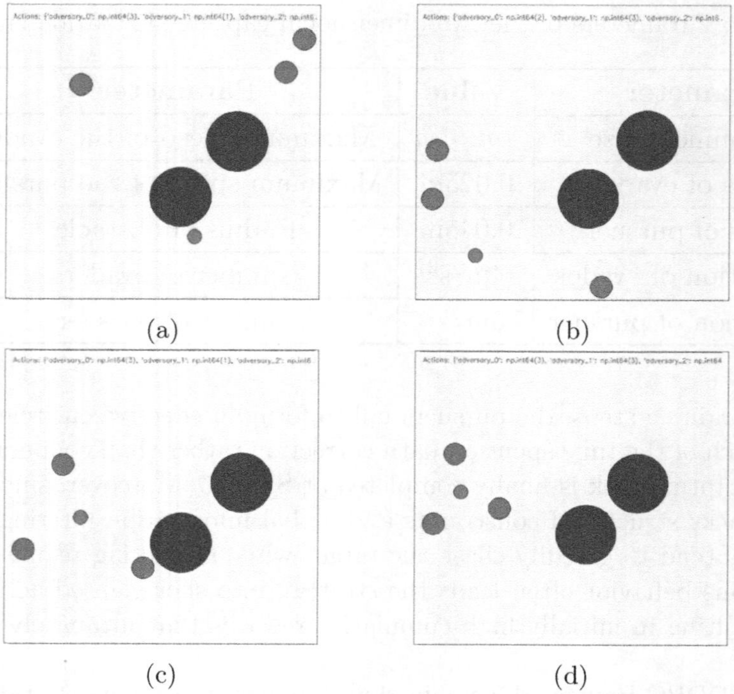


Figure 9.4: Baseline MADDPG capture process: (a) $Step = 0$; (b) $Step = 28$; (c) $Step = 39$; (d) $Step = 46$.

buffered Voronoi-based approach achieves faster capture efficiency, completing the capture task at Step 41. The training performance and pursuer reward curves are shown in Figs. 9.6 and 9.7, respectively. The buffered Voronoi constraint, through explicit spatial partitioning and structured action filtering, enables controlled oscillations and stable convergence during multi-robot cooperative capture. Since it forces the pursuers' movement endpoints to remain within their assigned regions, the robots must explore effective capture paths within limited spatial boundaries during training,

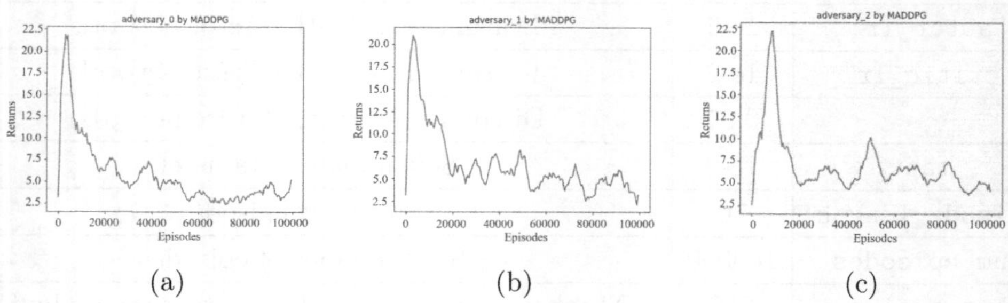


Figure 9.5: Cumulative rewards per pursuer under the baseline algorithm: (a) Cumulative reward of pursuer 0; (b) cumulative reward of pursuer 1; (c) cumulative reward of pursuer 2.

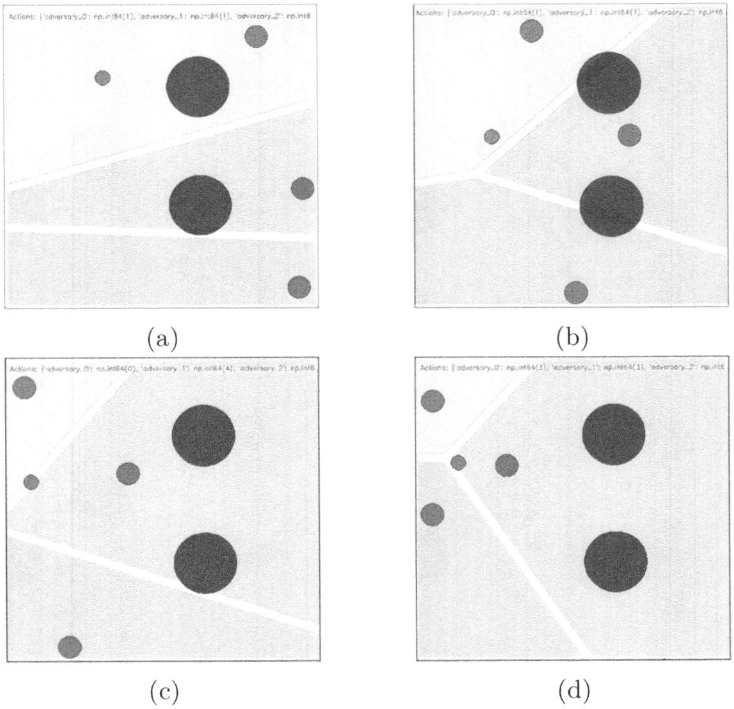

(a) (b) (c) (d)

Figure 9.6: MADDPG capture process under Buffered Voronoi constraints: (a) $Step = 0$; (b) $Step = 24$; (c) $Step = 33$; (d) $Step = 41$.

thereby increasing action selection complexity. The oscillations observed in the reward curves mainly stem from the trade-off between exploration efficiency and spatial restrictions. As training continues, the policy gradually converges, though occasional local adjustments and incomplete stabilization can still cause minor fluctuations. Ultimately, convergence is achieved through the synergistic effects of enforced regional collaboration and reward constraint mechanisms.

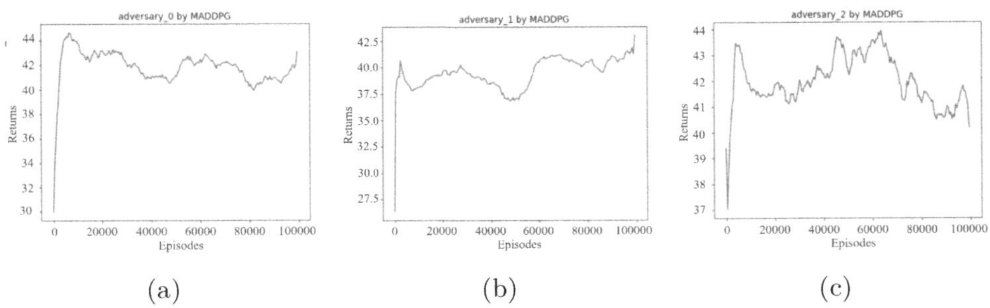


(a) (b) (c)

Figure 9.7: Cumulative rewards per pursuer under Buffered Voronoi constraints: (a) Cumulative reward of pursuer 0; (b) cumulative reward of pursuer 1; (c) cumulative reward of pursuer 2.

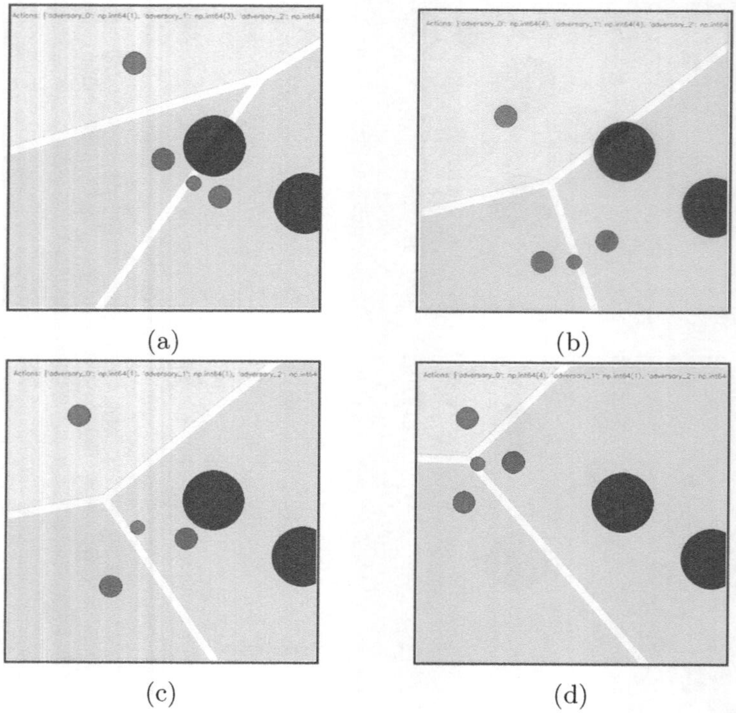


Figure 9.8: The capture process of MADDPG enhanced with attention mechanism and Buffered Voronoi constraints: (a) $Step = 0$; (b) $Step = 24$; (c) $Step = 28$; (d) $Step = 37$.

After incorporating the attention mechanism into the buffered Voronoi-based framework, the robots are able to effectively extract key information from complex and dynamic environments, thereby improving decision quality. This enhancement allows the robots to select the most efficient strategies within a limited time frame. From the capture states, it can be observed that by Step 24, the pursuers have already formed a clear capture trend, and by Step 37, the final capture task is completed—demonstrating higher capture efficiency compared to the buffered Voronoi case. The capture states and training curves are shown in Figs. 9.8 and 9.9, respectively. The attention-enhanced MADDPG achieves low oscillation and high convergence through deep coordination between dynamic information filtering and structured spatial constraints. The introduction of the attention mechanism requires the agents to continuously adjust their focus during the exploration phase, which leads to fluctuations in the reward curves. The final convergence of training indicates that the robots have successfully learned to optimize their capture strategies through the attention mechanism, with the reward values ultimately stabilizing at a higher level.

Building upon the aforementioned simulation analysis, this section further introduces other typical multi-agent reinforcement learning algorithms for comparative evaluation to validate the superiority of the proposed attention-enhanced MADDPG method under buffered Voronoi constraints. To ensure the scientific rigor and

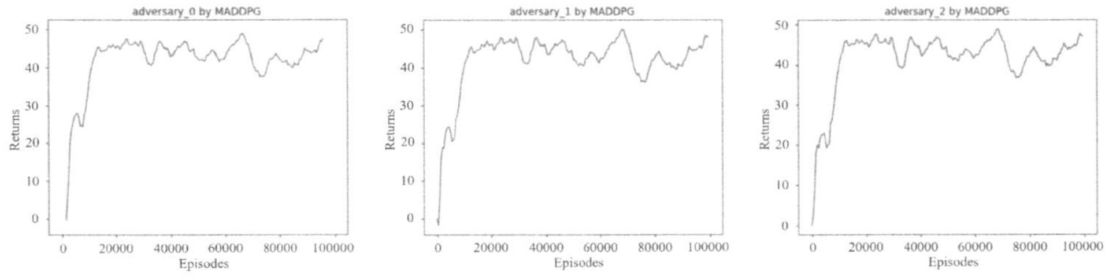


(a) (b) (c)

Figure 9.9: Cumulative rewards per pursuer under Buffered Voronoi constraints and attention enhancement: (a) Cumulative reward of pursuer 0; (b) cumulative reward of pursuer 1; (c) cumulative reward of pursuer 2.

reproducibility of the comparative experiments, this study will utilize the BenchMARL framework developed by Facebook Research to conduct a systematic evaluation of multi-agent reinforcement learning algorithms within the `simple_tag` environment of PettingZoo MPE. BenchMARL [26], as a standardized MARL training library, provides a unified interface that standardizes the training pipelines and evaluation criteria across different algorithms. It ensures high-performance computing and access to the latest algorithm implementations through its TorchRL backend, along with a flexible Hydra configuration system that supports parameter management and experimental reproducibility.

The capture success rate serves as the fundamental performance indicator for evaluating multi-agent cooperative pursuit missions. It quantitatively measures the effectiveness of pursuer teams in successfully encircling and capturing evaders within constrained operational parameters. In multi-agent cooperative pursuit tasks, the criterion for judging a successful capture primarily manifests as the effective encirclement of the target individual and the system reaching a stable cooperative state. A pursuit is considered successful when, within a limited number of time steps, all pursuers collaboratively adjust their positions through coordinated strategies to form a geometric encirclement around the evader. This requires that the distance between every pursuer and the evader is less than a predefined capture threshold, while the pursuers themselves maintain an approximately uniform angular distribution around the target, avoid collisions with each other, and sustain a relatively stable spatial configuration. This comprehensive criterion integrates distance constraints, spatial encircling properties, and angular equilibrium, effectively reflecting the multi-agent system's capability to form coordinated encirclements, restrict escape routes, and achieve collaborative capture in dynamic environments. The final success rate equals the number of successful pursuit episodes divided by the total number of test episodes.

In order to verify the effectiveness of the proposed method, this paper selects a variety of representative Multi-Agent Reinforcement Learning Algorithms for comparative experiments. It mainly includes: multi-agent deterministic policy gradient algorithm (MADDPG), multi-agent near end policy optimization algorithm (MAPPO), multi-agent trust domain policy optimization algorithm (MATRPO), multi-agent

Table 9.3: Comparison of characteristics of different training algorithms

| Algorithm | Learning paradigm | Policy/Action | Centralized critic |
|---|---|---|---|
| MADDPG [27] | *Off-policy* | Deterministic action | Inputs global state and joint actions |
| MASAC [28] | *Off-policy* | Maximizing entropy strategy | Double Q-network structure |
| MATD3 [29] | *Off-policy* | Deterministic action | Dual Critic + delayed update |
| MAPPO [30] | *On-policy* | Action probability distribution | Inputs global state or partial observation concatenation |
| MATRPO [31] | *On-policy* | Action probability distribution | Credit assignment based on neighboring policies |

extension algorithm based on soft actor critical (MASAC) and multi-agent dual delay deterministic policy gradient algorithm (MATD3). Maddpg, as a benchmark algorithm, can handle continuous action space tasks; Mappo uses the on policy update method, which can better track policy changes, but the sample utilization rate is low; MATRPO achieves stable policy updates through KL constraints, but it has insufficient exploration ability in complex environments; MASAC introduces the maximum entropy mechanism to enhance the exploration ability, which is suitable for high-dimensional continuous action space; While MATD3 alleviates the over estimation problem through double reviewers and delayed update mechanism, it is sensitive to implementation details. The main characteristics of different algorithms are shown in Table 9.3, and the hyper parameter settings of each algorithm are shown in Table 9.4. All algorithms are trained in the same environment, and the training rounds are set

Table 9.4: Hyperparameters used for SAC, PPO, DDPG, TRPO, and MADDPG

| Hyperparameter | MADDPG | MASAC | MATD3 | MAPPO | MATRPO |
|---|---|---|---|---|---|
| `Buffer_size` | 100000 | 100000 | 100000 | - | - |
| `Batch_size` | 256 | 2048 | 256 | 2048 | 2048 |
| `Actor_lr` | 1e-2 | 1e-2 | 1e-2 | 2.5e-4 | 2.5e-4 |
| `Critic_lr` | 1e-2 | 1e-2 | 1e-2 | 2.5e-4 | 2.5e-4 |
| Entropy coefficient ($\beta$) | - | - | - | 5e-2 | 1e-2 |
| Discount factor ($\gamma$) | 0.95 | 0.95 | 0.95 | 0.95 | 0.95 |
| Clip range ($\epsilon$) | - | - | - | 0.2 | 0.01 |
| Target update ($\tau$) | 0.005 | 0.005 | 0.005 | 0.005 | - |

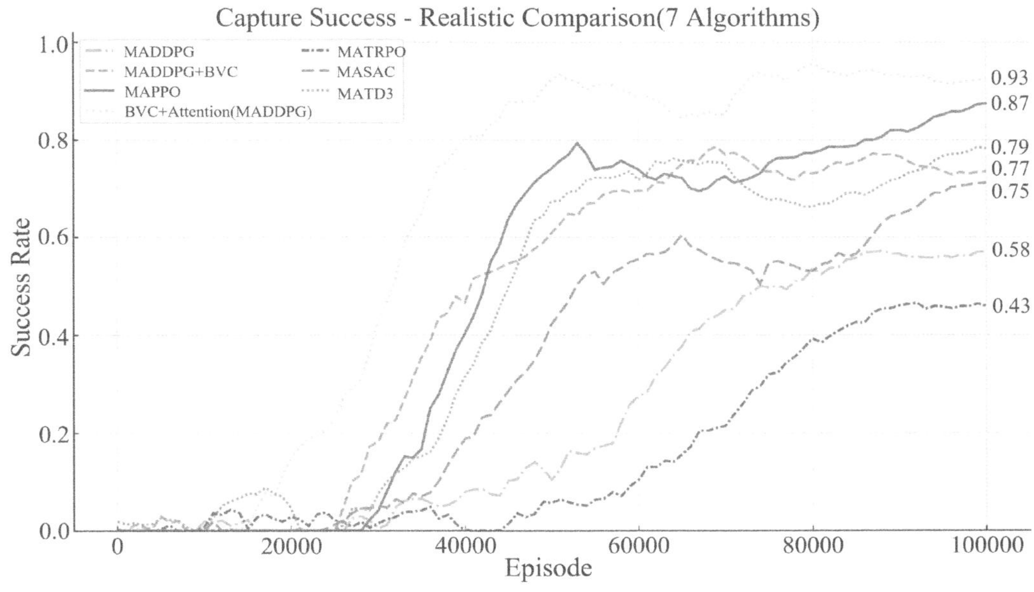


Figure 9.10: Comparison of capture success rates across different algorithms.

to 100000. One time energy evaluation is conducted every 500 rounds to ensure the fairness and reliability of the comparison results.

The comparison of success rates of different algorithms in the capture task is shown in Fig. 9.10, experimental results indicate that significant differences exist among various reinforcement learning algorithms in terms of convergence speed, stability, and final capture success rate. Among them, the BVC+Attention (MADDPG) method demonstrates the best overall performance. This approach integrates buffered Voronoi-based spatial partitioning and constraint mechanisms along with a self-attention mechanism into the MADDPG framework. Specifically, the buffered Voronoi method dynamically partitions the environment, assigns each agent an independent responsibility region, and introduces a buffer zone along the boundaries. This achieves geometric spatial division at the structural level and explicit spatial constraints at the control level. These constraints effectively limit the action range of agents, preventing inter-agent collisions and boundary violations, thereby ensuring the safety and stability of the policy during training. Meanwhile, this partitioning and constraint mechanism collectively facilitate spatial coordination and task allocation among multiple agents. The self-attention mechanism introduces dynamic feature selection at the information level, enabling the Critic network to automatically focus on the most critical state and action information for the current decision-making process. This significantly mitigates issues related to information redundancy and gradient instability caused by high-dimensional inputs. The synergistic effect of these two components allows the algorithm to achieve rapid and stable convergence after approximately 40,000 training episodes, ultimately reaching a capture success rate of 0.93, which is markedly superior to other methods.

As an on-policy policy optimization algorithm, MAPPO achieves stable policy iteration through trust region constraints and a clipped update mechanism, exhibiting high robustness and a final success rate of approximately 0.87. However, it is constrained by low sample efficiency and high training costs. MATD3 utilizes dual Critic networks and a delayed update mechanism to reduce Q-value overestimation, striking a balance between exploration and convergence stability, with a final success rate of 0.79. The MADDPG+BVC method, which incorporates buffered Voronoi tessellations, also benefits from its spatial constraints, significantly reducing collision risks in the early training stages and enhancing team collaboration. Nevertheless, due to the constraints limiting policy exploration initially, its convergence speed is slightly slower, resulting in a final success rate of 0.77. MASAC promotes policy diversity and exploration through entropy regularization, demonstrating strong adaptability in complex environments. However, the high stochasticity of the policy undermines team consistency, leading to significant oscillations in the training curve and a final success rate of approximately 0.75. The original MADDPG method, lacking both spatial partitioning constraints and an information filtering module, suffers from value estimation bias and gradient instability under high-dimensional joint state inputs. It exhibits slow convergence, large oscillations, and poor collaboration, with a final success rate of about 0.58. MATRPO ensures safe updates via KL divergence constraints, but its small update steps and insufficient exploration make it difficult to cope with dynamic pursuit-evasion scenarios, achieving a success rate of only 0.43.

In summary, the enhanced MADDPG framework, which combines buffered Voronoi spatial partitioning and constraint mechanisms with a self-attention-based feature focusing mechanism, exhibits optimal convergence efficiency and collaborative stability in multi-robot cooperative capture tasks. This method achieves efficient decision-making and global coordination among multiple agents through a dual-layer mechanism of "spatial constraints and information focus," fully validating the significant advantages of integrating geometric priors and attention-enhanced strategies in improving the performance of multi-agent systems.

## 9.7 Chapter Summary

This chapter presents a study on a multi-robot cooperative capture method that integrates Voronoi-based spatial partitioning with deep reinforcement learning (DRL). The aim is to improve coordination efficiency and safety in multi-agent pursuit tasks under dynamic and partially observable environments. A Voronoi-RL fusion framework is established, in which the geometric constraints of Buffered Voronoi Cells (BVC) are combined with the multi-agent deep deterministic policy gradient (MADDPG) algorithm. The Voronoi-based partitioning helps maintain safe distances between agents and reduce collisions, while reinforcement learning enables robots to adaptively optimize their pursuit behaviors through interaction with the environment. Simulation results show that the proposed method achieves more stable pursuit trajectories, higher capture success rates, and better cooperation compared with standard reinforcement learning algorithms. The introduction of attention mechanisms further enhances the ability of agents to process key environmental information and coordinate more effectively.

Overall, the combination of deep reinforcement learning and multi robot system is an important path toward swarm intelligence. In this chapter, DRL is enabled by the attention mechanism of structured prior (BVC) and cognitive heuristic, which initially shows its great potential in solving complex spatial decision-making problems. We firmly believe that with the continuous innovation of theoretical tools and the continuous evolution of computing platforms, the future of a highly autonomous and closely coordinated intelligent robot is bound to come.

## Bibliography


[1] Wenyu Cai, Hao Chen, and Meiyan Zhang. A survey on collaborative hunting with robotic swarm: Key technologies and application scenarios. *Neurocomputing*, 598:128008, 2024.

[2] Chen Tang, Ben AbbateMatteo, Jiaheng Hu, Rohan Chandra, Roberto Martín-Martín, and Peter Stone. Deep reinforcement learning for robotics: A survey of real-world successes. In *Proceedings of the AAAI Conference on Artificial Intelligence*, volume 39. AAAI Press, 2025.

[3] Rayadurga Gnana Prasuna and Sudharsana Rao Potturu. Deep reinforcement learning in mobile robotics–a concise review. *Multimedia Tools and Applications*, 83(28):70815–70836, 2024.

[4] Kai Hu, Ming Li, Zhen Song, Kai Xu, Qi Xia, Ning Sun, Peng Zhou, and Ming Xia. A review of research on reinforcement learning algorithms for multi-agents. *Neurocomputing*, 599:128068, 2024.

[5] James Orr and Ayan Dutta. Multi-agent deep reinforcement learning for multi-robot applications: A survey. *Sensors*, 23(7):3625, 2023.

[6] Hanna Kurniawati. Partially observable Markov decision processes and robotics. *Annual Review of Control, Robotics, and Autonomous Systems*, 5(1):253–277, 2022.

[7] Junyan Hu, Hanlin Niu, Joaquin Carrasco, Barry Lennox, and Farshad Arvin. Voronoi-based multi-robot autonomous exploration in unknown environments via deep reinforcement learning. *IEEE Transactions on Vehicular Technology*, 69(12):14413–14423, 2020.

[8] Ryan Lowe, Yi Wu, Aviv Tamar, Jean Harb, Pieter Abbeel, and Igor Mordatch. Multi-agent actor-critic for mixed cooperative-competitive environments. In *Advances in Neural Information Processing Systems*, volume 30. Curran Associates, Inc., 2017.

[9] Bo Li, Wenjia Zhao, Wei Zhang, Ying Zhang, Yuxin Tang, and Pengfei Liu. Multi-UAV roundup strategy method based on deep reinforcement learning cel-maddpg algorithm. *Expert Systems with Applications*, 245:123018, 2024.

[10] Zekun Bai, Zheng Tang, Xiaobo Wang, Yi Liu, Lei Zhang, and Zhiwei Li. Path planning of autonomous mobile robot in comprehensive unknown environment using deep reinforcement learning. *IEEE Internet of Things Journal*, 11(12):22153–22166, 2024.

[11] Zhilin Fan, Xin Wang, Shujing Li, Yifei Zhang, Liang Zhang, and Jian Wang. Reinforcement learning method for target hunting control of multi-robot systems with obstacles. *International Journal of Intelligent Systems*, 37(12):11275–11298, 2022.

[12] Yuhong Cao, Peng Xie, Ye Yuan, Hui Li, Lintao Liu, and YangQuan Chen. Deep reinforcement learning-based large-scale robot exploration. *IEEE Robotics and Automation Letters*, 9(5):4631–4638, 2024.

[13] Junchong Ma, Shuzhi Sam Ge, Chenguang Yang, Huayong Zhang, Zhiqiang Li, and Lin Cheng. Multi-robot target encirclement control with collision avoidance via deep reinforcement learning. *Journal of Intelligent & Robotic Systems*, 99(2):371–386, 2020.

[14] Derya Soydaner. Attention mechanism in neural networks: where it comes and where it goes. *Neural Computing and Applications*, 34(16):13371–13385, 2022.

[15] Kai Hu, Peng Zhou, Zhen Song, Ming Li, Ning Sun, Qi Xia, and Ming Xia. An overview: Attention mechanisms in multi-agent reinforcement learning. *Neurocomputing*, 598:128015, 2024.

[16] Shariq Iqbal and Fei Sha. Actor-attention-critic for multi-agent reinforcement learning. In *Proceedings of the 36th International Conference on Machine Learning (ICML)*, pages 2961–2970. Proceedings of Machine Learning Research (PMLR), 2019.

[17] Hangyu Mao, Zhengchao Liu, Jianye Hao, Chen Xiao, Ho-fung Leung, and Zhaopeng Meng. Modelling the dynamic joint policy of teammates with attention multi-agent ddpg. *arXiv preprint arXiv:1811.07029*, 2018.

[18] Chunjie Yang, Bodi Yuan, and Pengzhao Zhai. Actor-hybrid-attention-critic for multi-logistic robots path planning. *IEEE Robotics and Automation Letters*, 9(6):5559–5566, 2024.

[19] Chao Chen, Ming Li, Ning Sun, Peng Zhou, Kai Hu, Qi Xia, and Ming Xia. Global-and-local attention-based reinforcement learning for cooperative behaviour control of multiple uavs. *IEEE Transactions on Vehicular Technology*, 73(3):4194–4206, 2023.

[20] Kai Liu, Shuyue Zhang, Bin Wu, Xiaoxuan He, Lei Wang, and Jun Wang. Self-attention-based multi-agent continuous control method in cooperative environments. *Information Sciences*, 585:454–470, 2022.

[21] Lingli Yu, Peng Chen, Hanyue Zhang, Yuchen Wang, Liang Zhang, and Jian Wang. Hybrid attention-oriented experience replay for deep reinforcement learning and its application to a multi-robot cooperative hunting problem. *Neurocomputing*, 523:44–57, 2023.

[22] Aohong Nian, Xiaoping Chen, Jie Zhang, Haoran Liu, Yujie Wang, and Ming Li. Large-scale UAV swarm confrontation based on hierarchical attention actor-critic algorithm. *Applied Intelligence*, 54(4):3279–3294, 2024.

[23] Volodymyr Mnih, Adria Puigdomenech Badia, Mehdi Mirza, Alex Graves, Timothy Lillicrap, Tim Harley, David Silver, and Koray Kavukcuoglu. Asynchronous methods for deep reinforcement learning. In *International Conference on Machine Learning*, pages 1928–1937. PMLR, 2016.

[24] John Schulman, Filip Wolski, Prafulla Dhariwal, Alec Radford, and Oleg Klimov. Proximal policy optimization algorithms. *arXiv preprint arXiv:1707.06347*, 2017.

[25] Eric Jang, Shixiang Gu, and Ben Poole. Categorical reparameterization with gumbel-softmax. *arXiv preprint arXiv:1611.01144*, 2016.

[26] Matteo Bettini, Amanda Prorok, and Vincent Moens. Benchmarl: Benchmarking multi-agent reinforcement learning. *Journal of Machine Learning Research*, 25(217):1–10, 2024.

[27] Ryan Lowe, Yi Wu, Aviv Tamar, Jean Harb, Pieter Abbeel, and Igor Mordatch. Multi-agent actor-critic for mixed cooperative-competitive environments. In *Advances in Neural Information Processing Systems*, volume 30. Curran Associates, Inc., 2017.

[28] Shariq Iqbal, Bei Peng, and Fei Sha. Multi-agent soft actor-critic for partially observable environments. *Advances in Neural Information Processing Systems*, 34:17253–17265, 2021.

[29] Georgios Papoudakis, Filippos Christianos, Muhammad A Rahman, and Stefano V Albrecht. Addressing overestimation bias in multi-agent reinforcement learning with double critics and delayed policy updates. *arXiv preprint arXiv:2006.07869*, 2020.

[30] C. Yu, A. Velu, E. Vinitsky, Y. Wang, A. Bayen, and Y. Wu. The surprising effectiveness of mappo in cooperative multi-agent tasks. *arXiv preprint arXiv:2103.01955*, 2021.

[31] Jakub Kuba, Tim Allen, Hannah Roff, and Shimon Whiteson. Multi-agent trust region policy optimization. In *Proceedings of the International Conference on Autonomous Agents and Multiagent Systems (AAMAS)*, 2018.

CHAPTER 10

# Multi-Robot Physical Implementation Based on Voronoi Tessellations

## 10.1 Introduction

THE transition from simulation to physical hardware is a non-trivial and critical phase in robotics research. Simulations, while invaluable for initial algorithm development and validation, operate within an idealized world. They often neglect or oversimplify complex real-world phenomena such as friction, motor dynamics, sensor noise, processing delays, and unpredictable network latency. This discrepancy between the simulated environment and physical reality is known as the "Sim-to-Real Gap." Successfully deploying algorithms on physical robots requires addressing this gap by developing systems that are robust to these unmodeled factors. Therefore, the experiments presented in this chapter serve a dual purpose: not only to validate the fundamental correctness of the Voronoi-based strategies but also to rigorously test their performance and robustness under the inherent uncertainties and constraints of a real-world setting. The outcomes of these experiments provide crucial feedback for refining the theoretical models and control parameters, ultimately paving the way for reliable real-world applications.

The experimental part mainly focuses on multi-robot formation control, obstacle avoidance, and task execution strategies based on Voronoi tessellations. We use two ROS platform compatible robot hardware platforms—TurtleBot3 ground mobile robot, and Crazyflie quadcopter unmanned aerial vehicles, These are used to construct a heterogeneous multi-robot system experimental platform [1] [2]. ROS, as the mainstream open-source robot operating system today, has excellent modular design, powerful communication mechanism, and multilingual support, and has been widely used in education, scientific research, and industrial practice. With the powerful control and visualization capabilities of ROS, developers can efficiently achieve collaborative control, path planning, and state monitoring of different modules in multi-robot systems. At the same time, TurtleBot3 and Crazyflie are both standard

DOI: 10.1201/9781003714569-10

hardware platforms officially supported by ROS, with good openness and scalability, suitable for various needs from introductory teaching to advanced research. The TurtleBot3 is architected to readily accommodate a variety of external sensors, from RGB-D cameras like Intel's RealSense series for 3D environmental perception, to ultrasonic rangefinders for close-proximity obstacle avoidance. Similarly, the compact Crazyflie drone leverages a series of "expansion decks" to enhance its capabilities, enabling everything from GPS positioning and high-precision altitude hold to advanced functions like thermal imaging. The true elegance of this hardware ecosystem is that activating these new features often requires nothing more than installing the corresponding packages in the ROS, seamlessly translating hardware capabilities into accessible software interfaces.

This chapter will first introduce the basic architecture of the ROS system and the functions and configuration methods of related robot platforms. Then, it will present in detail the experimental design and implementation process of multi-robot systems in typical tasks such as formation control, formation switching, obstacle avoidance, target search and target capture. Through multiple rounds of validation of real robots in indoor environments, the stability, robustness, and operability of the collaborative algorithm based on Voronoi tessellations are comprehensively demonstrated, laying the foundation for the promotion of multi-robot systems to real-world application scenarios.

## 10.2 Overview of ROS and Multi-robot Platforms

The rapid development of hardware technology has not only promoted the rapid growth and complexity of the robotics field but also posed significant challenges to the software development of robot systems. The increasing richness of robot platforms and hardware devices has led to a stronger demand for software code reusability and modularity, while existing robot systems cannot adapt well to the requirements. To meet the enormous challenges faced by robot software development, developers and research institutions around the world have invested in the research and development of universal software frameworks for robots. In recent years, various excellent robot software frameworks have emerged, providing great convenience for software development work. One of the most outstanding software frameworks is the Robot Operating System (ROS) [3]. The ROS wiki website `http://wiki.ros.org/Robots`, shows a list of robot models that use ROS. Among them, the physical experiments in this book use, are TurtleBot3 mobile robots and Crazyflie drones. Next, we will provide a detailed introduction to ROS and the concepts and uses of the two robots mentioned above.

### 10.2.1 Introduction to the ROS Platform

Robot Operating System (ROS) is an open-source and flexible software framework that provides a range of open-source tools and libraries that facilitate developers in creating complex robot applications [4]. The functions provided by ROS to users include message passing between different nodes, abstract hardware description, and

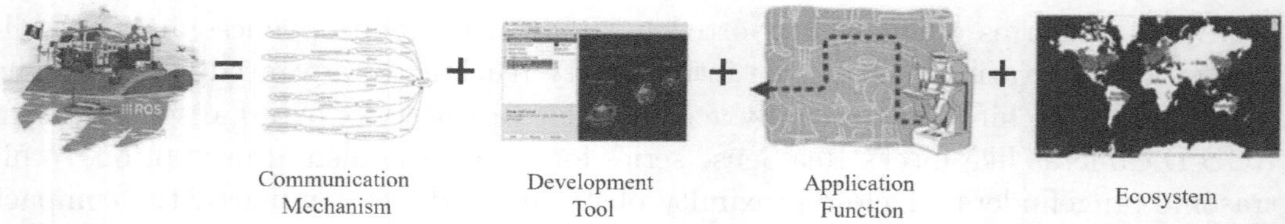


Figure 10.1: The four components of the ROS system.

underlying software drivers, greatly simplifying the task creation process of traditional robot control and providing stable control. ROS adopts a distributed architecture, where each functional module is independent, but can quickly couple different modules during program execution, making it easy to achieve hardware and software integration, sensor data processing, path planning, and control functions. Given the advantages and functions of ROS, it has been widely applied in many fields. For example, industrial automation, ROS is used to operate and control complex production lines; in agriculture, ROS is used to control agricultural robots for automated planting and harvesting; in the medical field, ROS is used to develop medical assistant robots; in drone technology, ROS is used to achieve flight control and mission planning; and in the field of autonomous vehicles, ROS is used to develop autonomous driving system. The main components of the ROS system include communication mechanisms, development tools, application functions, and ecosystem, as shown in Fig. 10.1.

The selection of the ROS as the foundational software framework was a deliberate design choice, critical for addressing the inherent complexities of multi-robot coordination. Far more than a mere collection of tools, ROS embodies a robust philosophy for building modular and distributable robotic software. Its architecture is particularly well-suited for the challenges posed by our heterogeneous multi-agent system. The successful orchestration of both ground and aerial robots in complex, dynamic tasks hinges on the advanced capabilities ROS provides for communication, spatial-temporal awareness, and system modularity.

At its core, the ROS framework is architected around a collection of fundamental components, including nodes, messages, topics, and services. These elements collectively form the ROS Computation Graph, which provides the foundational building blocks for creating complex, modular applications. This graph's power stems from its operation on a flexible, distributed publish-subscribe messaging paradigm. While a central `roscore` acts as a nameserver to facilitate initial discovery, all subsequent data transmission occurs directly between nodes in a peer-to-peer fashion. This architecture provides profound advantages for our Voronoi-based system. It ensures scalability, allowing new robots to be seamlessly integrated into the swarm by simply connecting their respective nodes to the same ROS Master without requiring architectural modifications. Furthermore, this design promotes a high degree of decoupling. Our primary Voronoi computation algorithm is encapsulated within a single, self-contained ROS node. This node subscribes to generalized robot state topics and publishes motion commands, remaining entirely agnostic to the specific hardware platforms generating or receiving this data. This abstraction was instrumental in enabling a single high-level controller to manage a diverse team of TurtleBot3s and Crazyflies.

Node is the fundamental processing unit in ROS, typically an executable program responsible for a single, specific task. For instance, one node might be responsible for controlling a robot's wheel motors, another for processing LiDAR data, and a third for executing the Voronoi-based path planning algorithm. In our experiments, the central control algorithm runs as a node on a host PC, while each robot runs its own set of nodes for sensing and actuation.

The ROS Master serves as a central discovery and registration point. It functions as a nameserver, allowing nodes to find and establish connections with one another. When a node starts, it registers with the Master, declaring the data it intends to publish or subscribe to. The Master then enables these nodes to establish direct, peer-to-peer connections for data exchange, without the data itself passing through the Master.

The primary method for asynchronous, continuous data flow is through topics. A topic can be thought of as a named bus for data. Nodes with information to share publish structured messages to a topic. Other nodes interested in this information can subscribe to that topic to receive the messages. For example, a LiDAR node continuously publishes '/sensor_msgs/LaserScan' messages to a '/scan' topic, which can be used by a localization node to build a map.

For situations requiring a synchronous, one-to-one interaction, ROS provides services. This communication model is analogous to a remote procedure call, where a client node sends a request to a server node and waits for a single response before continuing. A practical use case could be a service named '/start_formation_task' that, when called, triggers the multi-robot system to begin its mission and returns a success or failure message.

Finally, for long-running tasks where continuous feedback is needed, actions offer a more sophisticated alternative to services. An action client sends a goal to an action server, which begins executing the task. Unlike a service, the server is non-blocking and provides a stream of feedback to the client about its progress, and the client can preempt or cancel the goal at any time. This is ideal for tasks like navigating to a specific waypoint, allowing a control node to monitor the robot's progress toward its goal without halting its own operations.

Beyond basic messaging, two frequently overlooked yet mission-critical challenges in multi-robot systems are ensuring temporal consistency and maintaining a coherent spatial understanding across all agents. ROS provides robust, standardized solutions for both. Time synchronization is managed through a network-wide time source, typically governed by the Network Time Protocol (NTP), which guarantees that all message timestamps are consistent across the distributed system. This allows for the correct correlation of states and sensor readings from multiple agents, a fundamental prerequisite for any coherent group behavior. Simultaneously, the `tf2` library within ROS provides a dedicated and powerful system for managing the complex hierarchy of coordinate frames. It dynamically tracks the spatial relationships between a global `map` frame, each robot's `base_link`, and their various sensor frames. Our central control node leverages `tf2` to query the precise transformation of each robot relative to the global origin at any given moment, abstracting away the complex and error-prone task of manual geometric calculations and providing a unified spatial awareness for the entire swarm.

In terms of communication methods, the ROS system uses TCP/IP protocol to achieve communication between different nodes, and adopts asynchronous message passing mechanism to achieve efficient message exchange. In addition, the ROS system provides many advantages such as flexibility, scalability, and cross platform compatibility, making it an essential tool for research and practice in the field of robotics. The main features of ROS system include [5]:

(1) Openness and flexibility

The source code of ROS system is open source, and users can freely obtain, view and use the code. In addition, ROS provides many modules and tools that allow users to select and customize modules according to their needs and goals, while also facilitating integration with other software and hardware.

(2) Modularity and high integration

The unique modular characteristics of the ROS framework simplify the control of robot systems, avoiding the shortcomings of traditional robot system design confusion and difficulty in porting and secondary development. At the same time, the ROS system provides a variety of software libraries and algorithms for robot perception, navigation, and control, allowing users to directly utilize these ready-made functions to implement complex robot applications.

(3) Multi language support

The ROS system currently supports various programming languages such as C++, Python, Java, etc., reflecting the characteristics of weak language relevance, which can facilitate users to choose the appropriate programming language for development according to their own needs. In addition, the programming language in ROS system is very concise, and message interfaces can be created for communication through simple definitions, greatly improving user productivity.

Based on the analysis of the characteristics of the ROS system introduced above, it can be concluded that ROS, with its core advantages of open source, modularity, and multilingual support, not only solves the problems of poor reusability and integration difficulties in traditional robot software development but also provides efficient solutions for multi robot collaboration and complex task deployment through distributed architecture and flexible communication mechanisms. In the subsequent experiments, we will use the ROS framework and combine the TurtleBot3 mobile robot with the Crazyflie drone platform to complete the experimental verification of the previous algorithms.

There are two installation methods for ROS: software source installation and source code compilation installation [6]. The software source provides a huge application repository for the system, and compiling the source code requires solving complex software dependency relationships. Taking software source installation as an example, the corresponding Ubuntu operating system needs to be downloaded according to the ROS version, as shown in Table 10.1.

Table 10.1: The correspondence between Ubuntu and ROS versions

| Ubuntu | ROS 1.0 | ROS 2.0 |
|---|---|---|
| 16.04 | Kinetic LTS | Ardent |
| 18.04 | Melodic LTS | Dashing LTS |
| 20.04 | Noetic LTS | Foxy LTS |

After installing the Ubuntu system, it is necessary to configure the software source to allow three types of software sources: restricted, universal, and multiverse. Open the software source configuration interface in Ubuntu Software Center and check if all options match the image.

After completing the software source configuration, it is necessary to add the ROS software source. Open the terminal and enter the following command to add the official ROS software source image:

```
$ sudo sh -c 'echo "deb http://packages.ros.org/ros/ubuntu $(
    lsb_release -sc) main" > /etc/apt/sources.list.d/ros-latest.list'
```

Use the following command to add a key:

```
$ sudo apt-key adv --keyserver 'hkp://keyserver.ubuntu.com:80' --recv-
    key C1CF6E31E6BADE8868B172B4F42ED6FBAB17C654
```

Next, start installing ROS by using the following command to update the software source modifications:

```
$ sudo apt-get update
```

However, ROS has many versions, including desktop full version installation, desktop version installation, base version installation, and standalone feature package installation. The commands for installing different versions are as follows:

**Desktop Full Installation:** The most common installation version includes the basic features of ROS, as well as a rich set of robot universal function library packages, such as 2D/3D perception function, robot map modeling, autonomous navigation, etc., as well as `rviz` visualization tool, Gazebo simulation environment, `rqt` toolbox, etc.

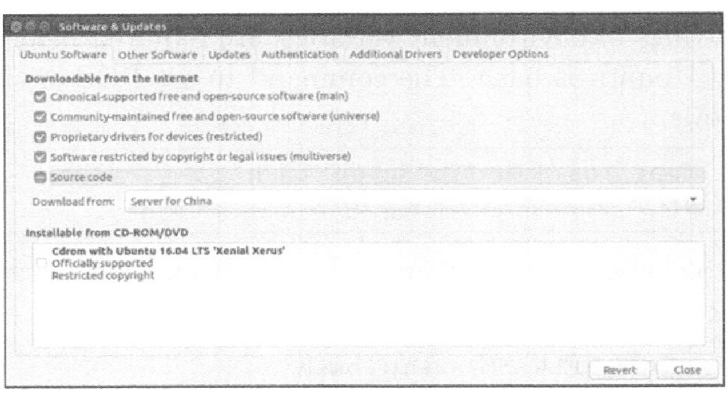


Figure 10.2: Setting up Ubuntu system software source.

The installation command is:

```
$ sudo apt-get install ros-kinetic-desktop-full
```

**Desktop installation:** This version is a fully installed, streamlined version that does not include robot feature packs or some tools. It only includes ROS basic features, robot universal function libraries, `rqt` toolbox, and `rviz` visualization tools.

The installation command is:

```
$ sudo apt-get install ros-kinetic-desktop
```

**ROS Base installation:** ROS Base simplifies the robot's general function library, feature packages, and tools, retaining only the basic features without any GUI (core featured packages, build tools, and communication mechanisms). Therefore, this version of the software has the smallest scale and is also the "smallest system" required for ROS. It is very suitable for direct installation on controllers with high performance and storage space requirements, providing the possibility for embedded systems to use ROS.

The installation command is:

```
$ sudo apt-get install ros-kinetic-ros-base
```

**Individual Package installation:** Regardless of which installation method is used, it is impossible to install all the feature packages in the ROS community on the computer. In later use, the following commands will be used to install independent feature packages according to specific needs

The installation word is:

```
$ sudo apt-get install ros-kinetiC-PACKAGE
```

Next, we need to initialize rosdep, which is a built-in tool in ROS. Its main function is to install system dependencies for certain featured packages, and it is also a necessary tool for some ROS core featured packages. After completing the above installation steps, the following commands are required for initialization and updates:

```
$ sudo rosdep init$ rosdep update
```

At this point, ROS has been successfully installed on the computer, defaulting to the/opt path. However, due to frequent use of terminal input commands in the future, simple settings of environment variables are required before use. The default terminal used by Ubuntu is bash. The command to set ROS environment variables in bash is as follows:

```
$ echo "source /opt/ros/kinetic/setup.bash">>~/.bashrc
$ source ~/.bashrc
```

If you want to change the environment variables currently used by the terminal, you can enter the following command:

```
$source /opt/ros/ROS-RELEASE/setup.bash
```

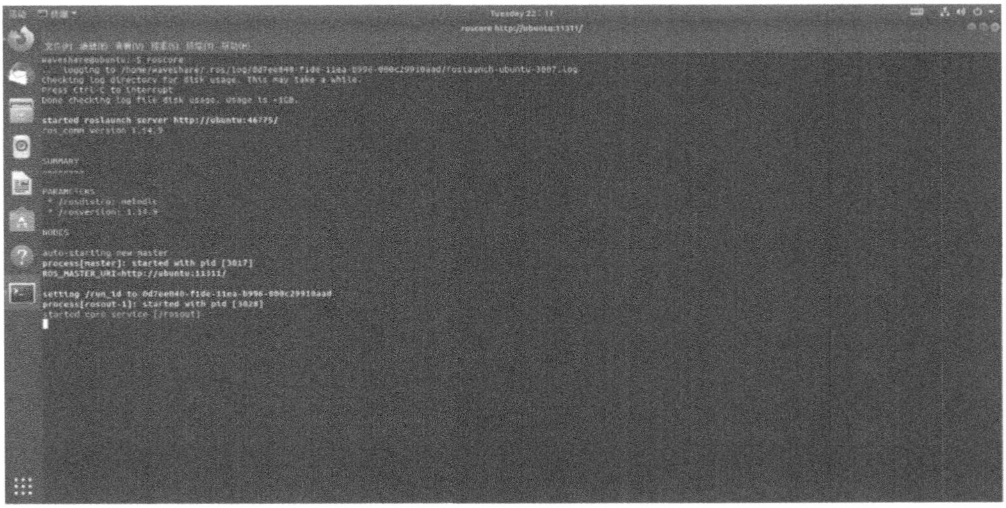

Figure 10.3: ROS installation successful.

Now open the terminal, enter roscore, as shown in Fig. 10.3, ROS has successfully run.

In Fig. 10.3, ROS-RELEASE represents the desired ROS version to be used (such as lunar, kinetic, indigo, hydro, Groovy, etc.).

### 10.2.2 Introduction to the TurtleBot3 Platform

The validation of the proposed Voronoi based algorithms within a two-dimensional space necessitated a physical testbed of ground mobile robots. For this purpose, the TurtleBot3 platform was selected as the foundational hardware for our multi-robot collaborative experiments. This decision was predicated on a strategic evaluation of its characteristics within the context of our research objectives. The TurtleBot3 [7] serves as an exemplary platform for bridging the simulation-to-reality gap due to its tight integration with the Robot Operating System (ROS), its embodiment of common robotic constraints, and its inherent suitability for multi-agent system research. Its native ROS compliance, in particular, provided a significant advantage by abstracting low-level hardware control through a standardized communication interface. This feature substantially accelerated the development cycle, permitting our research to concentrate on the high-level Voronoi control algorithm rather than on creating bespoke drivers and middleware.

A central objective of our research is the investigation of collective, rather than individual, robotic behavior. The TurtleBot3's low cost and compact form factor are instrumental in this regard, as these attributes facilitate the deployment of a sizeable fleet of robots. In our experiments, a team of five robots was assembled; the small size of each unit allows for a higher robot density within our constrained laboratory space. This ensures that inter-robot dynamics, e.g., collision avoidance and formation maintenance, are prominent and frequently tested aspects of the experiments.

Such scalability is a cornerstone for validating algorithms intended for multi-robot applications.

Each TurtleBot3 in the fleet functions as an independent, autonomous agent. Each TurtleBot3 is equipped with an onboard Raspberry Pi single-board computer, which possesses adequate computational power to execute the requisite ROS nodes for sensor processing, localization, and motion control. This self-contained processing capability allows each robot to receive high-level goals from our central Voronoi node and execute them autonomously. This architecture mirrors many real-world multi-robot deployments where individual agents must manage their own state and control loops while contributing to a collective task.

The physical subsystems of the TurtleBot3 introduce the very real-world challenges our algorithms are designed to overcome. Its primary sensor for environmental perception is a 360-degree LDS. This sensor serves a dual purpose in our experiments. First, it provides the raw data for the GMapping SLAM algorithm. This onboard process builds a map of the environment and continuously estimates the robot's pose, representing a critical departure from the perfect state information typically available in simulation. The performance of this SLAM algorithm directly impacts the quality of the input to our Voronoi controller; any localization error constitutes a disturbance that the formation control law must be robust enough to handle. Second, the LiDAR data is consumed by the ROS navigation stack for real-time local obstacle avoidance, a crucial capability for safe operation.

Furthermore, the robot's actuation mechanism imposes significant constraints that are absent in idealized models. The TurtleBot3 is built upon a differential drive, a classic example of a nonholonomic system. This kinematic constraint prevents the robot from moving laterally; to reach a target to its side, it must execute a turning maneuver. Consequently, a robot cannot instantaneously assume the position of its calculated Voronoi cell centroid but must instead follow a continuous, kinematically feasible path. This introduces a natural delay and tracking error between the ideal target position generated by the CVT algorithm and the robot's actual position. The experiments thus rigorously test the stability and performance of the control algorithm in the presence of motion constraints inherent to most ground vehicles.

There are two common models of TurtleBot3, namely TurtleBot3 Burger, TurtleBot3 WafflePi, and TurtleBot3 Labox. Each model provides different configurations and functions according to different usage requirements and application scenarios. Among them, TurtleBot 3-Burger is an entry-level version equipped with basic sensors and computing platforms, with a small size and low cost, making it very suitable for basic experiments and teaching in education and research; TurtleBot3 WafflePi is an upgraded version of TurtleBot3 Burger, integrating Raspberry Pi 3B+ as a computing platform, supporting more sensors and external devices, and meeting the needs of mid to high end users; The schematic diagrams of two types of robots are shown in Fig. 10.4:

This book will take TurtleBot3 Burger as an example to introduce the basic operation of the TurtleBot3 robot from the installation of the ROS environment library and related software, in preparation for experimental verification using the robot in the following sections.

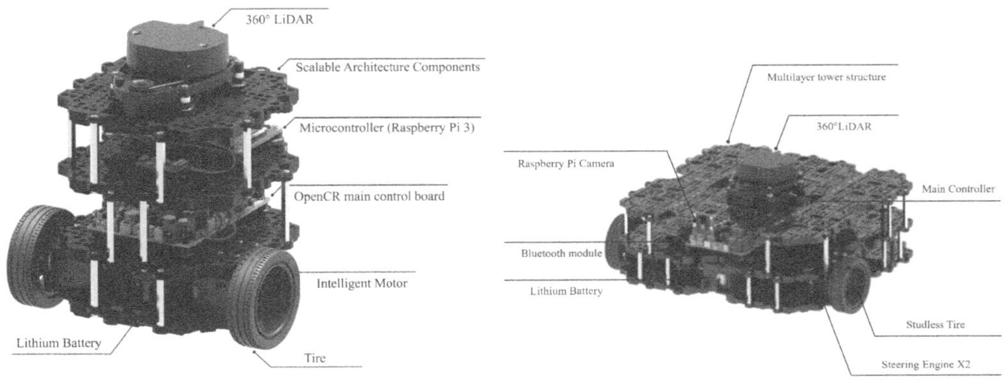


(a) (b)

Figure 10.4: TurtleBot3: (a) TurtleBot3 Burger; (b) Turtlebot3-WafflePi.

The TurtleBot3 dependency package in ROS needs to be installed separately. Use the following command to install dependencies:

```
sudo apt-get install ros-kinetic-joy ros-kinetic-teleop-twist-joy ros-
    kinetic-teleop-twist-keyboard ros-kinetic-laser-proc ros-kinetic-
    rgbd-launch ros-kinetic-depthimage-to-laserscan ros-kinetic-
    rosserial-arduino ros-kinetic-rosserial-python ros-kinetic-
    rosserial-server ros-kinetic-rosserial-client ros-kinetic-rosserial
    -msgs ros-kinetic-amcl ros-kinetic-map-server ros-kinetic-move-base
     ros-kinetic-urdf ros-kinetic-xacro ros-kinetic-compressed-image-
    transport ros-kinetic-rqt-image-view ros-kinetic-gmapping ros-
    kinetic-navigation ros-kinetic-interactive-markers
```

Clone the Turtlebot3 related source code to the workspace using the following command:

```
cd ~/catkin_ws/src/
git clone https://github.com/ROBOTIS-GIT/turtlebot3_msgs.git
git clone https://github.com/ROBOTIS-GIT/turtlebot3.git
cd ~/catkin_ws && catkin_make
```

Install the TurtleBot3 simulation package, which can be used to test the TurtleBot3 robot in a virtual environment:

```
cd ~/catkin_ws/src/
git clone https://github.com/ROBOTIS-GIT/turtlebot3_simulations.git
cd ~/catkin_ws && catkin_make
```

Install the multi-robot map merge package:

```
sudo apt-get install ros-kinetic-multirobot-map-merge
```

Install multi-robot code to support multi machine navigation operations:

```
git clone https://github.com/shunyiwang/ROS_MULTI/MULTI_ROBOT.git
```

Finally, set up the environment to ensure that the above software packages can be used normally:

```
echo "source ~/catkin_ws/devel/setup.bash" >> ~/.bashrc
```

If catkin_make is completed without any errors, use TurtleBot3 software to prepare for completion.

ROS needs to establish communication between Turtlebot3 and the remote PC. First, obtain the IP addresses of both: ifconfig

Edit the /. bashrc file of Turtlebot3 and set the ROS host and hostname: gedit /.bashrc

Add the following content to ensure that IP-OF_PC is replaced with the IP of PC, and IP-OF_ TURTLEBOT is replaced with the IP of TurtleBot3:

```
ROS_MASTER_URI = http://IP_OF_PC:11311
ROS_HOSTNAME   = IP_OF_TURTLEBOT
```

Similarly, configure ROS on the PC and add the following:

```
ROS_MASTER_URI = http://IP_OF_PC:11311
ROS_HOSTNAME   = IP_OF_PC
```

Finally, make the environment configuration effective:

```
source ~/.bashrc
```

At this point, all preliminary preparations have been completed, and we can start using TurtleBot3 for multi-robot collaborative mapping and navigation operations.

### 10.2.3 Introduction to the Crazyflie Platform

The Crazyflie 2.1 is a versatile open-source flying development platform equipped with low-latency/long-range radio and Bluetooth LE [8]. You can choose to download the app and use a mobile device as a controller, or use the Crazyradio 2.0 or Crazyradio-PA in combination with a computer to display data and fly using a game controller. The experiments in this book utilize the Crazyradio-PA to control the drone.

The flight control system of the Crazyflie is primarily developed based on open-source firmware, offering stable flight control capabilities. It allows users to flexibly adjust various flight parameters through a graphical interface or programming, making it suitable for experimental validation in path planning, swarm formation, target search, and more. As shown in Fig. 10.5, the Crazyflie's body consists of a Pilot Lamp, Wings, Battery, STM32 microcontroller, and the frame structure.

## 10.3 Experiments and Implementation

### 10.3.1 Multi-Robot Formation Control

This section conducted an indoor multi-robot formation experiment to verify the formation control algorithm proposed in Chapter 4 and simulate the automatic deployment of multiple robots at the scene of a hazardous accident. The experiment

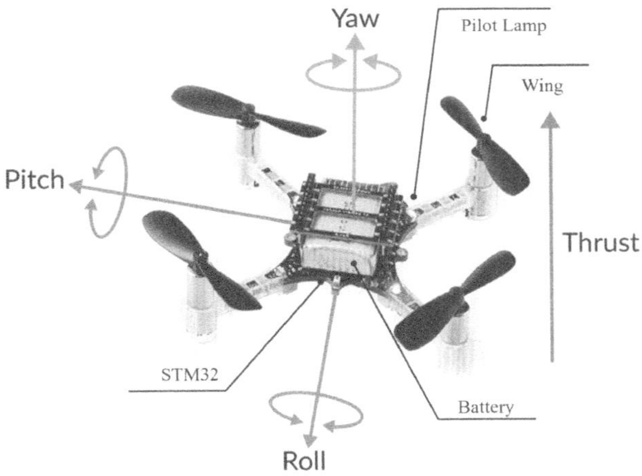


Figure 10.5: The Crazyflie platform.

used 5 TurtleBot3 Burger robots to form a formation and conducted formation experiments in both obstacle free and obstacle environments.

**Experimental Environment**

The experimental environment for multi-robot formation is shown in Fig. 10.6a, which is divided into barrier free environment and obstacle free environment. The task area of the robots is the gray ground cushion of 3.3m × 3.85m in the figure. The cylindrical obstacle in the center of the field in Fig. 10.6b simulates the hazardous accident tank that the robot formation needs to deal with. In this experiment, the robot used laser radar SLAM for positioning. Due to the low power of the robot's laser radar, it is difficult to cover the boundaries of the entire experimental environment. To achieve more accurate positioning, obstacles with auxiliary positioning were added to the

(a) (b)

Figure 10.6: Experimental environment: (a) Accessibility; (b) obstacle environment.

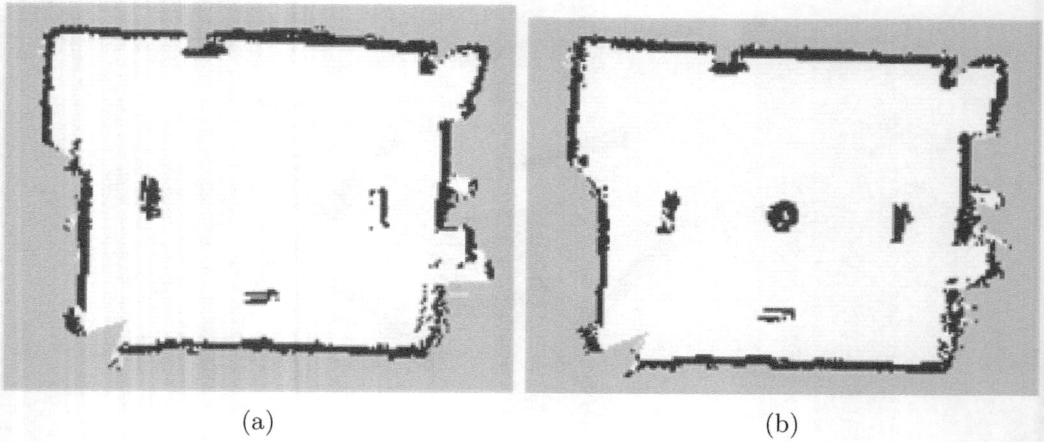


Figure 10.7: Experimental environment 2D grid map: (a) Accessibility; (b) obstacle environment.

edge of the robot's task area. The pre constructed two-dimensional grid map of the experimental environment by the robot is shown in Fig. 10.7.

This book uses a ROS host to simultaneously connect 5 TurtleBot3 Burgers to achieve collaborative control of multi-robot. As shown in Fig. 10.8, the robot and ROS host are in the same local area network. The ROS host controls the formation of multi-robot by sending the CVT algorithm's target point position for each iteration to the node of each robot. Meanwhile, the ROS host can also obtain the location information of each robot in the environment for visual display, as shown in Fig. 10.9.

**Multi-Robot Formation Obstacle Free Environment Experiment**

This section is an experiment on multi robot formation in an obstacle free environment to verify the effectiveness of the proposed formation algorithm.

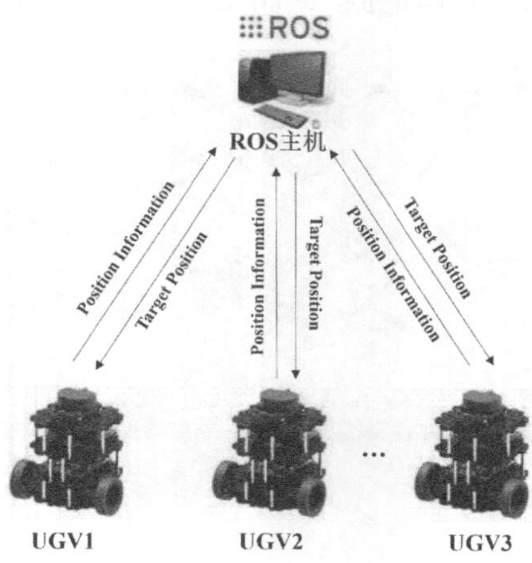


Figure 10.8: TurtleBot3 Burger connected to the ROS host.

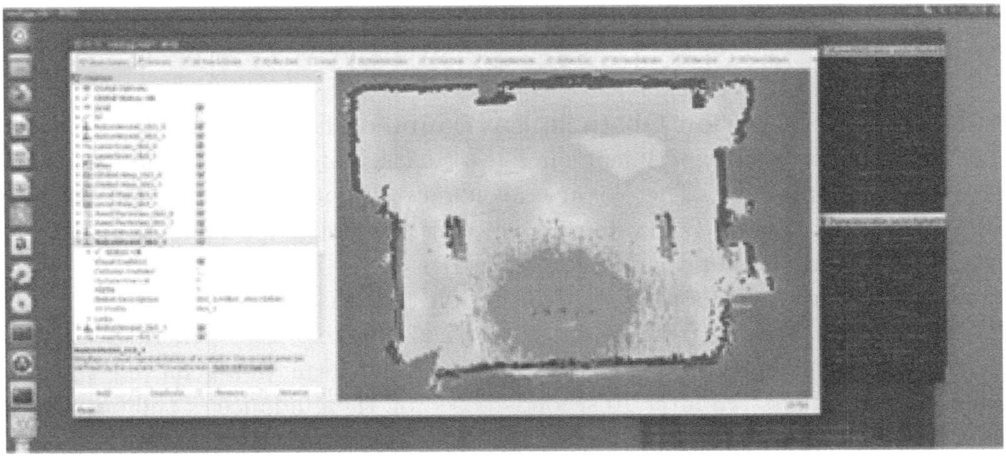

Figure 10.9: ROS host rviz visualization interface.

The process of generating a linear formation of multi-robot in an accessible environment is shown in Fig. 10.10a. The robot formation starts from directly below the experimental environment and generates a linear formation as shown in Figs 10.10d based on the CVT under the linear density function. During the experiment, the ROS host sent the target points generated by the CVT to each robot, and the formation

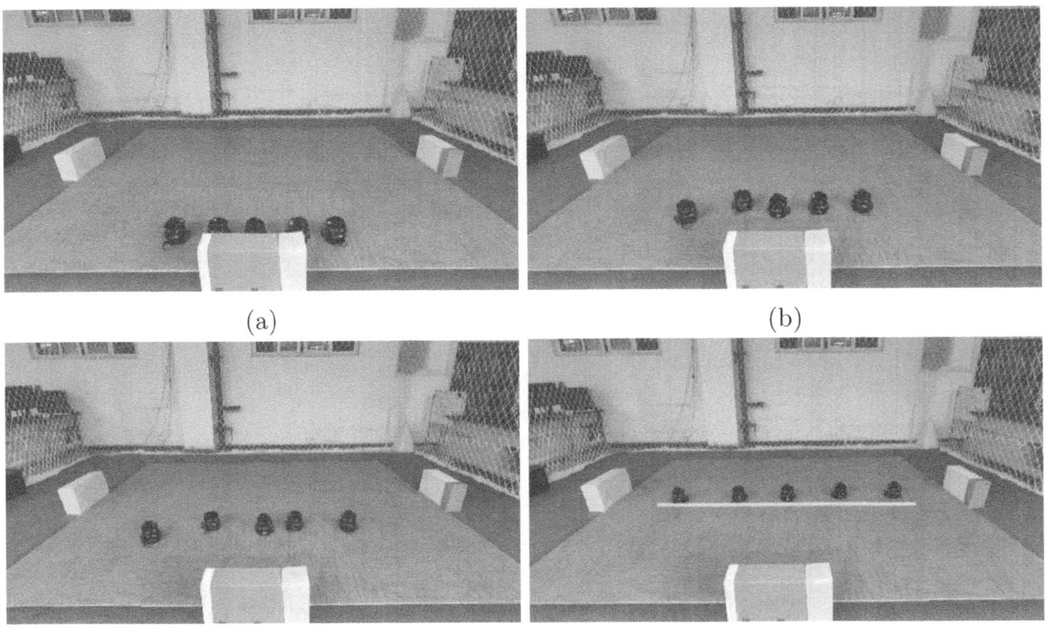

(a) (b) (c) (d)

Figure 10.10: Linear formation generation: (a) $t = 0s$; (b) $t = 3s$; (c) $t = 6s$; (d) $t = 10s$.

robots generated a predetermined formation based on the guidance of the CVT after 10 seconds.

**Multi-Robot Formation Obstacle Environment Experiment**

This section shows an experiment on the formation of a multi robot formation in an obstacle environment. The obstacle avoidance algorithm of the robot utilizes the DWA algorithm in the ROS system move_base navigation package to achieve local obstacle avoidance during the robot motion.

Shown in Fig. 10.11 is the robot formation task that starts from directly below the environment and, in front of the cylindrical obstacle, first generates a straight line formation to dispose of the formation. The robot formation achieved a predetermined formation pattern after 10 seconds based on the guidance of the target points generated by the CVT.

## 10.3.2 Multi-Robot Formation Switching

**Multi-Robot Formation Switching Experiment in Obstacle Free Environment**

(1) Experiment on switching linear formation to circular formation

The process of switching from a straight line formation to a circular formation in an obstacle free environment for a multi-robot formation is shown in Fig. 10.12. The

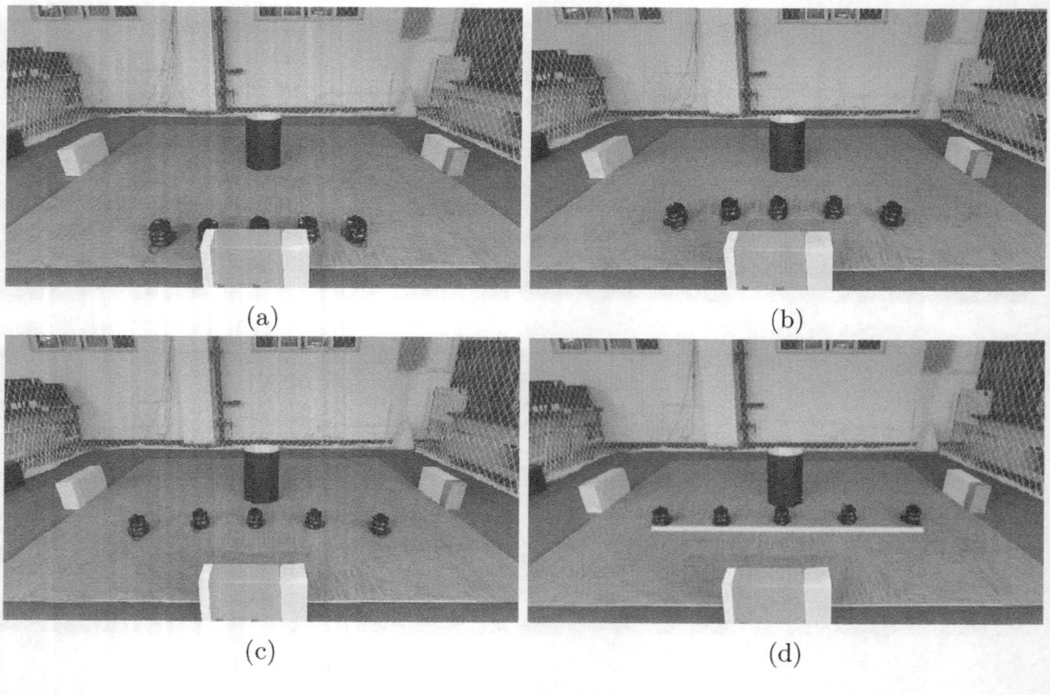

Figure 10.11: Generate a linear formation: (a) $t = 0s$; (b) $t = 3s$; (c) $t = 8s$; (d) $t = 10s$.

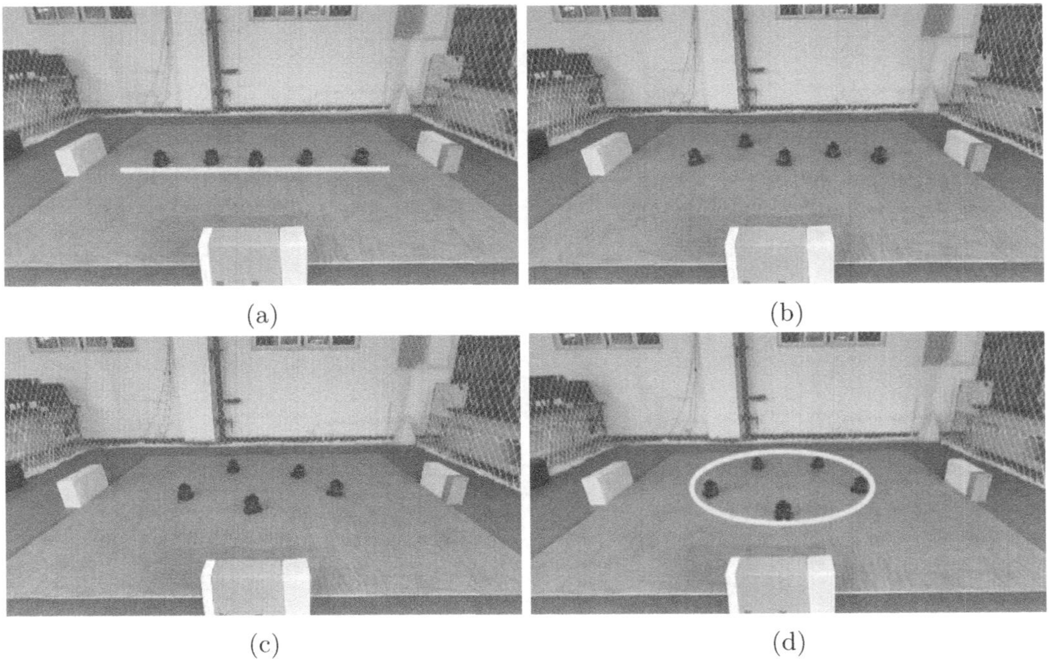

Figure 10.12: Switching from linear formation to circular formation: (a) $t = 0s$; (b) $t = 4s$; (c) $t = 8s$; (d) $t = 12s$.

initial formation of the robot formation is a linear formation, which switches to the circular formation shown in Fig. 10.12d based on the CVT division under different dynamic density functions at different times. The robot formation is guided by the target points generated by the CVT and switches to the predetermined formation after 12 seconds.

(2) Experiment on switching from circular formation to linear formation

The process of multi-robot formation switching from circular formation to linear formation in an accessible environment is shown in Fig. 10.13. The initial formation of the robot formation is a circular formation, which switches to a linear formation as shown in Fig. 10.13d based on the CVT division under different dynamic density functions at different times. The robot formation is guided by the target points generated by the CVT and switches to the predetermined formation after 14 seconds.

### Multi-Robot Formation Obstacle Environment Formation Switching Experiment

(1) Experimental study on switching capture formation in a zigzag formation

The process of switching from a multi-robot inline formation to a capture formation is shown in Fig. 10.14. The multi-robot team switch from the initial inline formation to the circular formation shown in Fig. 10.14d based on the CVT division under different dynamic density functions at different times. During the formation

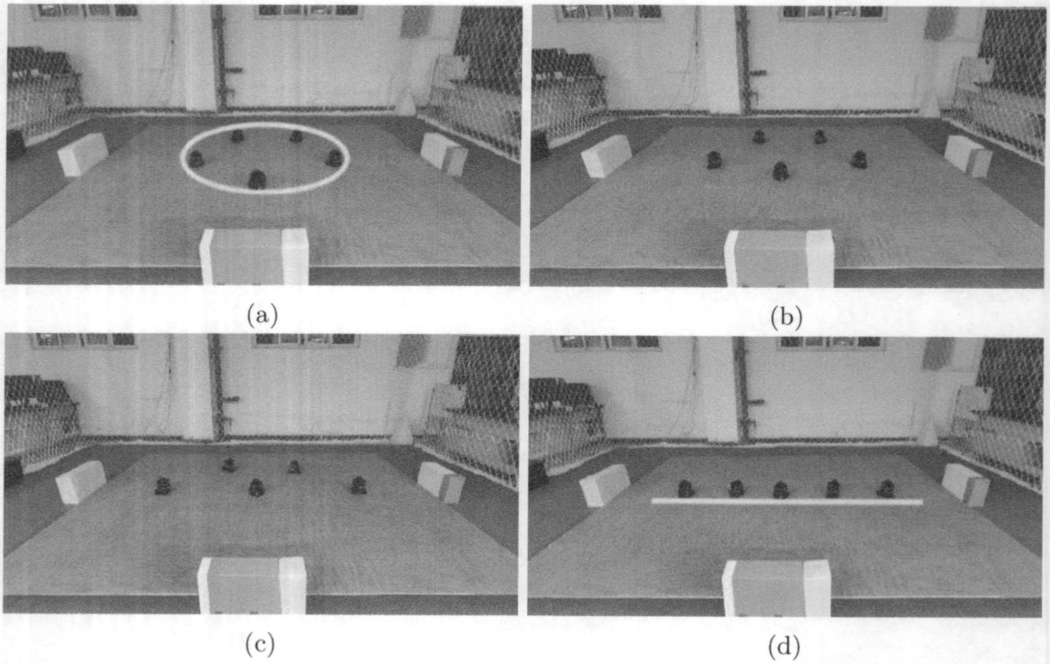

Figure 10.13: Switching from circular formation to linear formation: (a) $t = 0s$; (b) $t = 3s$; (c) $t = 8s$; (d) $t = 14s$.

switching process, the robot formation switches to the predetermined formation after 13 seconds based on the guidance of the target points generated by the CVT.

(2) capture formation switching line formation experiment

The process of switching from a multi-robot capture formation to a line formation is shown in Fig. 10.15. The multi-robot switch from the initial circular formation to the straight formation shown in Fig. 10.15d based on the CVT division under different dynamic density functions at different times. The robot formation is guided by the target points generated by the CVT algorithm and switches to the predetermined formation after 14 seconds.

### 10.3.3 Multi-Robot Formation Source Search

To further verify the effectiveness of the UCB-CVT algorithm and exploration strategy proposed in Chapter 6 of this book in locating pollution sources in practical scenarios, this section will use four TurtleBot3 robots equipped with photosensitive sensors and four Crazyflie drones to conduct experiments on reconstructing and locating the light source field. In this experimental environment, the task area size is set to (4.2m × 3.6m), with 4 TurtleBot3 robots having a physical size radius of 0.1m, 4 Crazyflie drones having a radius of 0.05m, and a positioning uncertainty radius of 0.2m.

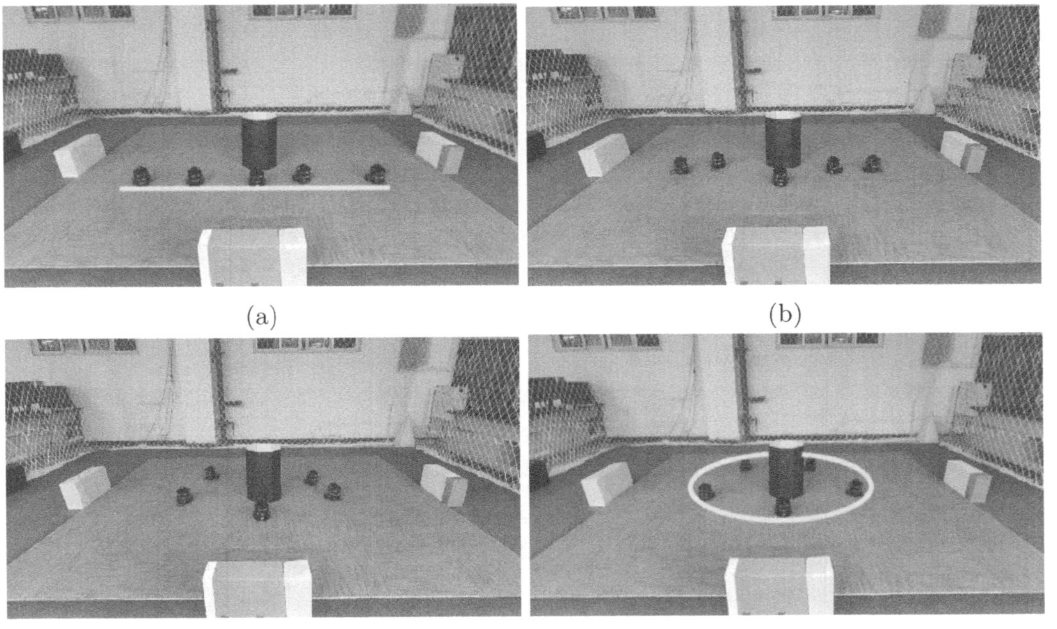

(a) (b) (c) (d)

Figure 10.14: One line formation switching capture formation: (a) $t = 0s$; (b) $t = 4s$; (c) $t = 8s$; (d) $t = 13s$.

**2D Environment TurtleBot3 Sourcing Seeking Experiment**

The TurtleBot3 robot in the experiment has the ability to sense light sources, as shown in Fig. 10.16. It uses a photosensitive sensor module to collect light field strength data. The sensor module uses the ESP8266-12E as the central processor, and sends the data to the base station as input for prior light field density information. In addition, SLAM technology is used to build an environmental map in advance to obtain real-time global position information of the robot, and the position information is sent to the base station through the ROS system as prior position information input.

The light source in this scene is located in the lower right corner, and four TurtleBot3 robots start from the upper left corner of the map. When the light intensity detected by the photosensitive sensor is less than 36Lux, the robot will be assigned to perform environmental exploration tasks. When the detected light intensity exceeds 36Lux, the robot's task will immediately switch to formation coverage. The experimental process is shown in Fig. 10.17, where the irregular polygons in red lines represent the Voronoi cells of different robots.

In the initial stage, as the collected illumination values approach 0, the mobile robot first randomly explores the task area. As the robot moves, the number of available prior sample points continues to increase, and the distribution of the predicted field becomes more accurate. At 33s, Robot 2 detected a light intensity value higher than 36Lux and began moving toward the predicted source position. But at this time, the remaining robots are still performing random exploration tasks until 87s when

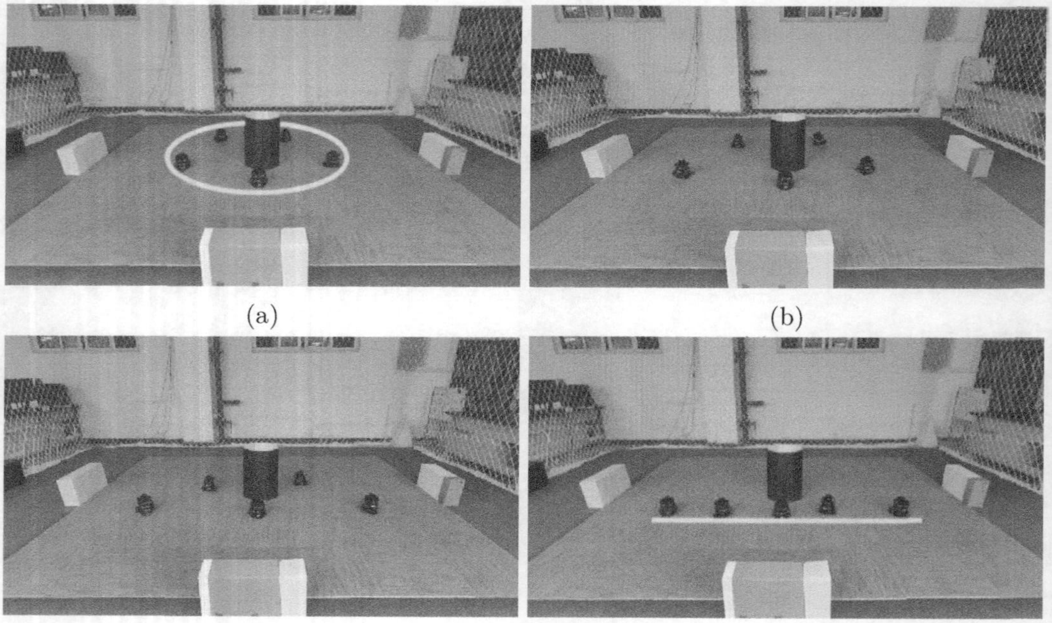

(a) (b)

(c) (d)

Figure 10.15: capture formation switching to line formation: (a) $t = 0s$; (b) $t = 5s$; (c) $t = 10s$; (d) $t = 14s$.

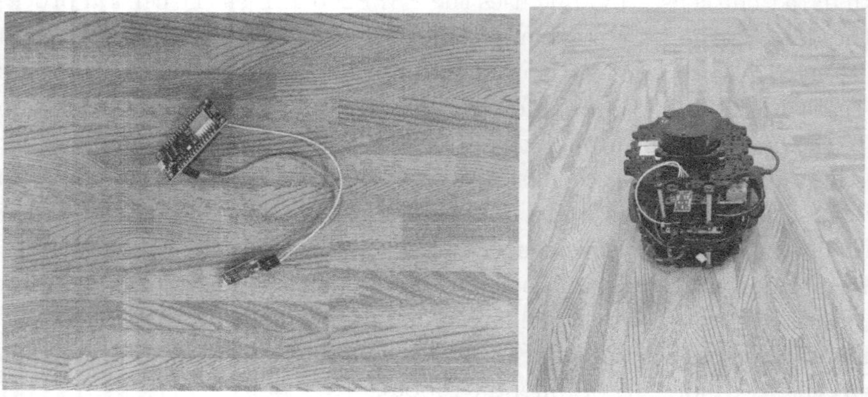

(a) (b)

Figure 10.16: Experimental environment setup: (a) Light source acquisition module; (b) TurtleBot3 robot.

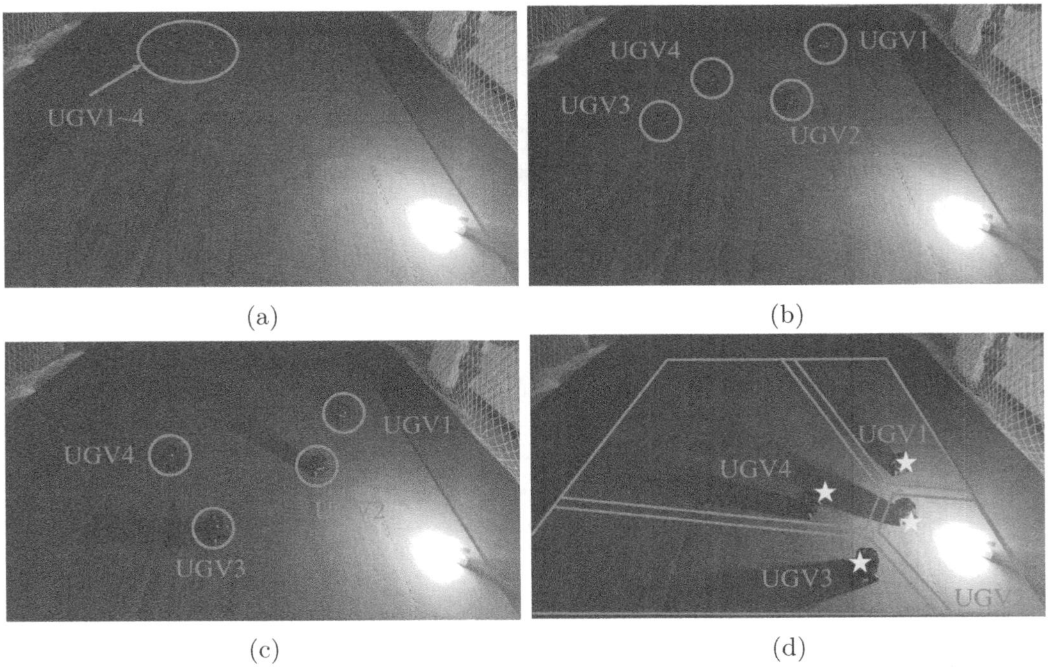


Figure 10.17: TurtleBot3 formation experiment: (a) $t = 0s$; (b) $t = 33s$; (c) $t = 87s$; (d) $t = 125s$.

robots 1 and 2 switch to formation coverage. Finally, at 125 seconds, the formation converged, achieving exploration of unknown environments and localization of the light source.

### 3D Environment Crazyflie Sourcing Seek Experiment

As shown in Fig. 10.18a, the Crazyflie drones used in the experiment have the capability of sensing light sources. First, the light-sensitive sensor collects data on light sources and sends the data to the base station via the ESP8266-12E module as the prior density information input. Additionally, retroreflective ball markers are attached to the Crazyflie drones, and their real-time position information is captured via Optitrack [7] [9] [10]. This information is then sent to the base station via the ROS system as the prior positional information input. The light sensor module can change the output voltage according to the ambient light intensity. The overall experimental environment is shown in Fig. 10.18b.

The overall process of the experiment is as follows. After capturing the motion status information of the Crazyflie and the light intensity information at the drone's current location, the obtained position and field strength information are transmitted to the ROS control terminal. The ROS control terminal analyzes and processes the data through MATLAB-ROS joint programming, ultimately obtaining the distribution of light sources and the next target position of the drone. Finally, the next target

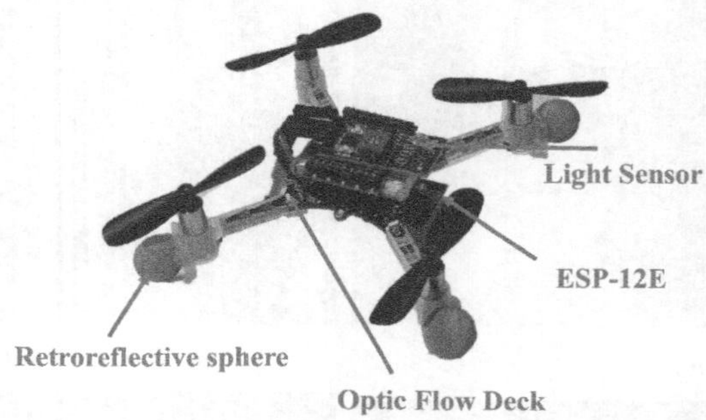


(a)

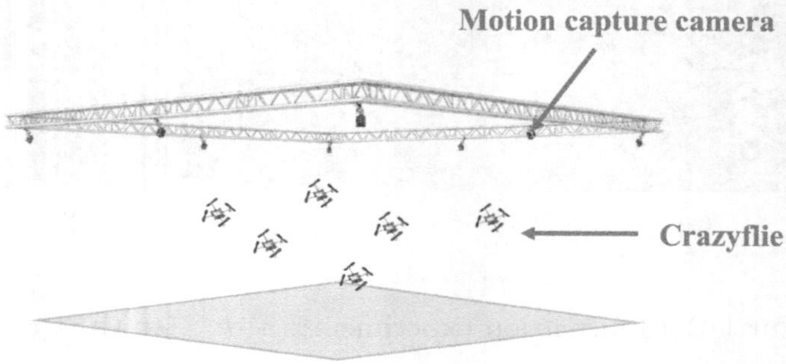


(b)

Figure 10.18: Experimental environment setup: (a) Crazyflie with attached photosensitive sensor; (b) schematic diagram of multiple Crazyflie formations.

position information is sent to the Crazyflie to achieve formation flight of multiple Crazyflie drones.

In the experiment, the gray rectangular area represents the task area. White nets are used around the task environment to ensure flight safety, preventing drones from flying out of the task area accidentally. The minimum and maximum flight heights of the Crazyflie drones in the experiment are set to 0.5m and 1.5m, respectively.

This experiment aims to validate the feasibility of the proposed coverage algorithm in a real environment. In contrast to the previous experiment, this section introduces the use of hanging light bulb sources to simulate pollution sources. In this experimental scenario, the light sources are located in the bottom right corner of the scene, and 3 Crazyflie drones depart from the top left corner of the map. Both the light intensity information obtained from light-sensitive sensors and the position information acquired from Optitrack are used as prior inputs to the algorithm. The distribution of the light source field is reconstructed by the movement of drones within the task area, ultimately pinpointing the location of the light source. The experimental process is illustrated in Fig. 10.19.

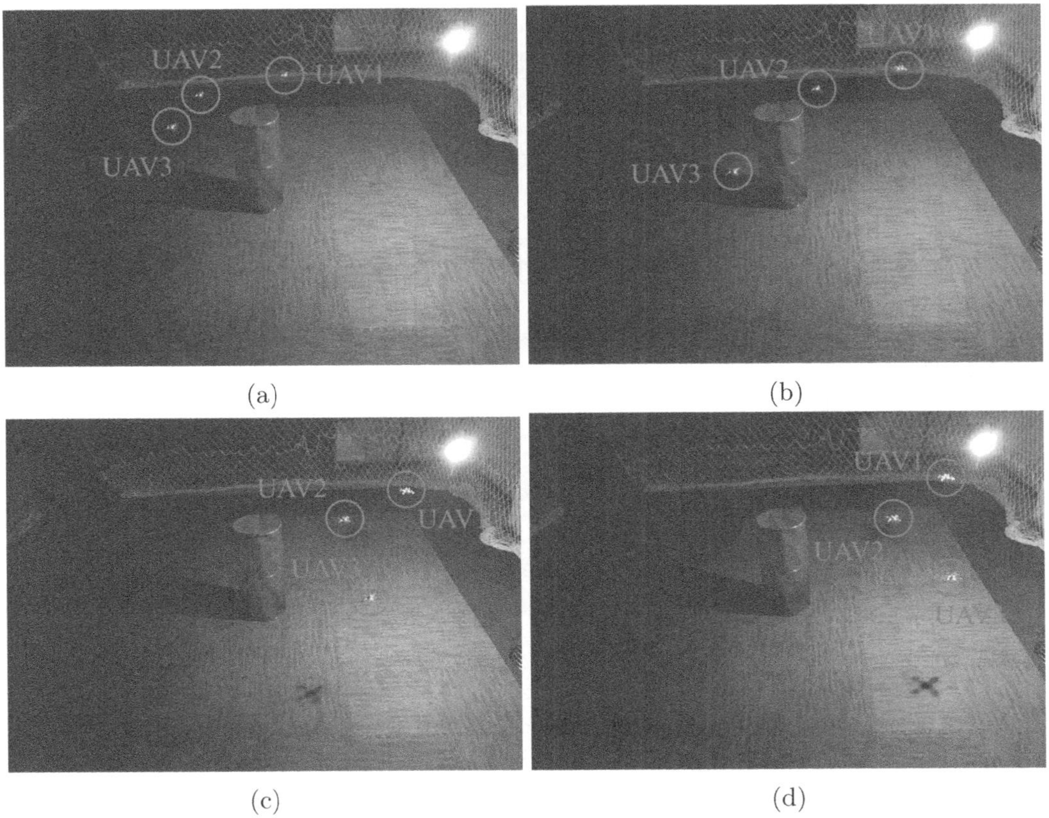


Figure 10.19: Formation coverage and source localization of 3 drones: (a) $t = 2s$; (b) $t = 7s$; (c) $t = 10s$; (d) $t = 12s$.

It can be seen that in the initial stages, due to the scarcity of light intensity information, the drones may head toward areas with greater uncertainty. However, as the number of available prior sample points increases, the predicted distribution of the field becomes more accurate. At this point, the drones tend to move toward areas where high density and uncertainty are balanced. Ultimately, the drone formation converges, achieving multi-drone online coverage and the localization of the source of light in environment. Due to the limitation of the scale of the experiment site, the UAV only needs a few iterations to reach the light source location, so the total iteration time in the experiment is relatively small. Whether the experimental scenario is simple or complex, the formation control principle is the same.

### 10.3.4 Multi-Robot Formation Capture

**Multi-Robot Formation Capture with Known Target Localization**

The experiment has three different environments: obstacle-free, multiple obstacles, and dynamic targets. In all three cases, the initial positions of the drones are located at the top left, and the target is positioned at the bottom right. The number of

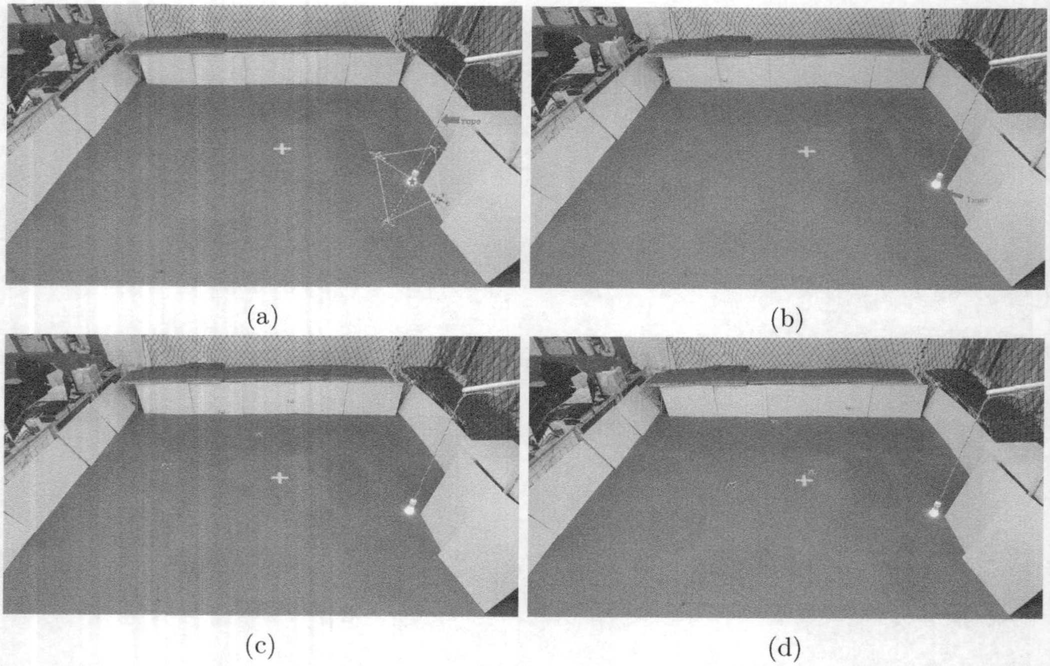


Figure 10.20: Multi-drone capture under an obstacle-free environment: (a) $t = 1s$; (b) $t = 5s$; (c) $t = 10s$; (d) $t = 14s$.

drones used in the experiment is four, and the velocity of the drones is 1.1 m/s. It is worth noting that, in order to avoid crashes caused by the turbulence generated when the distance between drones is too close, we did not minimize the volume of the target 3D-Voronoi cell; the safe distance between each drone is set as 0.3 m. At the beginning of our experiment, we assume that each drone has a good knowledge of the environment, including the positions of both the target and obstacles.

(1) Obstacle-Free Environment

In the obstacle-free environment, the drones were tasked with capturing a stationary glowing LED. The vertical yellow line connect the capture. The LED was suspended at the bottom right corner of the experimental scene at a position of (0.9 m, 0.9 m, 1.1 m). The capture process is illustrated in Fig. 10.20.

(2) Multiple Obstacles Environment

In the obstacle environment, five cylindrical obstacles were constructed. To avoid interfering with the tracking of the drones by the Optitrack system, the surfaces of the obstacles were covered with green cloth. The safe distance between drones and obstacles was set as 0.3 m. The capture process in this environment is shown in Fig. 10.21. From the final distribution in Figs. 10.20d and 10.21d, it can be observed that the drone formation has effectively captured the target.

(3) Dynamic Target Environment

In the dynamic target environment, three drones were designated as the capturing drones, and one drone served as the target. Throughout the entire experiment, the

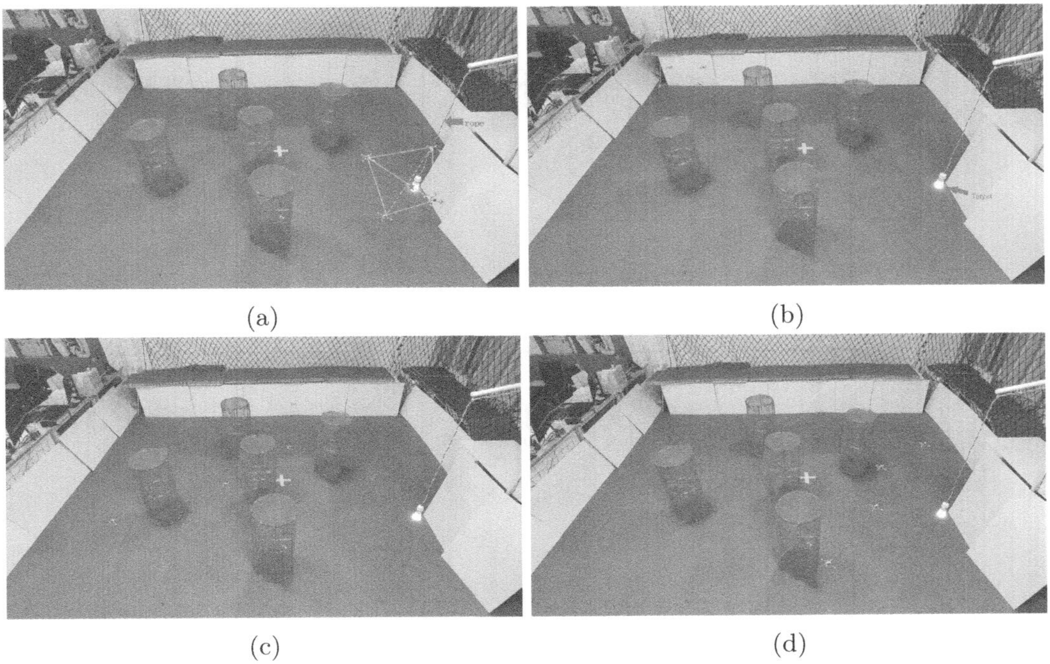

Figure 10.21: Multi-drone capture of under obstacles environment: (a) $t = 1s$; (b) $t = 7s$; (c) $t = 11s$; (d) $t = 16s$.

target did not employ any specific evasion strategy. It simply moved linearly from an initial point to another predetermined point. The capture process is illustrated in Fig. 10.22.

The final capture result in Fig. 10.22d is similar to the capture effect shown in 3a for $n = 3$. Through the experimental process, it can be observed that under the ODMV strategy, the drone formation is able to capture the target. This conclusion can be extended to capture experiments with a larger number of drones.

**Multi-Robot Formation Capture with Unknown Target Localization**

There are two different scenarios for this experiment in an obstacle environment, as shown in Fig. 10.23, which respectively capture a single target and multiple targets (with two targets). The targets in the experiment are all light sources. In the case of a single source, the initial position of the drone is located in the lower left corner, and the target light source is located in the upper right corner (0.9 m, 0.9 m, 1.1 m); In the case of multiple sources, the initial position of the drone remains unchanged, and a new light source located in the upper left corner (2.7 m, 0.2 m, 0.5 m) is added. The number of drones used in the experiment is 4, and the speed of the drones is 1.1 meters per second. In order to avoid collisions caused by turbulence when drones are too close, the safe distance between each drone is set to 0.3 meters. At the beginning of the experiment, each drone was given an environmental map, which included the location

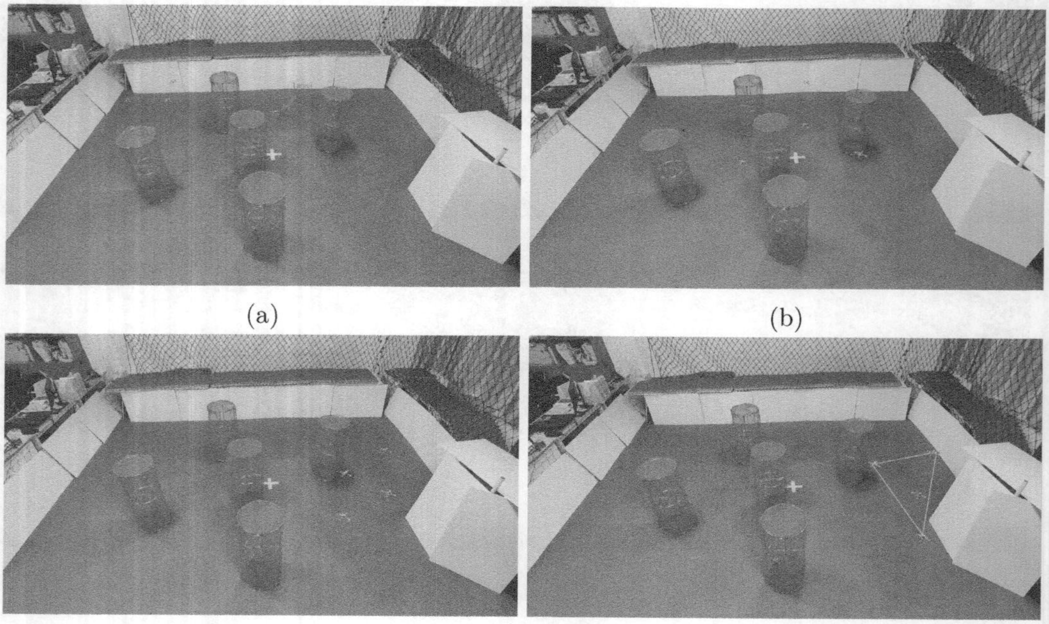

(a) (b) (c) (d)

Figure 10.22: Multi-drone capture under dynamic target environment: (a) $t = 1s$; (b) $t = 7s$; (c) $t = 12s$; (d) $t = 16s$.

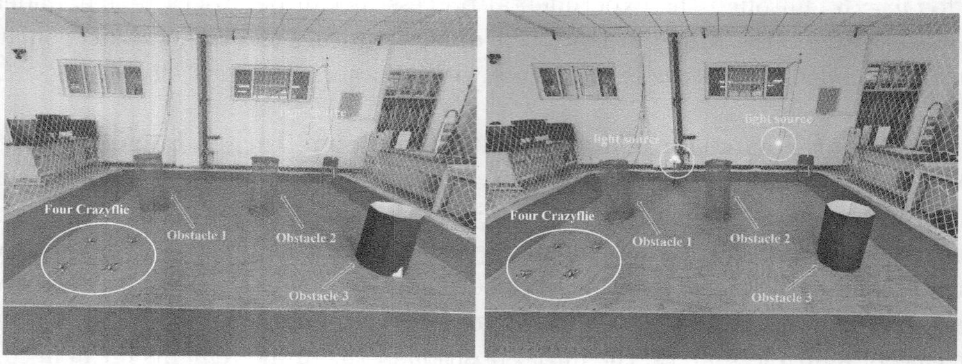


(a) (b)

Figure 10.23: Two different target source capture experiments: (a) Single-target source; (b) multi-target source.

of the target and obstacles. The relevant videos of this experiment can be viewed at the following link: Experiment of capture video: `https://youtu.be/1VzBe1976OE`

(1) Capture of a single target source in a multi-obstacle environment

Three cylindrical obstacles and one target light source were constructed in a single target source and multi-obstacle environment. The light source bulb is suspended in the upper right corner (0.9m, 0.9m, 1.1m) of the experimental scene, and four Crazyflie drones are used to conduct a capture experiment on the target source. To avoid interfering with the OptiTrack system tracking of the drone, a green mesh material is covered on the surface of the intermediate obstacle. The safe distance between the drone and the obstacle is set to 0.3 meters, and the ultimate goal of the experiment is to surround the target light source within a circular area composed of four Crazyflie drones. The partial state of the capture process in this environment is shown in Fig. 10.24. From the final distribution in Fig. 10.24d, it can be observed that the drone formation uniformly surrounded the target light source at 17s.

(2) Capture of multi-target sources in a multi-obstacle environment

On the basis of a single target source, an additional target light source has been added, located in the upper left corner (2.7 m, 0.2 m, 0.5 m), forming a multi-target

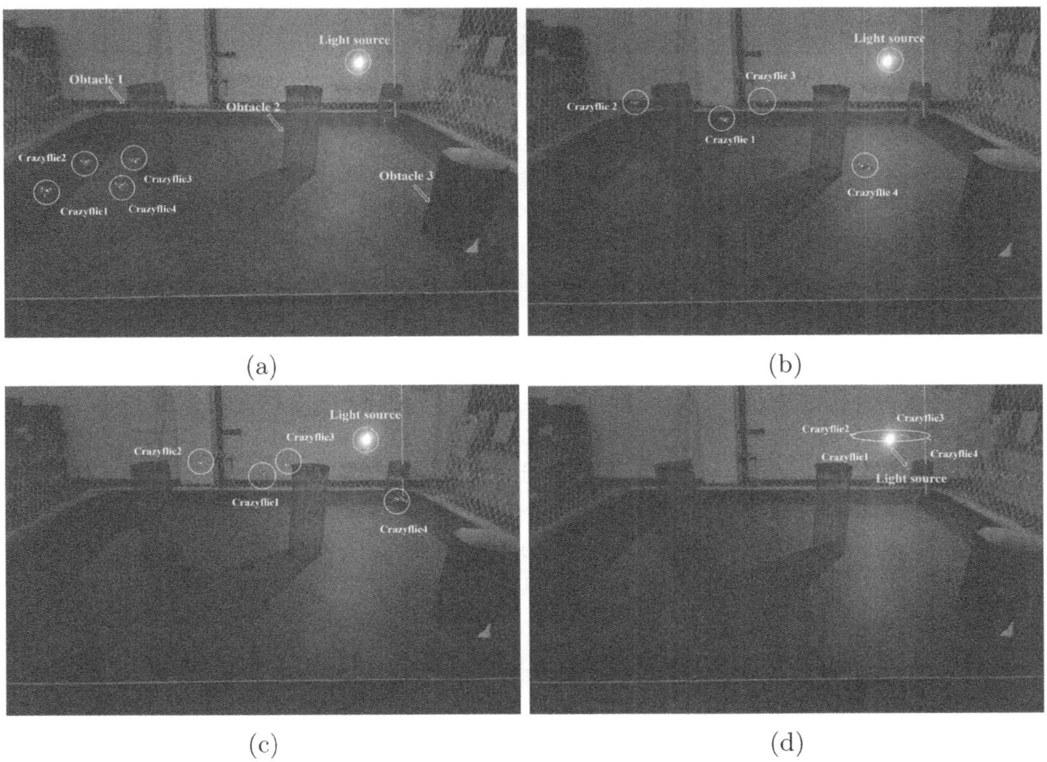


Figure 10.24: Multi-UAV capture single-source experiment: (a) The initial capture process ($t = 1\ s$); (b) during the capture process ($t = 8\ s$); (c) during the capture process ($t = 12\ s$); (d) complete uniform capture ($t = 17\ s$).

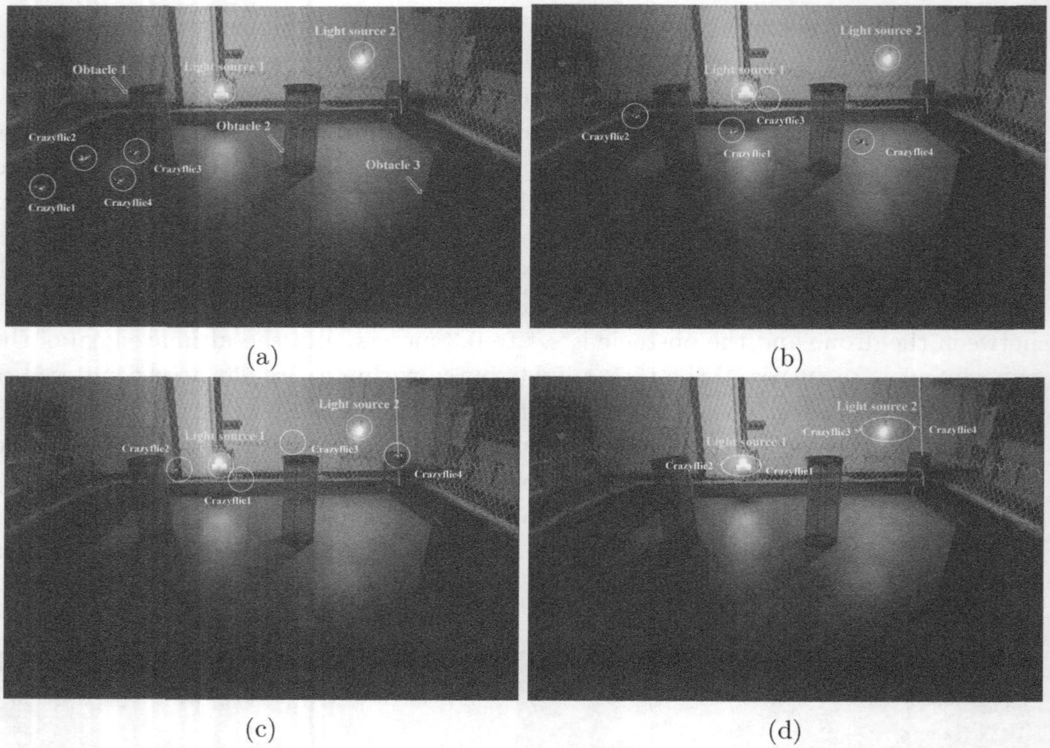


Figure 10.25: Multi-UAV capture multi-source experiment: (a) The initial capture process ($t = 1\ s$); (b) during the capture process ($t = 9\ s$); (c) during the capture process ($t = 14\ s$); (d) complete uniform capture ($t = 18\ s$).

source and multi-obstacle environment. Due to the limited number of drones, four Crazyflie drones were used for the experiment, with two drones assigned to each target source for capture. The safe distance between the drones and obstacles was set to 0.3 meters. The capture process is shown in Fig. 10.25. From the final distribution in Fig. 10.25d, it can be observed that the drone formation completed the effective capture of two target sources at 18 seconds. Due to the limitations of the experimental site, the drone only needs about 5 iterations to reach the light source position, so the total iteration time in the experiment is relatively less than the simulated.

## 10.4 Chapter Summary

This chapter shows a complete cycle from theoretical modeling to practical validation by constructing an experimental platform for multi-robot systems and applying Voronoi-based cooperative formation algorithms in real-world environments. Leveraging the modular and distributed architecture of the ROS platform, the experiments integrated TurtleBot3 ground mobile robots and Crazyflie drones to conduct multiple rounds of validation across several typical tasks, including formation control, obstacle avoidance, source search, and target capture. The powerful communication and visualization capabilities of ROS significantly enhanced coordination among different robotic modules, enabling real-time control and state monitoring of the system.

Through the aforementioned indoor physical experiments, the proposed Voronoi-based algorithms demonstrated strong stability, robustness, and operability in dynamic and constrained environments. The experimental results not only verified the feasibility of the previously introduced algorithms but also laid a solid foundation for their application in real-world scenarios such as industry, emergency response, and smart cities. This chapter highlights the crucial importance of deeply integrating theoretical models with practical robotic platforms, marking a key step toward advancing multi-robot collaboration from laboratory research to real-world deployment.

# Bibliography


[1] M Theeng Tamang, D Maheriya, S Sharif, and T Sutharssan. Autonomous navigation for TurtleBot3 robots in Gazebo simulation environment. In *2024 International Conference on Innovation and Intelligence for Informatics, Computing, and Technologies (3ICT)*, pages 568–574. IEEE, 2025.

[2] Christian Llanes, Zahi Kakish, Kyle Williams, and Samuel Coogan. Crazysim: A software-in-the-loop simulator for the Crazyflie nano quadrotor. In *2024 IEEE International Conference on Robotics and Automation (ICRA)*, pages 12248–12254. IEEE, 2024.

[3] Anis Koubaa et al. *Robot Operating System (ROS)*, volume 1. Springer, 2017.

[4] Steven Macenski, Tully Foote, Brian Gerkey, Chris Lalancette, and William Woodall. Robot operating system 2: Design, architecture, and uses in the wild. *Science Robotics*, 7(66):eabm6074, 2022.

[5] Michael Cashmore, Maria Fox, Derek Long, Daniele Magazzeni, Bram Ridder, Arnau Carrera, Narcis Palomeras, Natalia Hurtos, and Marc Carreras. ROSplan: Planning in the robot operating system. In *Proceedings of the International Conference on Automated Planning and Scheduling*, volume 25, pages 333–341, 2015.

[6] Ramkumar Gandhinathan and Lentin Joseph. *ROS Robotics projects: build and control robots powered by the Robot Operating System, machine learning, and virtual reality*. Packt Publishing Ltd, 2019.

[7] Joshua S Furtado, Hugh HT Liu, Gilbert Lai, Herve Lacheray, and Jason Desouza-Coelho. Comparative analysis of optitrack motion capture systems. In *Advances in Motion Sensing and Control for Robotic Applications: Selected Papers from the Symposium on Mechatronics, Robotics, and Control (SMRC'18)-CSME International Congress 2018, May 27-30, 2018 Toronto, Canada*, pages 15–31. Springer, 2019.

[8] Wojciech Giernacki, Mateusz Skwierczyński, Wojciech Witwicki, Paweł Wroński, and Piotr Kozierski. Crazyflie 2.0 quadrotor as a platform for research and education in robotics and control engineering. In *2017 22nd International Conference on Methods and Models in Automation and Robotics (MMAR)*, pages 37–42. IEEE, 2017.

[9] B Castillo, C Riascos, JM Franco, J Marulanda, and P Thomson. Assessing spatiotemporal behavior of human gait: a comparative study between low-cost smartphone-based MoCap and Optitrack systems. *Experimental Techniques*, 49(1):3–13, 2025.

[10] Marek Retinger, Jacek Michalski, Piotr Kozierski, and Wojciech Giernacki. Toward improving tracking precision in motion capture systems. In *2023 International Conference on Unmanned Aircraft Systems (ICUAS)*, pages 919–925. IEEE, 2023.

CHAPTER 11

# Conclusions and Future Prospects

## 11.1 Concluding Remarks

THIS book systematically explores the applications of Voronoi Tessellation in multi-robot systems. Starting from geometric modeling, it delves into the theoretical foundations, algorithmic design, and engineering implementation pathways for multi-robot coordination. The book covers a wide range of topics, from constructing Voronoi diagrams in 2D, 3D, and non-convex environments, to their applications in formation control, dynamic formation switching, collision avoidance, source search, target capture strategies, health management, and ROS-based hard ware implementation. These demonstrate the strong capabilities and broad applicability of Voronoi Tessellations in spatial allocation and distributed intelligence.

The book begins by introducing the basic geometric properties of Voronoi Tessellations and extends them from two-dimensional space to three-dimensional and non-convex environments. Then, by incorporating CVT algorithms, a distributed formation strategy for multi-robot coordination is constructed. A density-function-guided dynamic switching method further enhances the system adaptability to changing tasks. For safety concerns, strategies such as buffered Voronoi cells and separating hyperplanes are proposed to address collision avoidance in complex environments. For task-oriented applications, this book integrates sparse Gaussian process regression with Voronoi-based methods in pollution source search and coverage control tasks, proposing a novel perception and decision-making mechanism that combines confidence bounds with geometric properties. In terms of capture and encirclement control, various geometry-driven strategies are developed for both known and unknown target locations. Furthermore, a Voronoi-based health management mechanism is introduced to enable adaptive task load balancing with improved system robustness. To bridge theory and engineering practice gap, the book presents implementation schemes based on ROS, TurtleBot3, and Crazyflie platforms. As discussed in Chapters 3–5 and Chapter 10 offer a progressive modeling-simulation-deployment workflow tailored to practice-oriented readers, significantly lowering the barrier for beginners to understand and implement Voronoi-based control frameworks.

DOI: 10.1201/9781003714569-11

In summary, this book builds a comprehensive control architecture centered on Voronoi Tessellations, integrating theoretical modeling, algorithmic development, and practical deployment. It provides a rigorous geometric foundation while offering generalizability and scalability for real-world task applications, serving as a solid reference for further research and applications in the field.

## 11.2 Future Research Directions and Challenges

Although Voronoi tessellations has demonstrated considerable potential in multi-robot systems, its broader application still faces several challenges. Future research may focus on the following directions, each of which we will explore in detail.

### 11.2.1 Dynamic Environments and Real-Time Adaptability

The majority of existing methods based on Voronoi diagrams remain largely predicated on the assumption of a static or quasi-static environment [1]. In contrast, real-world applications—including urban search and rescue, dynamic target capture, and cooperative logistics—are intrinsically highly dynamic and uncertain. It is therefore imperative for future research to transcend this foundational assumption and tackle the problem of real-time Voronoi reconstruction amidst asynchronous sensing and communication.

1) Event-Triggered Local Update Mechanism: Globally synchronous recomputation of the Voronoi diagram incurs prohibitively high communication and computational costs in large-scale swarms, which is often unnecessary. Future algorithms should shift toward an event-driven paradigm [2]. For instance, local communication and diagram updates should only be triggered when a robot detects significant changes within its adjacent Voronoi cell, such as obstacle intrusion or neighbor movement beyond a set threshold. This necessitates research into defining effective event-triggering conditions and designing efficient local information exchange protocols to minimize communication overhead while maintaining the spatial consistency of the overall Voronoi diagram.

2) Asynchronous Communication and Latency: In practical systems, communication between robots is prone to delays, packet loss, and even interruption. Future Voronoi-based control algorithms must therefore possess robustness against such asynchronous and unreliable communication. A promising direction involves algorithms based on a "predictor-corrector" mechanism, where each robot predicts its own Voronoi cell using its motion model and received information, and subsequently performs a Voronoi reconstruction upon receiving new data.

3) Rapid Response in Dynamic Environments: The system must react swiftly to suddenly appearing dynamic obstacles, such as pedestrians or other vehicles. Research can focus on integrating Voronoi tessellations with fast local obstacle avoidance algorithms. For instance, the Voronoi diagram can be responsible for macro-level, strategic area allocation, while upon detecting an imminent collision risk, a robot switches to a reactive control mode based on local sensors. It would then seamlessly revert to the optimized Voronoi-based path once the threat has passed.

### 11.2.2 Integration with Learning-Based Control

When dealing with dynamic uncertainties in unknown environments or task variability, control strategies based purely on geometric models may prove inadequate. Integrating Voronoi tessellations with cutting-edge machine learning methods represents a critical pathway toward developing smarter, data-driven strategy optimization.

1) Reinforcement Learning for Optimizing Voronoi Parameters: In weighted Voronoi diagrams, weights are key determinants of region allocation. Reinforcement learning is well-suited for online learning of optimal weight allocation strategies to optimize global objectives, such as coverage efficiency and task completion time [3,4]. Robots can treat environmental features—such as target density and terrain complexity—as states, and adjustments to their own weights as actions. Through team reward signals, they learn when to increase or decrease their influence in specific scenarios.

2) Graph Neural Networks and Topological Relationship Learning: The dual graph of a Voronoi diagram—the Delaunay triangulation—naturally defines the communication topology among robots. Owing to their strong capability in processing topological data, Graph Neural Networks (GNNs) are particularly suitable for learning complex collaborative strategies over such a topology. GNNs can be used to predict the movement intentions of neighboring robots, assess link stability, and even dynamically optimize the generation rules of the Voronoi diagram itself. For instance, they can make cells more compact in areas requiring close collaboration, and sparser in regions where rapid traversal is needed.

3) Meta-Learning for Rapid Adaptation: In applications that require frequent task switching, the system must quickly adapt to new environmental dynamics. Meta-learning can train a prior model capable of rapid adaptation to new tasks. This model takes a small number of samples from a new task as input and quickly adjusts the parameters of the Voronoi control, thereby significantly reducing the tuning time in novel scenarios.

### 11.2.3 Multi-Objective Tasks and Heterogeneous Coordination

In real-world multi-robot systems, robots are often heterogeneous—equipped with varying sensors, mobility levels, and load capacities, and must balance multiple, often conflicting, objectives.

1) Generalized Voronoi Models Supporting Heterogeneity: Standard Voronoi diagrams assume homogeneous robots. Weighted Voronoi diagrams offer a preliminary approach to handle heterogeneity [5], but more sophisticated models are needed for future applications. For instance, capability-aware Voronoi diagrams incorporate not only geometric distances but also functional attributes of robots into their metric. This requires defining new metric spaces that integrate robots' physical states and capability vectors into distance calculations.

2) Game-Theoretic Modeling for Multi-Objective Optimization: In multi-objective problems—such as simultaneously maximizing coverage area, minimizing energy consumption, and balancing communication load—the system can be modeled as a cooperative game [6]. Each robot acts as a player, and its strategy involves choosing

behaviors under the Voronoi-based control law. The global objective of the system corresponds to the potential function of the game. Research can focus on designing utility functions such that, as robots pursue their individual utilities, the resulting Nash equilibrium approximates the globally optimal solution for the multi-objective problem.

3) Hierarchical Control Architecture: For complex heterogeneous systems, a hierarchical control architecture can be employed. At the top level, a Voronoi diagram-based task planner allocates macro-level task regions to distinct robot sub-teams. At the bottom level, each sub-team employs homogeneous Voronoi control laws for internal coordination. This layered structure helps manage the complexity of the system.

### 11.2.4 Human-Robot Coexistence and Explainable Collaboration

In fields such as healthcare, education, and disaster response, robots are no longer isolated autonomous units but partners that require close collaboration with humans [7,8]. Introducing Voronoi tessellations into human-robot collaborative systems to construct an interpretable, intervenable, and consensus-driven control structure presents a significant challenge.

1) Voronoi as an Intuitive Human-Robot Interface: The spatial decomposition provided by a Voronoi diagram possesses intuitive geometric meaning, making it an excellent tool for shared situational awareness in mixed teams. The operational areas of humans and robots can be clearly visualized through their respective Voronoi cells. This allows humans to quickly comprehend robot intentions and plans, thereby enhancing the interpretability of collaboration.

2) Control Authority Handover: The system must support a seamless transfer of control. Under normal circumstances, robots operate autonomously within their Voronoi cells. When a human operator deems intervention necessary, they can directly "edit" the Voronoi diagram—for instance, by manually adjusting the boundary of a cell or reassigning a cell to a different robot. The system should be able to interpret such high-level intent and automatically generate the underlying control sequences for execution. This approach realizes "What-level" programming, freeing the operator from "How-level" micromanagement.

3) Voronoi-based Trust and Safety Model: Safety rules can be established based on Voronoi. For example, protected Voronoi cells can be defined around humans, the boundaries of which must not be crossed by robots. Robot path planning must respect these protected cells. By adjusting the size of these "safety cells" the system's trust level toward human partners can be dynamically expressed.

### 11.2.5 Topologically Constrained Complex Spatial Partitioning

In complex domains such as urban underground pipe networks, air traffic systems, and marine operations, the environments are often non-Euclidean, multi-layered, or subject to topological constraints [9,10], Future studies should explore extended models.

1) Voronoi tessellations on Manifolds: For robot swarms operating on curved surfaces, Voronoi tessellations must be defined on two-dimensional manifolds. This requires changing the distance metric from Euclidean distance to geodesic distance. The research focus will lie on how to efficiently compute and update Voronoi diagrams on manifolds and how to integrate them with robot motion control on surfaces.

2) Multi-Layer and 3D Voronoi tessellations: In scenarios such as urban air mobility, the workspace is a complex three-dimensional structure. This necessitates research into true 3D Voronoi tessellations, taking into account multi-layer constraints. Furthermore, generalized Voronoi diagrams can be utilized by incorporating obstacles like buildings as generating points, thereby naturally planning safe flight corridors in three-dimensional space.

3) Topological-Navigation Voronoi tessellations: For large-scale, structured environments, an abstract topological map can be precomputed, where nodes represent key locations and edges represent connecting pathways. Based on this, a topology-based Voronoi diagram can be defined, whose "distance" metric is the shortest-path cost within the graph. This tessellations method is particularly well-suited for task allocation and patrol route planning in large-scale environments.

### 11.2.6 Algorithmic Realization under Hardware Constraints

Many current Voronoi algorithms may overlook practical hardware limitations, such as localization errors, communication latency, and motion resolution. Future research should incorporate perception noise modeling and bandwidth limitations into Voronoi diagrams algorithms to achieve software-hardware co-design and more robust real-world deployment [11,12].

1) Uncertainty-Aware Voronoi Modeling: The core challenge lies in mitigating the impact of localization errors and sensor noise on partition stability. Key research directions should include probabilistic Voronoi diagrams, which integrate the uncertainty of robot positions directly into the calculation of cell boundaries, thereby defining the confidence level of these boundaries. An alternative approach is conservative robust optimization, which constructs worst-case Voronoi cells to guarantee that a robot remains within its assigned cell despite its localization uncertainty, thus algorithmically preventing cell ownership conflicts caused by perceptual noise.

2) Distributed Algorithm Design under Computational Constraints: To address limitations in bandwidth and processor performance, lightweight asynchronous algorithms must be developed. This involves researching quantized communication protocols that transmit only critical positional updates or low-precision data to drastically reduce bandwidth consumption. Concurrently, an event-triggered local update mechanism should be established to replace global synchronous computation. This allows robots to recompute their local Voronoi diagrams only when their own state or a neighbor's state changes beyond a set threshold, thereby significantly reducing the overall communication and computational load on the system.

3) Integration of Software and Hardware at the Control Level: To address the mismatch between theoretical control laws and robot kinematics/dynamics, a deep integration of Voronoi tessellations with advanced control strategies is required. A

primary direction is adopting a hierarchical architecture: the top layer utilizes Voronoi tessellations for regional planning and task allocation, while the bottom layer employs controllers such as Model Predictive Control (MPC) to translate Voronoi-based objectives into smooth, executable trajectories that adhere to motion constraints. Furthermore, Sim-to-Real transfer techniques are crucial. By embedding high-fidelity hardware models in simulation, the algorithms can be validated and optimized to ensure their robustness in real-world deployment.

In summary, with its geometric clarity and distributed adaptability, Voronoi tessellations are increasingly becoming a key modeling tool for multi-robot system coordination and intelligent decision-making. Looking ahead, their development will exhibit the following trends: Algorithmically, the focus will shift from static, synchronous, and deterministic models toward dynamic, asynchronous, data-driven, and robust hybrid models. In terms of application, they will extend from relatively structured homogeneous systems to open, heterogeneous, and human-robot mixed complex scenarios. Theoretically, interdisciplinary integration with game theory, machine learning, topology, and other fields will deepen, forming a more solid theoretical foundation.

With the continuous advancement of algorithms, computing power, and hardware, Voronoi tessellations will become further embedded in autonomous systems, pervasive robotics, and human-robot symbiotic frameworks. They are not merely a computational geometry tool but are likely to evolve into a "spatial grammar" for swarm intelligence, defining the relationship between individuals and space to manifest collective intelligence, ultimately driving the development of next-generation intelligent collaborative ecosystems.

This notion resonates with Fei-Fei Li's vision of the "spatial world model," in which intelligence is grounded in understanding, perceiving and generating structured three-dimensional spaces. By integrating the geometric constructs of Voronoi diagrams into such a world-model framework, one can bridge space-aware reasoning with multi-agent coordination, thereby enabling both distributed spatial cognition and efficient cooperative behavior in swarm systems.

The simulation code and dataset supporting each chapter of this book have been uniformly hosted on the GitHub platform (Link: `https://github.com/storybox126/Robotic-Swarming-Control-via-Voronoi-Tessellations.git`) Readers can obtain complete resources through this repository.

# Bibliography

[1] Qingchen Bi, Xuebo Zhang, Jian Wen, Zhangchao Pan, Shiyong Zhang, and Runhua Wang. CURE: A hierarchical framework for multi-robot autonomous exploration inspired by centroids of unknown regions. *IEEE Transactions on Automation Science and Engineering*, 21(3):3773–3786, 2023.

[2] Jiayu Zou, Haitao Zhang, Chao Zhai, Ning Xing, Yong Ma, and Xingjian Liu. Cooperative hierarchical coverage control of multi-agent systems with

non-holonomic constraints in poriferous environments. *IEEE Transactions on Control of Network Systems*, 12(2):1512–1520, 2025.

[3] Junyan Hu, Hanlin Niu, Joaquin Carrasco, Barry Lennox, and Farshad Arvin. Voronoi-based multi-robot autonomous exploration in unknown environments via deep reinforcement learning. *IEEE Transactions on Vehicular Technology*, 69(12):14413–14423, 2020.

[4] Yufei Yang, Changsheng Zhang, Yi Liu, Jiaxu Ning, and Ying Guo. Deep reinforcement learning assisted novelty search in Voronoi regions for constrained multi-objective optimization. *Swarm and Evolutionary Computation*, 91:101732, 2024.

[5] Hao Cheng, Bailing Tian, Xuewei Zhang, and Hongming Shen. A distributed algorithm for multi-robot task allocation via weighted buffered Voronoi partition. *IEEE Transactions on Industrial Electronics*, 71(12):16098–16107, 2024.

[6] Maulik Bhatt, Yixuan Jia, and Negar Mehr. Strategic decision-making in multi-agent domains: A weighted constrained potential dynamic game approach. *IEEE Transactions on Robotics*, 41:2749–2764, 2025.

[7] S. Secil and M. Ozkan. Minimum distance calculation using skeletal tracking for safe human-robot interaction. *Robotics and Computer-Integrated Manufacturing*, 73:102253, 2022.

[8] Yankun Wang, Yong Fan, Dongyang Wang, Xiaoming Li, Yang Wang, and Renzhong Guo. Near relationship enhanced multisourced data fusion method for voice-interactive indoor positioning. *IEEE Internet of Things Journal*, 11(4):5912–5919, 2023.

[9] Oscar de Groot, Laura Ferranti, Dariu M. Gavrila, and Javier Alonso-Mora. Topology-driven parallel trajectory optimization in dynamic environments. *IEEE Transactions on Robotics*, 41:110–126, 2024.

[10] Rahul Kumar Padhy, Krishnan Suresh, and Aaditya Chandrasekhar. Voroto: multiscale topology optimization of Voronoi structures using surrogate neural networks. *Engineering with Computers*, 41(2):1123–1139, 2025.

[11] Erick J. Rodríguez-Seda, Xiaotian Xu, Josep M. Olm, Arnau Dòria-Cerezo, and Yancy Diaz-Mercado. Self-triggered coverage control for mobile sensors. *IEEE Transactions on Robotics*, 39(1):223–238, 2022.

[12] Mingxin Wei, Lanxiang Zheng, Ying Wu, Han Liu, and Hui Cheng. Safe learning-based control for multiple UAVs under uncertain disturbances. *IEEE Transactions on Automation Science and Engineering*, 21(4):7349–7362, 2023.

# Index

For Product Safety Concerns and Information please contact our
EU representative GPSR@taylorandfrancis.com Taylor & Francis
Verlag GmbH, Kaufingerstraße 24, 80331 München, Germany